T0319368

Control Theory Applications for Dynamic Production Systems

Control Theory Applications for Dynamic Production Systems

Time and Frequency Methods for Analysis and Design

Neil A. Duffie
University of Wisconsin
Madison, Wisconsin

This edition first published 2022

Registered Office
John Wiley & Sons, Inc., 111 River Street, Hoboken, NJ 07030, USA

Editorial Office
111 River Street, Hoboken, NJ 07030, USA

For details of our global editorial offices, customer services, and more information about Wiley products visit us at www.wiley.com.

Wiley also publishes its books in a variety of electronic formats and by print-on-demand. Some content that appears in standard print versions of this book may not be available in other formats.

A catalogue record for this book is available from the Library of Congress

Hardback ISBN: 9781119862833; ePub ISBN: 9781119862857; ePDF ISBN: 9781119862840;
Obook ISBN: 9781119862864

Cover Image: Courtesy of John Miller, grayjaygraphics.com
Cover Design: Wiley

To Hans-Peter Wiendahl (1938–2019)

Contents

Preface

Production planning, operations, and control are being transformed by digitalization, creating opportunities for automation of decision making, reduction of delays in making and implementing decisions, and significant improvement of production system performance. Meanwhile, to remain competitive, today's production industries need to adapt to increasingly dynamic and turbulent markets. In this environment, production engineers and managers can benefit from tools of control system engineering that allow them to mathematically model, analyze, and design dynamic, changeable production systems with behavior that is effective and robust in the presence of turbulence. Research has shown that the tools of control system engineering are important additions to the production system engineer's toolbox, complementing traditional tools such as discrete event simulation; however, many production engineers are unfamiliar with application of control theory in their field. This book is a practical yet thorough introduction to the use of transfer functions and control theoretical methods in the modeling, analysis, and design of the dynamic behavior of production systems. Production engineers and managers will find this book a valuable and fundamental resource for improving their understanding of the dynamic behavior of modern production systems and guiding their design of future production systems.

This book was written for a course entitled Smart Manufacturing at the University of Wisconsin-Madison, taught for graduate students working in industry. It has been heavily influenced by two decades of industry-oriented research, mainly in collaboration with colleagues in Germany, on control theory applications in analysis and design of the dynamic behavior of production systems. Motivated by this experience, the material in this book has been selected to

- explain and illustrate how control theoretical methods can be used in a practical manner to understand and design the dynamic behavior of production systems
- focus application examples on production systems that can include production processes, machines, work systems, factories, communication, and production networks
- present both time-based and frequency-based analytical and design approaches along with illustrative examples to give production engineers important new perspectives and tools as production systems and networks become more complex and dynamic

- apply control system engineering software in examples that illustrate how dynamic behavior of production systems can be analyzed and designed in practice
- address both open-loop and closed-loop decision-making approaches
- present discrete-time and continuous-time theory in an integrated manner, recognizing the discrete-time nature of adjustments that are made in the operation of many production systems and complementing the integrated nature of supporting tools in control system engineering software
- recognize that delays are ever-present in production systems and illustrate modeling of delays and the detrimental effects that delays have on dynamic behavior
- show in examples how information acquisition, information sharing, and digital technologies can improve the dynamic behavior of production systems
- "bridge the gap" between production system engineering and control system engineering, illustrating how control theoretical methods and control system engineering software can be effective tools for production engineers.

This material is organized into the following chapters:

Chapter 1 *Introduction.* The many reasons why production engineers can benefit from becoming more familiar with the tools of control system engineering are discussed, including the increasingly dynamic and digital environment for which current and future production systems must be designed. Several examples are described that illustrate the opportunities that control theoretical time and frequency perspectives present for understanding and designing the dynamic behavior of production systems and their decision-making components.

Chapter 2 *Continuous-Time and Discrete-Time Models of Production Systems.* Methods for modeling the dynamic behavior of production systems are introduced, both for continuous-time and discrete-time production systems and components. The result of modeling is differential equations in the continuous-time case or difference equations in the discrete-time case. These describe how the outputs of a production system and its components vary with time as a function of their time-varying inputs. The concepts of linearizing a model around an operating point and linearization using piecewise approximations also are presented.

Chapter 3 *Transfer Functions and Block Diagrams.* Use of the Laplace transform and Z transform to convert continuous-time differential equation models and discrete-time difference equation models, respectively, into relatively more easily analyzed algebraic models is introduced. The concept of continuous-time and discrete-time transfer functions is introduced, as is their use in block diagrams that clearly illustrate dynamic characteristics, cause–effect relationships between the inputs and outputs of production systems and their components, delay, and closed-loop topologies. Transfer function algebra is reviewed along with methods for defining transfer functions in control system engineering software.

Chapter 4 *Fundamental Dynamic Characteristics and Time Response.* Fundamental dynamic characteristics of production system and component models are defined including time constants, damping ratios, and natural frequencies. The significance of the roots of characteristic equations obtained from transfer functions is reviewed, including using the roots to assess stability. Methods are presented for using continuous-time and discrete-time transfer functions to calculate the response of production systems as a function of time and determine characteristics such as settling time and overshoot in oscillation, with practical emphasis on use of control system engineering software.

Chapter 5 Frequency Response. Methods are presented for using transfer functions to calculate the response of production systems and their components to sinusoidal inputs that represent fluctuations in variables such as demand. Characteristics of frequency response that are important in analysis and design are defined including bandwidth, zero-frequency magnitude, and magnitude and phase margins. Theoretical foundations are presented, with practical emphasis on using control system engineering software to calculate and analyze frequency response.

Chapter 6 Design of Decision Making for Closed-Loop Production Systems. Approaches for design of decision making for closed-loop production systems using time response, transfer functions, and frequency response are introduced. Design for common closed-loop production system topologies is reviewed, and approaches such as PID control, feedforward control, and cascade control are introduced. Challenges and options for decision making in systems with significant time delays are addressed, and the use of control system engineering software in design is illustrated with examples.

Chapter 7 Application Examples. Examples are presented in which analysis and design of the dynamic behavior is of higher complexity, requiring approaches such as use of matrices of transfer functions and modeling using multiple sampling rates. The examples illustrate analysis and design from both the time and frequency perspectives. In the first application example, the potential for improving performance by using digital technologies to reduce delays in a replanning cycle is explored. Other application examples then are presented that illustrate analysis and design production systems with multiple inputs and outputs, networks of production systems with information sharing, and production systems with multiple closed loops.

After becoming familiar with the material presented in this book, production engineers can expect to be able to apply the basic tools of control theory and control system engineering software in modeling, analyzing, and designing the dynamic behavior of production systems, as well as significantly contribute to control system engineering applications in production industries.

Acknowledgments

I am grateful to many former graduate students and international research associates in my laboratory for the fruitful discussions and collaboration we have had on topics related to this book. I am particularly indebted to Professor Hans-Peter Wiendahl (1938–2019) for his inspiring encouragement of the research that culminated in this book, which is dedicated to him; he is greatly missed. Professor Katia Windt provided indispensable feedback regarding the contents of this book and its focus on production systems, and I owe much to collaborations with her and Professors Julia Arlinghaus, Michael Freitag, Gisela Lanza, and Bernd Scholz-Reiter. I thank the Department of Mechanical Engineering of the University of Wisconsin-Madison for the environment that made this book possible and, above all, I am deeply indebted to my wife Colleen for her companionship and her unwavering support of my research and the writing of this book.

Acknowledgments

1

Introduction

To remain competitive, today's industries need to adapt to increasingly dynamic and turbulent markets. Dynamic production systems[1] and networks need to be designed that respond rapidly and effectively to trends in demand and production disturbances. Digitalization is transforming production planning, operations, control, and other functions through extensive use of digitized data, digital communication, automatic decision-making, simulation, and software-based decision-making tools incorporating AI algorithms. New sensing, communication, and actuation technologies are making new types of measurements and other data available, reducing delays in decision-making and implementing decisions, and facilitating embedding of models to create more "intelligent" production systems with improved performance and robustness in the presence of turbulence in operating conditions.

In this increasingly dynamic and digital environment, production engineers and managers need tools that allow them to mathematically model, analyze, and design production systems and the strategies, policies, and decision-making components that make them responsive and robust in the presence of disturbances in the production environment, and mitigate the negative impacts of these disturbances. Discrete event simulation, queuing networks, and Petri nets have proved to be valuable tools for modeling the detailed behavior of production systems and predicting how important variables vary with time in response to specific input scenarios. However, these are not convenient tools for predicting fundamental dynamic characteristics of production systems operating under turbulent conditions. Large numbers of experiments, such as discrete event simulations with random input scenarios, often must be used to draw reliable conclusions about dynamic behavior and to subsequently design effective decision rules. On the other hand, measures of fundamental dynamic characteristics can be obtained quickly and directly from control theoretical models of production systems. Dynamic characteristics of interest can include

- time required for a production system to return to normal operation after disturbances such as rush orders or equipment failures (settling time)
- difference between desired values of important variables in a production system and actual values (error)

1 Production systems include the physical equipment, procedures, and organization needed to supply and process inputs and deliver products to consumers.

Control Theory Applications for Dynamic Production Systems: Time and Frequency Methods for Analysis and Design, First Edition. Neil A. Duffie.

- tendency of important variables to oscillate (damping) or tendency of decision rules to over adjust (overshoot)
- whether disturbances that occur at particular frequencies cause excessive performance deviations (magnification) or do not significantly affect performance (rejection)
- over what range of frequencies of turbulence in operating conditions the performance of a production system is satisfactory (bandwidth).

Unlike approaches such as discrete event simulation in which details of decision rules and the physical progression of entities such as workpieces and orders through the system often are modeled, control theoretical models are developed using aggregated concepts such as the flow of work. The tools of control system engineering can be applied to the simpler, linear models that are obtained, allowing decision-making to be directly designed to meet performance goals that are defined using characteristics such as those listed above. Experience has shown that the fidelity of this approach often is sufficient for understanding the fundamental dynamic behavior of production systems and for obtaining valuable, fundamentally sound, initial decision-making designs that can be improved with more detailed models and simulations.

Production engineers can significantly benefit from becoming more familiar with the tools of control system engineering because of the following reasons:

- The dynamic behavior of production systems can be unexpected and unfavorable. For example, if AI is incorporated into feedback with the expectation of improving system behavior, the result instead might be unstable or oscillatory. If a control theoretical model is developed for such a system, even though it is an approximation, it can be an effective and convenient means for understanding why such a system behaves the way it does. A control theoretical analysis can replace a multitude of simulations from which it may be difficult to draw fundamental conclusions and obtain initial guidance for design and implementation of decision-making.
- Many useful decision-making topologies already have been developed and are commonly applied in other fields but are unlikely to be (re)invented by a production engineer who is unfamiliar with control system engineering. Well-known practical design approaches arising from control theory can guide production engineers toward systems that are stable, respond quickly, avoid oscillation, and are not sensitive to day-to-day variations in system operation and variables that are difficult to characterize or measure.
- Delays and their effects on a production system can be readily modeled and analyzed. While delay often is not significant in design of electro-mechanical systems, delay can be very significant in production systems. The implications of delay need to be well understood, including the penalties of introducing delay and the benefits of reducing delay.
- Analysis and design using frequency response is an important additional perspective in analysis and design of dynamic behavior. Production systems often need to be designed to respond effectively to lower-frequency fluctuations such as changes in demand but not respond significantly to higher-frequency fluctuations such as irregular arrival times of orders to be processed. Analysis using frequency response is not a separate theory; rather, it is a fundamental aspect of basic control theory that complements and augments analysis using time response. Production engineers,

who are mostly familiar with time domain approaches such as results of discrete-event simulation, can significantly benefit from this alternative perspective on dynamic behavior and analysis and design using frequency response.

In this book, emphasis is placed on analysis and examples that illustrate the opportunities that control theoretical time and frequency perspectives present for understanding and designing the behavior of dynamic production systems. The dynamic behavior of the components of these systems and their interactions must be understood first before decision-making can be designed and implemented that results in favorable overall dynamic behavior of the production system, particularly when the structure contains feedback. In the replanning system with the topology in Figure 1.1, control theoretical modeling and analysis reveal that relationships between the period between replanning decisions and delays in making and implementing decisions can result in undesirable oscillatory behavior unless these relationships are taken into account in the design of replanning decision-making. Benefits of reducing delays using digital technologies can be quantified and used to guide replanning cycle redesign. In the production capacity decision-making approach shown in Figure 1.2, modeling and analysis from a frequency perspective can be used to guide design of the decision rules used to adjust capacity provided by permanent, temporary, and cross-trained employees, but also reveals that these decision rules can work at cross-purposes unless phase differences are explicitly considered.

In the planning and scheduling system shown in Figure 1.3, failure to understand the interactions between backlog regulation and work-in-progress (WIP) regulation when designing their decision rules can lead to unexpected and adverse combined dynamic behavior. Design guided by modeling and analysis achieves system behavior that reliably meets goals of effective backlog and WIP regulation. In the four-company production network shown in Figure 1.4, modeling and analysis of interactions

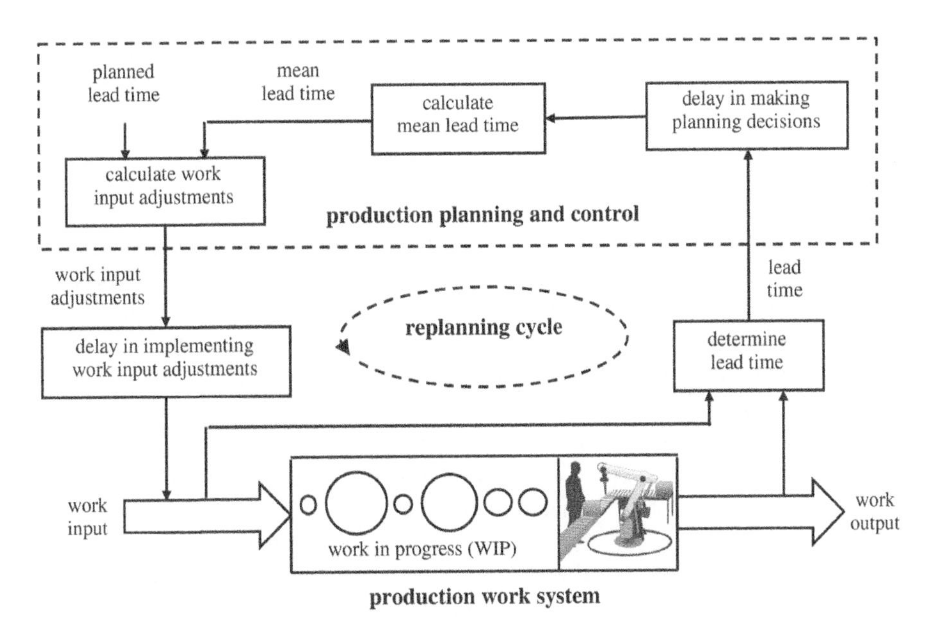

Figure 1.1 Replanning cycle with significant delays.

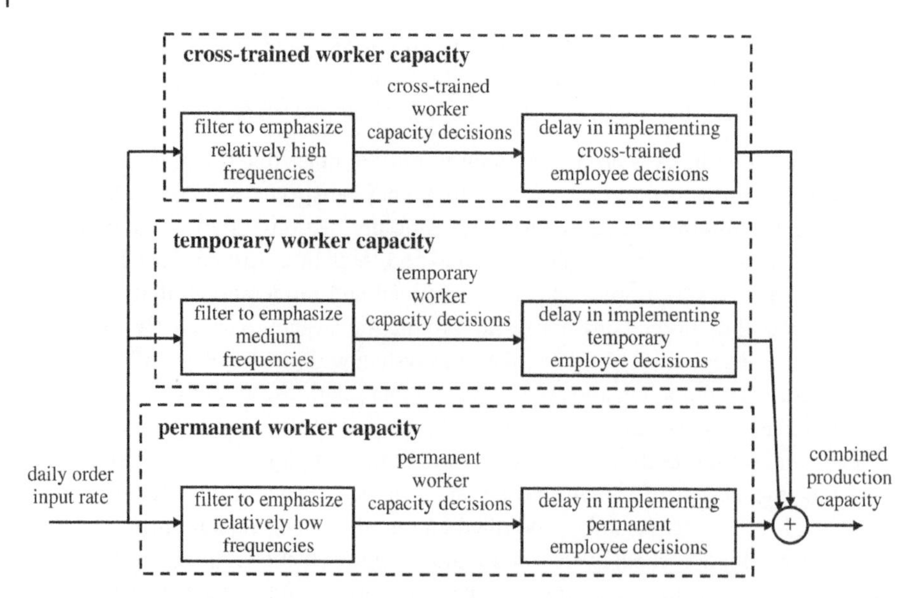

Figure 1.2 Adjustment of permanent, temporary, and cross-trained employee capacity based on frequency content of variation in order input rate.

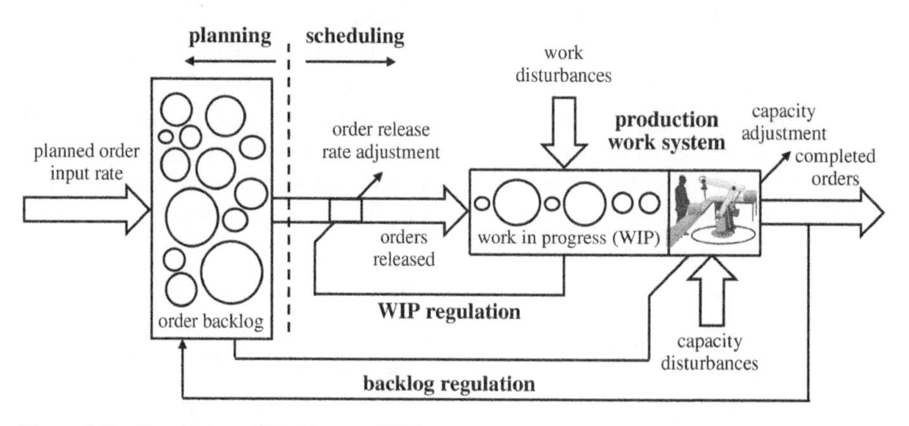

Figure 1.3 Regulation of backlog and WIP.

between companies allows decision rules to be designed for individual companies that result in favorable combined dynamic behavior. Benefits and dynamic limitations of information sharing between companies can be quantified and used in evaluating the merits and costs of information sharing and designing the structure in which it should be implemented. In the production operation shown in Figure 1.5, control theoretical modeling and analysis of the interacting components enables design of control components that together result in favorable, efficient behavior.

There has been considerable research in the use of control theoretical methods to improve understanding of the dynamics behavior of production systems and supply chains [1–4], but many production engineers are unfamiliar with the application of the tools of control system engineering in their field, tools that are well-developed and used extensively by electrical, aerospace, mechanical, and chemical engineers for mathematically modeling, analyzing, and designing control of electro-mechanical

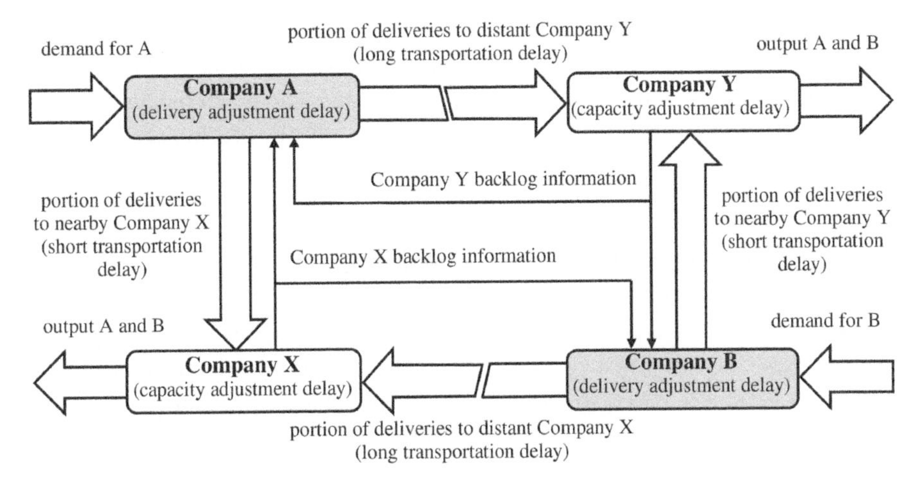

Figure 1.4 Adjustment of deliveries based on feedback of backlog information.

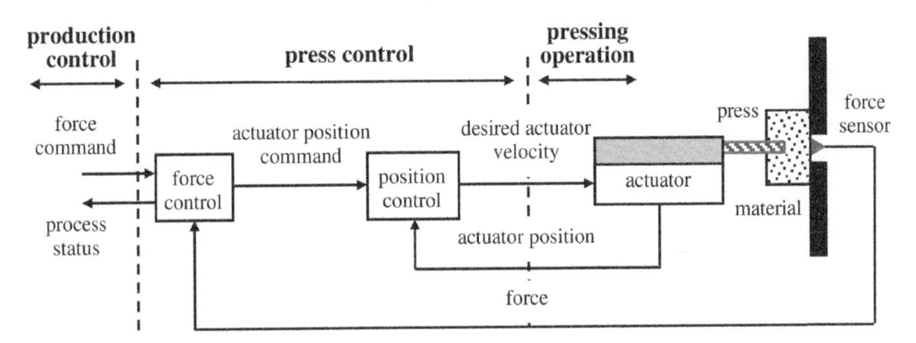

Figure 1.5 Control of force and position in a pressing operation.

systems and chemical processes. The tools of control system engineering include a daunting variety of mathematical approaches, but even the most basic control theoretical methods for modeling, analysis, and design can be important additions to the productions system engineer's toolbox, complementing tools such as discrete event simulation. The content of this book has been chosen to be immediately relevant to practicing production engineers, providing a fundamental understanding of both continuous-time and discrete-time control theory while avoiding unnecessary material. Some aspects of control theory covered in traditional texts are omitted here; for example, the principles of obtaining discrete-time models from continuous-time models are discussed, but the variety of mathematical methods for doing so are not because practicing production engineers rarely or never use these methods; instead, practicing production engineers need to obtain results quickly with the aid of control system engineering software. Similarly, practicing production engineers rarely or never need to find explicit solutions for differential and difference equations, and such solutions are only discussed in this book when they support important practical developments. Straightforward examples are presented that illustrate basic principles, and software examples are used to illustrate practical computation and application. The goal throughout this book is to provide production engineers and managers with valuable and fundamental means for improving their understanding of the dynamic behavior of modern production systems and guiding their design of future production systems. A brief biography is included at the end of this book for readers who are interested in further study including additional theoretical derivations, alternative methods of analysis and design, other application areas, and advanced topics in the ever-evolving field of control system engineering.

1.1 Control System Engineering Software

Control system engineering software is an essential tool for control system designers. MATLAB® and its Control System Toolbox™ from The MathWorks, Inc.[2] is one of the more widely used, and MATLAB® programs have been included in many of the examples in this book to illustrate how such software can be used to obtain practical results quickly using transfer functions and control theoretical methods.[3] Computations that would be very tedious to perform by hand can be performed by such software using a relatively small number of statements, and numerical and graphical results can be readily displayed. Programming control system engineering calculations on platforms other than MATLAB® often uses functions and syntax that are similar to those in the Control System Toolbox™. For purposes of brevity and compatibility between platforms, some programming details are omitted in the examples in this book.

References

1 Ortega, M. and Lin, L. (2004). Control theory applications to the production–inventory problem: a review. *International Journal of Production Research* 42 (11): 2303–2322.

2 Sarimveis, H., Patrinos, P., Tarantilis, C., and Kiranoudis, C. (2008). Dynamic modeling and control of supply chain systems: a review. *Computers & Operations Research* 35 (11): 3530–3561.

3 Ivanov, D., Dolgui, A., and Sokolov, B. (2012). Applicability of optimal control theory to adaptive supply chain planning and scheduling. *Annual Reviews in Control* 36 (1): 73–84.

4 Duffie, N., Chehade, A., and Athavale, A. (2014). Control theoretical modeling of transient behavior of production planning and control: a review. *Procedia CIRP* 17: 20–25. doi: 10.1016/j.procir.2014.01.099.

---

2 MATLAB® and Control System Toolbox™ are trademarks of The MathWorks, Inc. The reader is referred to the Bibliography and documentation available from The MathWorks as well as many other publications that address the use of MATLAB® and other software tools for control system analysis and design.

3 Other software such as Simulink®, a trademark of The MathWorks, Inc., facilitates modeling and time-scaled simulations. While such tools are commonly used by control system engineers, production engineers often find that discrete-event simulation software is more appropriate for detailed modeling of production systems. The reader is referred to the Bibliography and many publications that describe discrete-event and time-scaled simulation.

2

Continuous-Time and Discrete-Time Modeling of Production Systems

The dynamic behavior of a production system is the result of the combined dynamic behavior of its components including the decision-making components that implement decision rules. Production system behavior is not simply the sum of component behaviors, and it only can be understood and modeled by considering the structure of the production system, the nature of interconnections between individual components, and dynamic behavior that results from these interactions. In this chapter, methods for control theoretical modeling of the dynamic behavior of production systems are introduced, both for continuous-time and discrete-time production systems and their components. Then, in subsequent chapters, methods will be introduced that can be used to combine models of production system components and design control components and decision rules that result in desired production system dynamic behavior.

A control theoretical dynamic model of a production system or component is, in the continuous-time case, a set of differential and algebraic equations or, in the discrete-time[1] case, a set of difference and algebraic equations that describe how the time-varying outputs of the production system or component are related to its time-varying inputs. The mathematical methods that will be introduced in subsequent chapters are valid for linear models, and development of these models is the focus of this chapter. Because many production system components have behavior that is at least to some extent nonlinear, linearization of models around operating points and linearization using piecewise approximation are described. Some important dynamic attributes of models of production systems and components are introduced including delay, integration, and time constants.

Steps in control theoretical modeling of a production system and its components can be summarized as follows:

- *Make appropriate assumptions*: The level of detail with which production systems and their components can be modeled is limited by practical and theoretical considerations. Aggregated models may, for example, focus on the flow of orders through a production system. The amount of work that has been done may be represented,

1 The term "discrete-time" differentiates this type of model from discrete-event simulation models.

Control Theory Applications for Dynamic Production Systems: Time and Frequency Methods for Analysis and Design, First Edition. Neil A. Duffie.

but not the handling and processing of individual orders. This can facilitate understanding of fundamental behavior, which may be important in the initial design of a production system and its decision-making components. There is a tradeoff: a complex model often can make it difficult to recognize and design fundamental dynamic behavior, but the fidelity of a model is directly affected by the assumptions and simplifications that have been made.
- *Understand the physics of the production system and its components*: Mathematical relationships need to be found that describe input–output relationships in a production system (production process, machine, work system, factory, production network, etc.) and its components. Inputs, outputs, and internal variables need to be identified. Fundamental principles (logistical, mechanical, electrical, chemical, thermal, etc.) or experiments can be used to obtain these relationships.
- *Linearize relationships if necessary*: The models that are obtained must be linear to enable subsequent analysis using control theoretical methods. Linearization is often performed using selected operating points, and care must be taken in using the linearized models that are obtained: they are relatively accurate when variables are in the vicinity of the operating points and are relatively inaccurate when variables deviate from the operating points.
- *Develop linear algebraic equations and differential or difference equations*: These equations can be transformed using methods described in Chapter 3 to facilitate combining models of components into a model of an entire production system and selecting and designing control actions that are implemented in decision-making components.
- *Simplify the model*: Often, the fundamental dynamic behavior of a production system can be adequately described by a subset of the mathematical elements in a dynamic model that has been obtained. For example, it may be possible to eliminate insignificant time delays to reduce complexity. Again, there is a tradeoff: even though relatively complex models can be handled both theoretically and by control system engineering software, relatively simple models often provide better insight for guiding the design of control components and their decision rules.
- *Verify model fidelity and modify if necessary*: A model that has been obtained should be thoroughly examined to ensure that it adequately represents the important aspects of the dynamic behavior of the production system. Both numerical and physical experiments can be helpful in comparing modeled behavior to actual behavior. Model linearization and simplification may adversely affect the fidelity of the model outside a given range of some variables or at smaller time scales. The model may need to be improved to enable successful subsequent use in design of decision-making and prediction of resulting system dynamic behavior.

The model obtained is highly dependent on the physical and logistical nature of the production system and the components being modeled. The utility of the model is highly dependent on the nature of the analyses that subsequently will be performed and the decision-making components that are to be designed as a result. For this reason, past experience in modeling, analysis and design plays a significant role in anticipating the model that is required. Furthermore, models usually need to be modified during subsequent steps of analysis and design: additional features may need to be added, additional simplifications may need to be made, and fidelity may need to be improved. Additionally, the nature of the decisions that are made in production systems, as well as the structure of the information and communication that is required to support decision-making may change as the result of understanding gained during analysis and design; this can require further model modification.

2.1 Continuous-Time Models of Components of Production Systems

Variables in continuous-time models have a value at all instants in time. Many physical variables in production systems are continuous variables even though they may change abruptly. Examples include work in progress (WIP), lead time, demand, and rush orders. Continuous-time modeling results in differential and algebraic equations that describe input–output relationships at all instants in time t. Although the specific output function of time that results from a specific input function of time often is of interest, the goal is to obtain models that are valid for any input function of time or at least a broad range of input functions of time because the operating conditions for production systems and their components can be varying and unpredictable.

Example 2.1 Continuous-Time Model of a Production Work System with Disturbances

For the production work system illustrated in Figure 2.1 it is desired to obtain a continuous-time model that predicts work in progress (WIP) $w_w(t)$ hours as a function of the rate of work input to the work system $r_i(t)$ hours/day, the nominal production capacity $r_p(t)$ hours/day, WIP disturbances $w_d(t)$ hours and capacity disturbances $r_d(t)$ hours/day. WIP disturbances can be positive or negative due to rush orders and order cancellations, while capacity disturbances usually are negative because of equipment failures or worker absences. Units of hours of work are chosen rather than orders or items, and units of time are days.

The rate of work output by the production system $r_o(t)$ hours/day is

$$r_o(t)=r_p(t)+r_d(t)$$

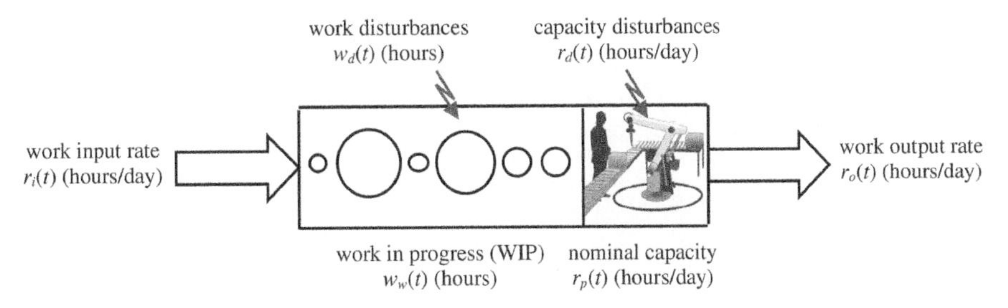

Figure 2.1 Continuous variables in a continuous-time model of a production work system.

The WIP is

$$w_w(t) = w_w(0) + w_d(t) + \int_0^t \left(r_i(t) - r_o(t)\right) dt$$

where $w_w(0)$ hours is the initial WIP. Integration often is an element of models of production systems and their components. The corresponding differential equation is

$$\frac{dw_w(t)}{dt} = \frac{dw_d(t)}{dt} + r_i(t) - r_p(t) - r_d(t)$$

The dynamic behavior represented by this continuous-time model is illustrated in Figure 2.2 for a case where there is a capacity disturbance $r_d(t)$ of −10 hours/day that starts at time $t = 0$ and lasts until $t = 3$ days. The initial WIP is $w_w(t) = 30$ hours for $t \leq 0$ days. The rate of work input is the same as the nominal production capacity, $r_i(t) = r_p(t)$ hours/day, and there are no WIP disturbances: $w_d(t) = 0$ hours. The response of WIP to the capacity disturbance is shown in Figure 2.2 and is calculated using Program 2.1 for $0 \leq t \leq 3$ days using the known solution[2]

$$w_w(t) = w_w(0) - r_d(0)t$$

This model represents the production work system using the concept of work flows rather than representing the processing of individual orders. Numerous aspects of real work system operation are not represented such as setup times, operator skills and experience, reduction in actual capacity due to idle times when the work in progress is low, and physical limits on variables. Also, WIP cannot be negative and often cannot be greater than some maximum due to buffer size. Capacity cannot be negative and cannot be greater than some maximum that is determined by physical characteristics such as the number of workers, number of shifts, available equipment, and available product components or raw materials.

Program 2.1 WIP response calculated using solution of differential equation

```
ww0=30;  % initial WIP (hours)
rd0=-10;  % capacity disturbance (hours/day)

t(1)=-2; rd(1)=0; ww(1)=ww0;  % initial values
t(2)=0; rd(2)=rd0; ww(2)=ww(1);  % disturbance starts
t(3)=3; rd(3)=0; ww(3)=ww(2)-3*rd0;  % solution of differential equation
t(4)=6; rd(4)=0; ww(4)=ww(3);  % disturbance ends

stairs(t,rd); hold on  % plot disturbance and WIP vs time - Figure 2.2
plot(t,ww); hold off
xlabel('time t (days)')
legend ('capacity disturbance r_d(t) (hours/day)','WIP w_w(t) (hours)')
```

2 Here and in subsequent examples, well-known solutions for the differential equations obtained are not derived. The tools of control system engineering that are presented in subsequent chapters generally make it unnecessary in practice for production engineers to find such solutions.

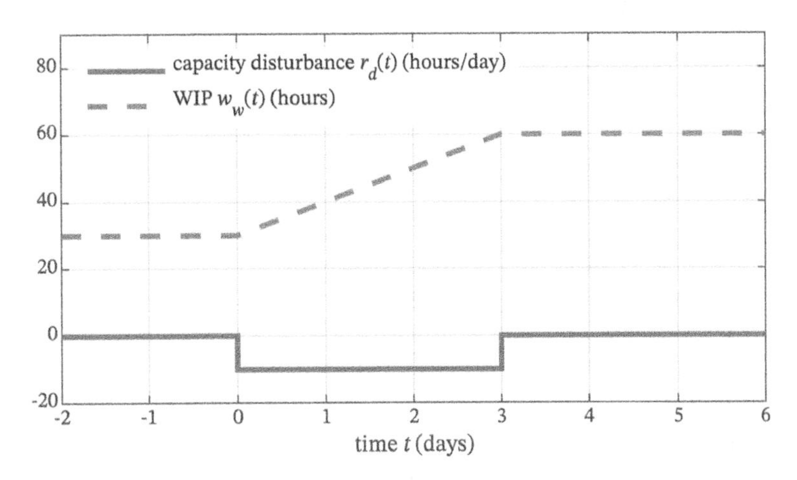

Figure 2.2 Response of WIP to 3-day capacity disturbance.

Example 2.2 Continuous-Time Model of Backlog Regulation in the Presence of Rush Orders and Canceled Orders

A continuous-time model is needed for the production system illustrated in Figure 2.3 that would facilitate design of an order release rate adjustment decision rule that tends to maintain backlog at a planned level. This type of decision-making can be referred to as backlog regulation, and it could operate manually or automatically. There are fluctuations in order input rate and disturbances in the form of rush orders and canceled orders that must be responded to by adjusting the order release rate in a manner that tends to eliminate deviations of actual backlog from planned backlog.

There are two main components: accumulation of orders in the backlog, and order release rate decision-making. The differential equation that describes backlog is similar to that developed in Example 2.1:

$$\frac{dw_b(t)}{dt} = \frac{dw_d(t)}{dt} + r_i(t) - r_o(t)$$

where $w_b(t)$ orders is the order backlog, $w_d(t)$ orders represents disturbances such as rush orders and order cancelations, $r_i(t)$ orders/day is the order input rate, and $r_o(t)$ orders/day is the order release rate.

There are many possible decision rules that can be used to implement backlog regulation. One straightforward option is to adjust the order release rate $r_o(t)$ orders/day as a function of the both difference between planned backlog and actual backlog and the integral of that difference. This decision rule is described by

$$r_o(t) = K_1\left(w_b(t) - w_p(t)\right) + K_2 \int_0^t \left(w_b(t) - w_p(t)\right) dt$$

or

$$\frac{dr_o(t)}{dt} = K_1 \frac{d\left(w_b(t) - w_p(t)\right)}{dt} + K_2\left(w_b(t) - w_p(t)\right)$$

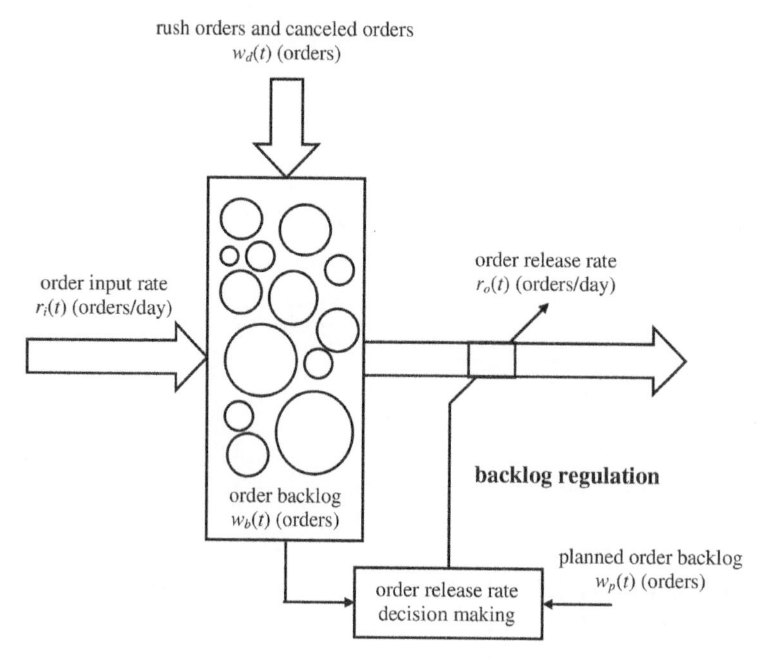

Figure 2.3 Backlog regulation in the presence of rush orders and canceled orders.

where $w_p(t)$ orders is the planned order backlog and K_1 days^{-1} and K_2 days^{-2} are backlog regulation decision parameters for which values are selected that result in favorable closed-loop backlog regulation dynamic behavior. Knowledge of order input rate $r_i(t)$ orders/day and work disturbances $w_d(t)$ orders is not required in this decision rule; instead, work disturbances and fluctuations in order input rate cause backlog $w_b(t)$ orders to increase or decrease, and deviations from planned backlog $w_p(t)$ orders subsequently are responded to using the decision rule.

Combining these component models to obtain a complete system model that relates backlog to the various inputs would be straightforward in this case, but this is easily and generically done using the transformation methods described in Chapter 3. Analysis using this combined model would allow selection of a combination of values of decision rule parameters K_1 and K_2 that satisfy requirements for dynamic behavior such as quick return of backlog to plan after a rush order or canceled order disturbance. Other options for the form of the decision rule could result in significantly different and possibly improved regulation of backlog.

Example 2.3 Continuous-Time Model of Mixture Temperature Regulation using a Heater

Experimental results can be used to obtain component models. Consider a portion of a production process in which it is necessary to deliver a mixture at a desired temperature. A heater at the outlet of a pipe is used to raise the temperature of the mixture flowing in the pipe to the desired level. Predicting the heater voltage required to deliver the mixture at the correct temperature is unlikely to be successful because of uncertainty in mixture inlet temperature. Therefore, as shown in Figure 2.4, a closed-loop

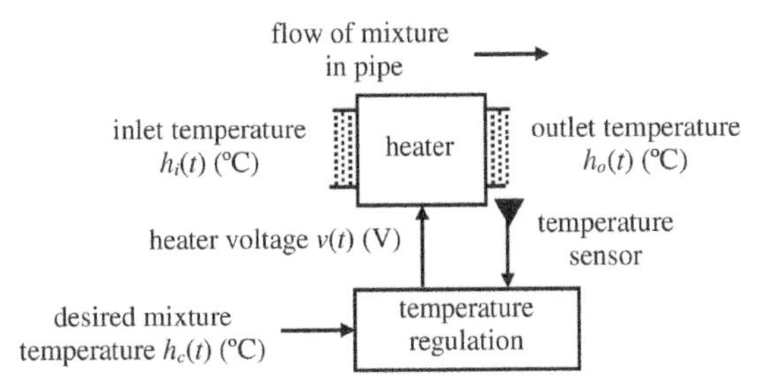

Figure 2.4 Mixture outlet temperature regulation in which a heater is used to raise the temperature of a mixture to a desired temperature.

temperature regulation approach is used in which a temperature sensor is placed at the end of the pipe and feedback from this sensor is used to adjust the heater voltage until the desired mixture temperature is obtained.

The experimental results in Figure 2.5 show how the temperature of the mixture at the outlet of the pipe changes as a function of heater voltage and time. The temperature of the mixture at the inlet of the pipe is assumed to be constant throughout the test: $h_i(t) = h_i(0)$ °C for all t seconds. Initially, the heater voltage has been $v(t) = 0$ V for a long enough time to ensure that the temperature of the mixture at the outlet of the pipe $h_o(t)$ °C is the same as the temperature of the mixture at the inlet: $h_o(0) = h_i(0)$ °C. A constant heater voltage $v(t) = 50$ V is applied for time $0 \le t$ seconds, and the temperature of the mixture at the outlet of the pipe is measured for the period of 400 seconds. The change in temperature[3] of the mixture $\Delta h(t)$ °C then is calculated where

$$\Delta h(t) = h_o(t) - h_i(t)$$

The temperature of the mixture changes relatively rapidly at the beginning of the experiment as shown Figure 2.5, but has reached a final value at the end of the experiment. This behavior can be characterized by the relationship

$$\tau \frac{d\Delta h(t)}{dt} + \Delta h(t) = K_h v(t)$$

where time constant τ seconds characterizes how quickly temperature difference $\Delta h(t)$ °C changes in response to heater voltage $v(t)$ V and constant of proportionality K_h °C/V relates the final temperature difference to the applied heater voltage; K_h can be referred to as the mixture heating parameter.

The known solution of this differential equation for constant input $v(t) = v(0)$ V and initial condition $\Delta h(t) = 0$ is

$$\Delta h(t) = K_h \left(1 - e^{-t/\tau}\right) v(0)$$

---

3 It often is convenient to use relative change as a variable in dynamic models of the components of production systems.

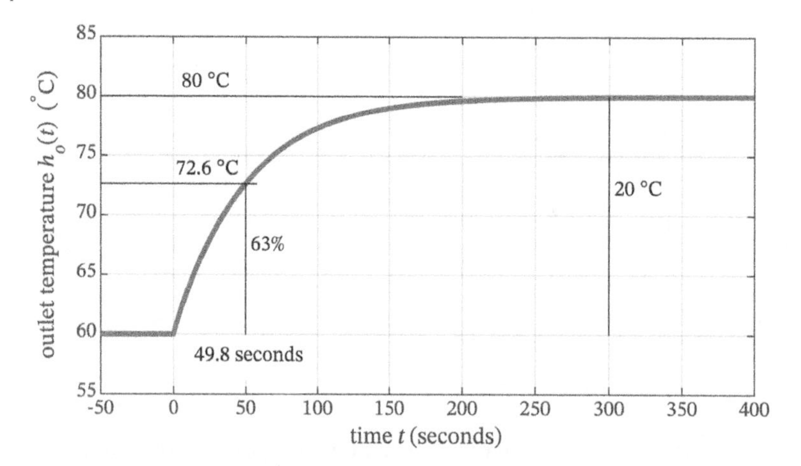

Figure 2.5 Experimental results obtained by applying a constant heater voltage $v(t)$ = 50 V starting at time t = 0 seconds and measuring the outlet temperature $h_o(t)$ °C when the inlet temperature is constant $h_i(t)$ = 60°C.

and when $t = \tau$ seconds,

$$\Delta h(\tau) = K_h\left(1 - e^{-1}\right)v(0) = 0.632K_h v(0)$$

The estimated time constant $\tau = 49.8$ seconds therefore can be obtained by noting the time in Figure 2.5 when approximately 63% of the final change in temperature is reached.

The estimate of the value of mixture heating parameter K_h can be obtained from the ratio of the constant final change in temperature in Figure 2.5 to the constant voltage applied to the heater: $K_h = 20/50 = 0.4$°C/V. The model of mixture heating then is approximately

$$49.8\frac{d\Delta h(t)}{dt} + \Delta h(t) = 0.4v(t)$$

One option for the decision rule used in the mixture temperature regulation component is

$$\frac{dv(t)}{dt} = K_c\left(h_c(t) - h_o(t)\right)$$

where $h_c(t)$ °C is the desired temperature of the mixture at the outlet of the pipe and K_c (V/second)/°C is a voltage adjustment decision parameter. This decision rule causes the heater voltage to continually change until $h_o(t) = h_c(t)$ °C. Choosing a relatively large value of K_c causes heater voltage and outlet temperature to change more quickly but tends to require higher heater voltages and can result in oscillatory outlet temperature behavior. Choosing a relatively small value of K_c causes the heater voltage to change less quickly and when the inlet temperature of the mixture fluctuates, longer-lasting deviations in outlet temperature can result. The analysis and design methods described in subsequent chapters facilitate selection of a value for temperature regulation parameter K_c and assessment of the appropriateness of this decision rule.

2.2 Discrete-Time Models of Components of Production Systems

Variables in discrete-time models have a value only at discrete instants in time separated by a fixed time interval T. While many physical variables in production systems are fundamentally continuous, they often are sampled, calculated, or changed periodically. Examples include work in progress (WIP) measured manually or automatically at the beginning of each day, mean lead time calculated at the end of each month, and production capacity adjusted at the beginning of each week. Discrete-time modeling results in difference and algebraic equations that describe input–output relationships and represent the behavior of a production system at times kT where k is an integer.

Example 2.4 Discrete-Time Model of a Production Work System with Disturbances

The production work system illustrated in Figure 2.6 can be represented by a discrete-time model that predicts work in progress (WIP) at instants in time separated by fixed period T. This period could, for example, be one week (T = 7 days), one day (T = 1 day), one shift (T = 1/3 day) or one hour (T = 1/24 day). The modeled values of work in progress then are $w_w(kT)$ hours. If it is assumed that work input rate $r_i(t)$ hours/day, nominal capacity $r_p(t)$ hours/day, and capacity disturbances $r_d(t)$ hours/day are constant (or nearly constant) over each period $kT \le t < (k+1)T$ days, the work output rate at time kT days then is

$$r_o(kT) = r_p(kT) + r_d(kT)$$

and the WIP is

$$w_w((k+1)T) = w_w(kT) - w_d(kT) + w_d((k+1)T) + T(r_i(kT) - r_p(kT) - r_d(kT))$$

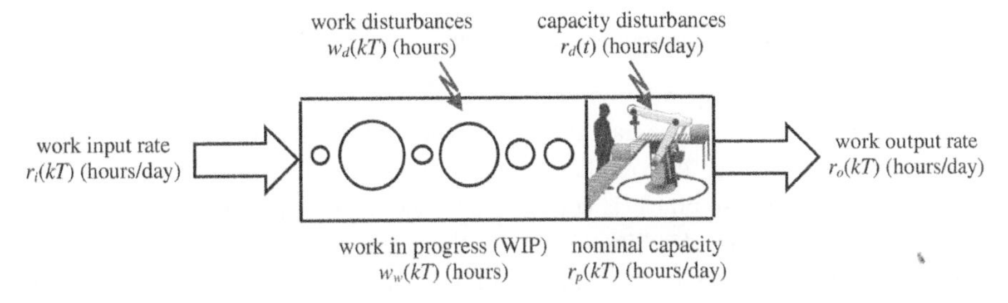

Figure 2.6 Discrete variables in a discrete-time model of a production work system.

This discrete-time model only represents the values of WIP $w_w(kT)$ at instants in time separated by period T; values between these instants are not represented. When the inputs are constant during period $kT \le t < (k+1)T$, it is clear from the model obtained in Example 2.1 that $w_w(t)$ increases or decreases at a constant rate over that period; however, this information is not contained in the discrete-time model.

The dynamic behavior represented by this discrete-time model is illustrated in Figure 2.2 for a case where $T = 1$ day and there is a capacity disturbance $r_d(kT)$ of −10 hours/day that starts at time $kT = 0$ and lasts until $kT = 3$ days. The initial WIP is $w_w(kT) = 30$ hours for $kT \le 0$ days. The rate of work input is the same as the nominal production capacity, $r_i(kT) = r_p(kT)$, and there are no WIP disturbances: $w_d(kT) = 0$. In this case, the difference equation for WIP can be written as

$$w_w(kT) = w_w((k-1)T) - Tr_d((k-1)T)$$

The response of WIP to the capacity disturbance, with the other inputs constant, is shown in Figure 2.7 and calculated recursively in Program 2.2 using the this difference equation. In Figure 2.7, the responses are denoted by the discrete values at times kT as well as by a staircase plot. The latter is used by convention to indicate that no information is present in the discrete-time model regarding the WIP between the instants in time separated by period T.[4]

Program 2.2 WIP Response calculated recursively using difference equation

```
T=1;   % discrete period (days)
kT=[-2,-1,0,1,2,3,4,5,6];   % times kT (days)
rd=[0,0,-10,-10,-10,0,0,0,0];   % capacity disturbance at times kT
(hours/day)

ww(1)=30;   % initial WIP at time kT=-2 days (hours)
for k=-2:5   % instants in time kT between kT=-2 and kT=5 days
    ww(k+4)=ww(k+3)-T*rd(k+3);   % next WIP
end

stairs(kT,rd); hold on   % disturbance and WIP vs kT - Figure 2.7
stairs(kT,ww); hold off
xlabel('time kT [days]')
legend ('capacity disturbance r_d(t) (hours/day)','WIP w_w(t) (hours)')
```

Example 2.5 Discrete-Time Model of Planned Lead Time Decision-Making

The decision-making component shown in Figure 2.8 is used to calculate periodic adjustments that increase or decrease the lead time used to plan operations in a production system. Lateness of order completion can negatively affect production operations and customer satisfaction, and it is common practice to increase planned lead times when the trend is to miss deadlines. On the other hand, planned lead times can be decreased when the trend is to complete orders early; this can be a competitive

4 Henceforth, staircase plots will be presented without explicitly denoting discrete values at times kT.

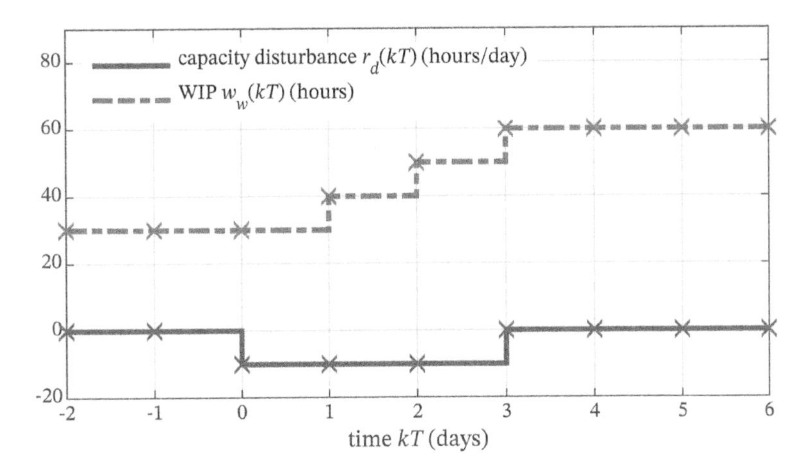

Figure 2.7 Response of WIP to a 3-day capacity disturbance; each discrete value is denoted with an X.

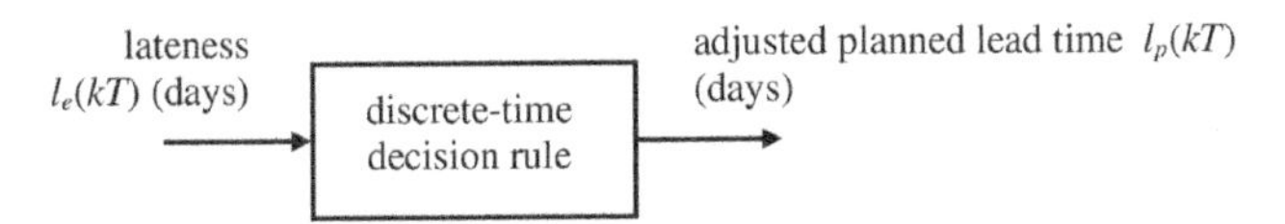

Figure 2.8 Discrete-time decision-making component for adjusting planned lead time in a production system as a function of lateness of order completion.

advantage because earlier due dates can be promised when customers are placing or considering placing orders.

An example of a discrete-time decision rule that could be used periodically to adjust planned lead time is

$$l_p(kT) = l_p((k-1)T) + \Delta l_p(kT)$$

$$\Delta l_p(kT) = l_e(kT) + K_l\left(\frac{l_e(kT) - l_e((k-1)T)}{T}\right)$$

where $l_p(kT)$ days is the planned lead time, $\Delta l_p(kT)$ days is the change in planned lead time, $l_e(kT)$ days is a measure of lateness that could be obtained statistically from recent order due date and completion time data, and K_l weeks is a decision-making parameter that needs to be designed to obtain favorable dynamic behavior of the production system into which the decision-making component is incorporated. T weeks is the period between adjustments. This decision rule both increases planned lead time when orders are late and also increases planned lead time when lateness is increasing; the contribution of the latter is governed by the choice of parameter K_l.

The dynamic behavior of the component is illustrated in Figure 2.9b for the case shown in Figure 2.9a where lateness $l_e(kT)$ increases to 8 days over a period of 6 weeks. The response of change in planned lead time to this increase in lateness is shown in

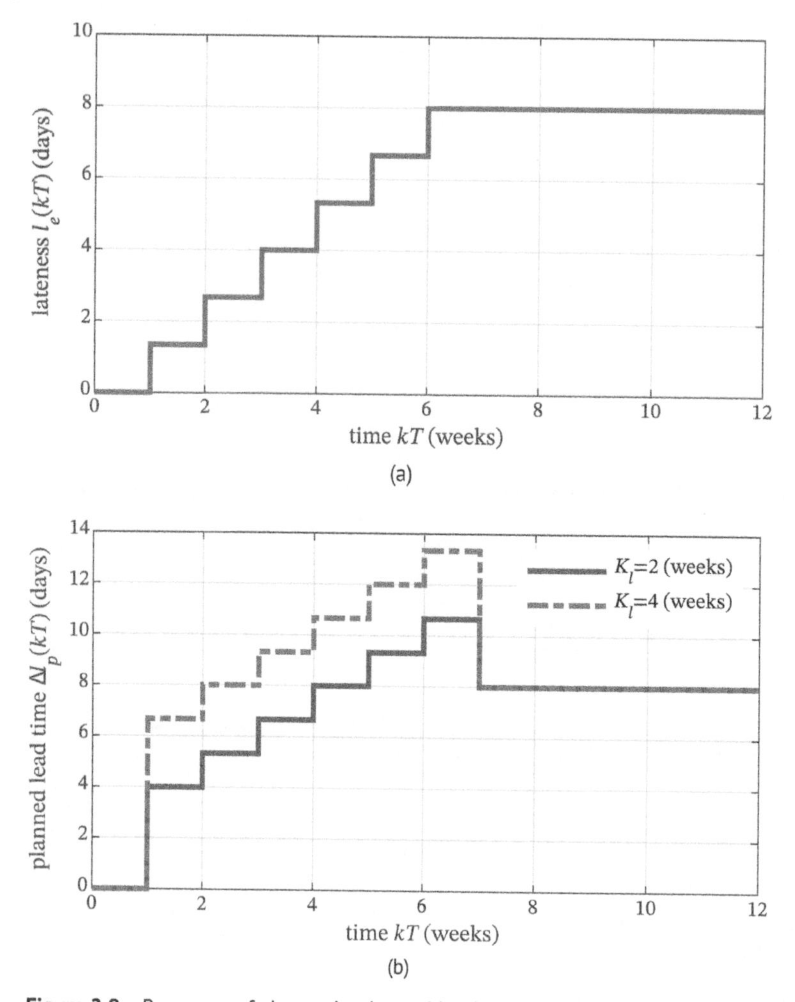

Figure 2.9 Response of change in planned lead time to lateness in order completion.

Figure 2.9b for $K_l = 2$ weeks and $K_l = 4$ weeks. The period between adjustments is $T =$ 1 week. The response was calculated recursively using the above difference equation in a manner similar to that shown in Program 2.2. As expected, the larger value of K_l results in larger adjustments when lateness is increasing, a stronger response to this trend in lateness.

Example 2.6 Exponential Filter for Number of Production Workers to Assign to a Product

The exponential filter shown in shown in Figure 2.10 is used in a component of a production system to make periodic decisions regarding the workforce that should be assigned to a product when there are fluctuations in demand for the product. The exponential filter has a weighting parameter $0 < \alpha \leq 1$ that determines how significantly the amplitudes of higher-frequency fluctuations in number of workers are reduced with respect to the amplitudes of fluctuations in demand. This reduction is

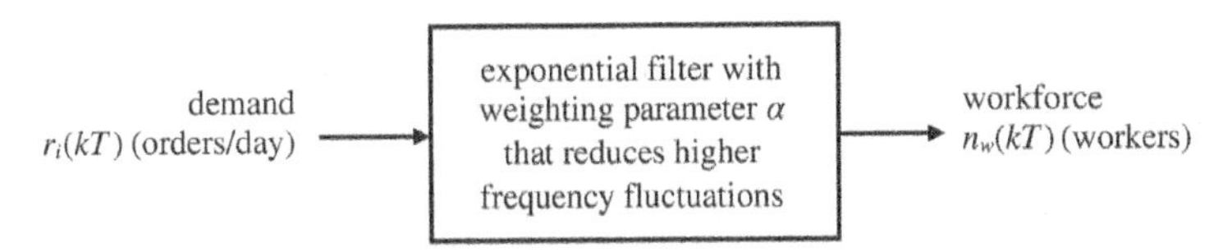

Figure 2.10 Exponential filter for smoothing demand to determine the number of production workers to assign to a product.

important because making rapid, larger amplitude changes in the number of workers is likely to be costly and logistically difficult.

The discrete-time equation for the filter is

$$n_w(kT) = (1-\alpha) n_w((k-1)T) + \alpha K_w r_i(kT)$$

where $n_w(kT)$ is the number of workers, $r_i(kT)$ orders/day is the demand, K_w workers/(orders/day) is the fraction of a worker's day required for an order, and T days is the period between calculations of the number of workers to assign to the product.

For weighting parameter $\alpha = 0.1$, $K_w = 1/8$ workers/(orders/day), and $T = 10$ days the response of number of workers is shown in Figure 2.11b for the fluctuation in demand shown in Figure 2.11a. The response was calculated recursively using the above difference equation in a manner similar to that shown in Program 2.2. The relatively small value of α results in significant smoothing of the number of workers with respect to the fluctuations in demand.

2.3 Delay

Delays are common in production systems and sources of delay include data gathering and communication, decision-making and implementation, setup times, processing times, and buffers. For example, decisions may not be made until sometime after relevant information is obtained, and, for logistical reasons, implementation of decisions may not be immediate. Disturbances may not have immediate effects, and these effects may not be detected until they have propagated through a production system. Delays often are detrimental and limit achievable performance; therefore, it is important to include delays in models when they are significant.

Example 2.7 Continuous-Time Model of Delay in a Production System

In the example illustrated in Figure 2.12, Company A obtains unfinished orders from a supplier, distant Company B, and then performs the work required to finish them. Both companies are assumed to process orders at the same rate as they are received. Unfinished orders are shipped from Company B to Company A, and the time between an order leaving Company B and arriving at Company A is a constant D days. Companies A and B have lead times L_A and L_B days, respectively, which is the time between when the company receives an order and when the company has completed processing the order; lead times are assumed to be constant and can be modeled as delays.

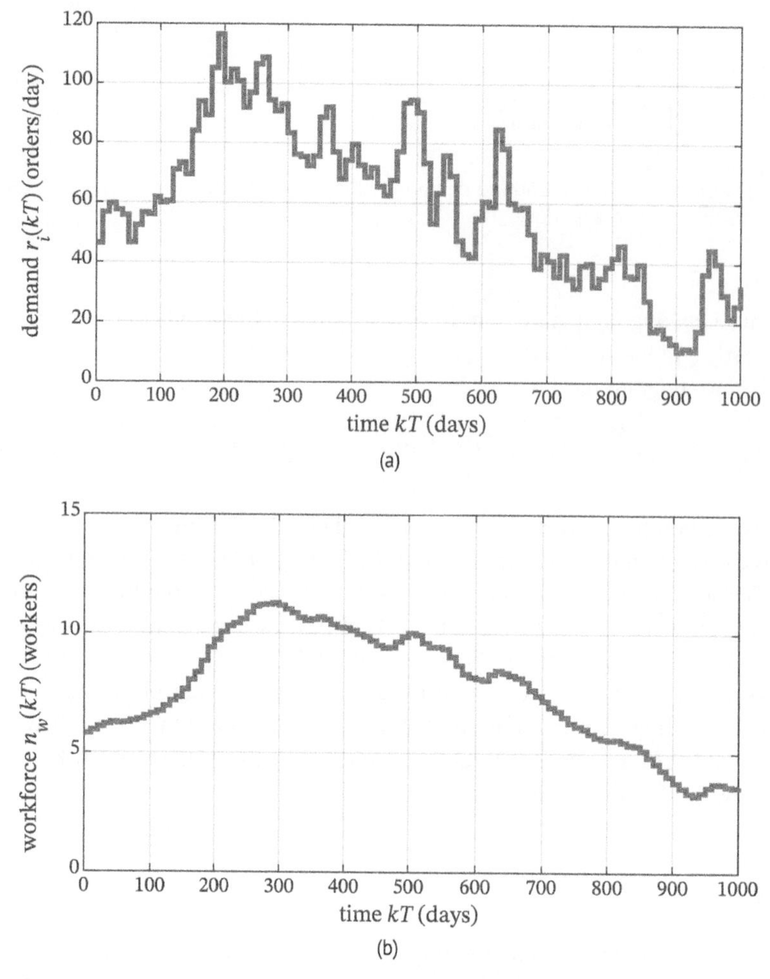

Figure 2.11 Response of desired workforce to fluctuations in demand.

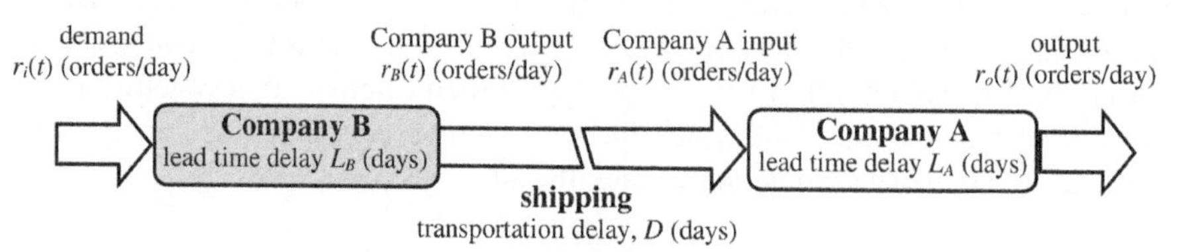

Figure 2.12 Lead time and transportation delays in a two-company production system.

If the order input rate to Company B is demand $r_i(t)$ orders/day and the order input rate to Company A is $r_A(t)$ orders/day, the order output rates from Companies B and A, $r_B(t)$ and $r_o(t)$, respectively, are

$$r_B(t) = r_i(t - L_B)$$

$$r_o(t) = r_A(t - L_A)$$

Shipping is described by

$$r_A(t) = r_B(t - D)$$

Combining the delays, the relationship between demand and the completed order output rate of Company A is

$$r_o(t) = r_i(t - L_B - D - L_A)$$

Example 2.8 Discrete-Time Model of Assignment of Production Workers with Delay

The rate of orders input to a production system often fluctuates and it is necessary to adjust production capacity to follow this order input rate. The use of a discrete-time exponential filter in decision-making to smooth fluctuations was described in Example 2.6 and as shown in Figure 2.13, an exponential filter is used in a similar manner in this example to determine the portion of the production capacity to be provided by permanent workers; this portion cannot be adjusted quickly and should not be adjusted at high frequencies. The remaining portion is provided by cross-trained workers; this portion can be adjusted immediately.

Order input rate $r_i(kT)$ orders/day is measured regularly with a period of T days, weekly for example, and the portion of production capacity provided by permanent workers $r_p(kT)$ orders/day is adjusted; however, because of logistical issues in hiring and training, there is a delay of dT days in implementing permanent worker adjustment decisions where d is a positive integer. The exponential filter is used to focus adjustments in permanent worker capacity on relatively low frequencies:

$$r_f(kT) = \alpha r_i(kT) + (1 - \alpha) r_f((k-1)T)$$

where $0 < \alpha \leq 1$. A relatively high value of weighting parameter α results models relatively rapid adjustment of permanent worker capacity, whereas a relatively low value

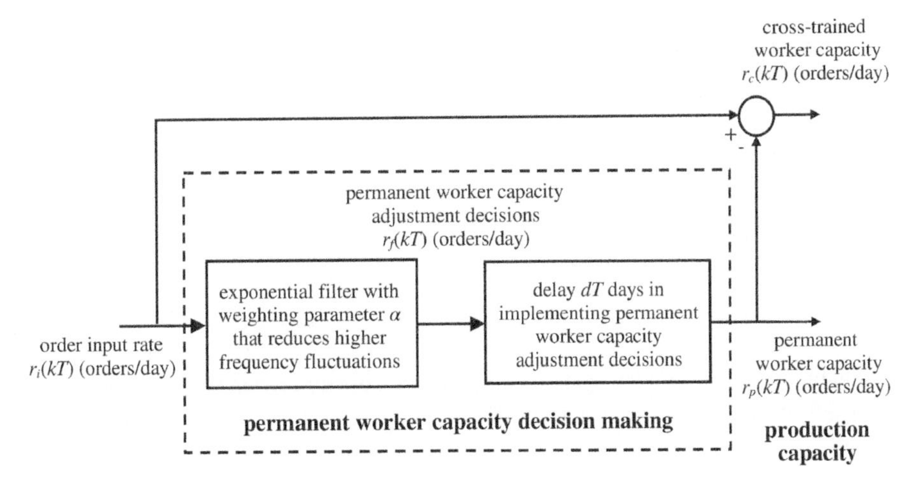

Figure 2.13 Adjustment of permanent and cross-trained worker capacity in a production system to match fluctuating order input rate.

of weighting parameter α models significant smoothing and relatively slow adjustment of permanent worker capacity.

The portion of production capacity provided by permanent workers is

$$r_p\left((k+d)T\right)=r_f\left(kT\right)$$

where dT days is the delay in implementing permanent worker capacity adjustments. Hence, the portions of fluctuating order input that are addressed by permanent worker capacity $r_p(kT)$ orders/day and cross-trained capacity $r_c(kT)$ orders/day are

$$r_p\left((k+d)T\right)=\alpha r_i\left(kT\right)+\left(1-\alpha\right)r_p\left((k-1+d)T\right)$$

$$r_c\left(kT\right)=r_i\left(kT\right)-r_p\left(kT\right)$$

2.4 Model Linearization

A component behaves in a linear manner if input x_1 produces output y_1, input x_2 produces output y_2, and input $x_1 + x_2$ produces output $y_1 + y_2$. The following are examples of linear relationships:

$$y\left(t\right)=Kx\left(t\right)$$

$$\frac{dy\left(t\right)}{dt}=Kx\left(t\right)$$

$$y\left((k+1)T\right)=Kx\left(kT\right)$$

The following are examples of nonlinear relationships:

$$y\left(t\right)=Kx\left(t\right)v\left(t\right)$$

$$\frac{dy\left(t\right)}{dt}=Kx\left(t\right)^2$$

$$y\left((k+1)T\right)=Kx\left(kT\right)\left(1-x\left((k-1)T\right)\right)$$

$$\left\{\begin{array}{c} Kx\left(t\right)\geq y_{max} : y\left(t\right)=y_{max} \\ y_{min}<Kx\left(t\right)<y_{max} : y\left(t\right)=Kx\left(t\right) \\ Kx\left(t\right)\leq y_{min} : y\left(t\right)=y_{min} \end{array}\right\}$$

In reality, most production system components have nonlinear behavior, but often the extent of this nonlinearity is insignificant and can be ignored, with care, when a model is formulated. On the other hand, behavior that is significantly nonlinear often can be modeled in a simpler but sufficiently accurate manner using approximate linear models obtained using approaches such as those described in the following subsections.[5]

2.4.1 Linearization Using Taylor Series Expansion – One Independent Variable

A nonlinear function $f(x)$ of one variable x can be expanded into an infinite sum of terms of that function's derivatives evaluated at operating point x_o:

$$f(x)=f(x_o)+\frac{1}{1!}(x-x_o)\left.\frac{df}{dx}\right|_{x_o}+\frac{1}{2!}(x-x_o)^2\left.\frac{d^2f}{dx^2}\right|_{x_o}+\cdots \quad (2.1)$$

x_o is the operating point about which the expansion made. Over some range of $(x - x_o)$ higher-order terms can be neglected, and the following linear model in the vicinity of the operating point is a sufficiently good approximation of the function:

$$f(x)\approx f(x_o)+K(x-x_o) \quad (2.2)$$

where

$$K=\left.\frac{df}{dx}\right|_{x_o} \quad (2.3)$$

Such an approximation is illustrated in Figure 2.14.

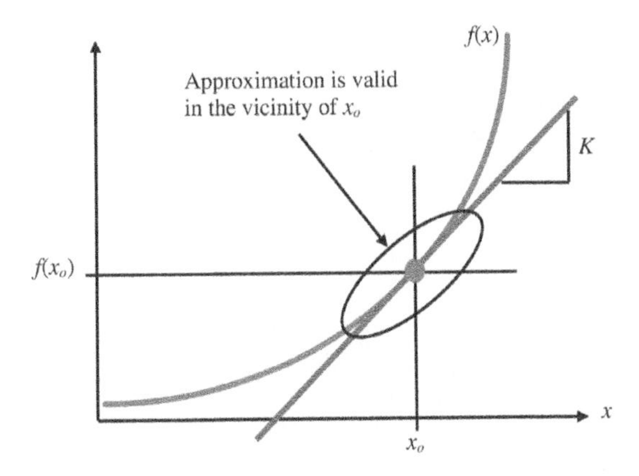

Figure 2.14 Linear approximation of function $f(x)$ at operating point x_o.

---

5 The reader is referred to the Bibliography and many other publications on nonlinear dynamics and nonlinear control theory for other approaches to modeling nonlinear behavior.

Example 2.9 Production System Lead Time when WIP Is Constant and Capacity Is Variable

A production work system such as that illustrated in Figure 2.15 has constant work in progress (WIP) w hours and variable production capacity $r(t)$ hours/day. The lead time $l(t)$ hours then is approximately

$$l(t) \approx \frac{w}{r(t)}$$

The relationship between lead time and capacity is nonlinear; however, a linear approximation of this relationship in the vicinity of operating point r_o can be obtained using Equations 2.2 and 2.3:

$$\left.\frac{dl}{dr}\right|_{r_o} = -\frac{w}{r_o^2}$$

$$l(t) \approx \frac{w}{r_o} - \frac{w}{r_o^2}\left(r(t) - r_o\right)$$

The percent error in lead time calculated using the linear approximation due to deviation of actual capacity $r(t)$ from the chosen capacity operating point r_o is shown in Figure 2.16 and calculated using

$$e_l(t) = 100 \times \left(\frac{\frac{w}{r(t)} - \left(\frac{w}{r_o} - \frac{w}{r_o^2}\left(r(t) - r_o\right) \right)}{\frac{w}{r(t)}} \right)$$

Clearly, capacity should not deviate significantly from the operating point if this approximation is used in a model. If, for example, lead time is to be regulated by adjusting capacity, capacity might vary significantly from the operating point that was used to design lead-time regulation decision rules. An option[6] in this case could be to

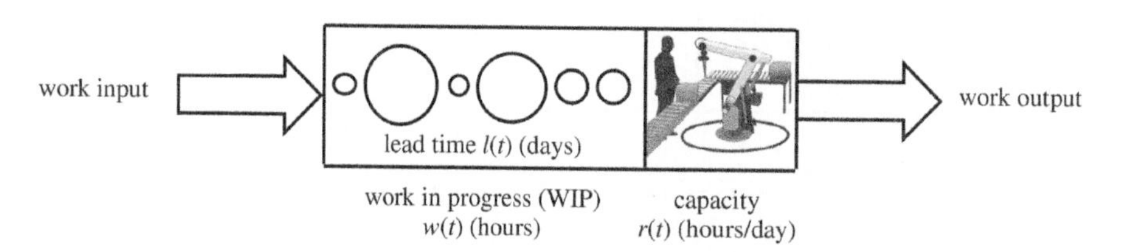

Figure 2.15 Production work system with variable capacity.

6 Other options include choosing to regulate a variable other than lead time (work in progress, due date deviation, etc.) and designing non-linear decision rules.

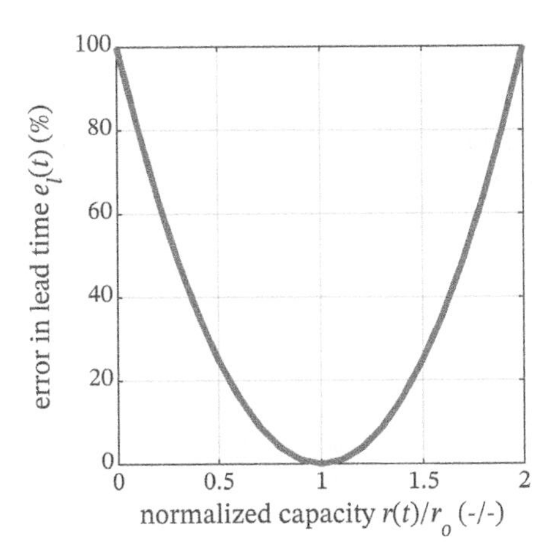

Figure 2.16 Percent error in lead time due to deviation of actual capacity from capacity operating point chosen for linear approximation.

- calculate the parameters for a linearized model for each of several capacity operating points
- design lead time regulation decision rules for each operating point using the model for that operating point
- switch between decision rules as operating conditions vary.

2.4.2 Linearization Using Taylor Series Expansion – Multiple Independent Variables

A nonlinear function $f(x,y,...)$ of several variables x, y, ... can be expanded into an infinite sum of terms of that function's derivatives evaluated at operating point x_o, y_o, ...:

$$\begin{aligned} f(x,y,...) = f(x_o,y_o,...) &+ \frac{1}{1!}(x-x_o)\left.\frac{\partial f}{\partial x}\right|_{x_o,y_o,...} + \frac{1}{2!}(x-x_o)^2\left.\frac{\partial^2 f}{\partial x^2}\right|_{x_o,y_o,...} + \cdots \\ &+ \frac{1}{1!}(y-y_o)\left.\frac{\partial f}{\partial y}\right|_{x_o,y_o,...} + \frac{1}{2!}(y-y_o)^2\left.\frac{\partial^2 f}{\partial y^2}\right|_{x_o,y_o,...} + \cdots \end{aligned} \tag{2.4}$$

Over some range of $(x - x_o)$, $(y - y_o)$, ... higher-order terms can be neglected and a linear model is a sufficiently good approximation of the nonlinear model in the vicinity of the operating point:

$$f(x,y,\ldots) \approx f(x_o,y_o,\ldots) + K_x(x-x_o) + K_y(y-y_o) + \cdots \tag{2.5}$$

where

$$K_x = \left.\frac{\partial f}{\partial x}\right|_{x_o,y_o,\ldots} \quad K_y = \left.\frac{\partial f}{\partial y}\right|_{x_o,y_o,\ldots} \quad \cdots \tag{2.6}$$

Example 2.10 Production System Lead Time when WIP and Capacity are Variable

In the case where the production work system illustrated in Figure 2.15 has variable work in progress (WIP) $w(t)$ hours and variable production capacity $r(t)$ hours/day, the lead time is

$$l(t) \approx \frac{w(t)}{r(t)}$$

For work in progress operating point w_o and capacity operating point r_o, an approximating linear function for lead time in the vicinity of operating point w_o, r_o, can be obtained using Equations 2.5 and 2.6:

$$l(t) \approx \frac{w_o}{r_o} + K_r\left(r(t) - r_o\right) + K_w\left(w(t) - w_o\right)$$

where

$$K_r = \left.\frac{\partial l}{\partial r}\right|_{w_o, r_o} = -\frac{w_o}{r_o^2}$$

$$K_w = \left.\frac{\partial l}{\partial w}\right|_{w_o, r_o} = \frac{1}{r_o}$$

2.4.3 Piecewise Approximation

In practice, variables in models of production systems may have a limited range of values. Maximum values of variables such as work in progress and production capacity cannot be exceeded, and these variables cannot have negative values. In many cases, operating conditions where limits have been reached may not be of primary interest when analyzing and designing the dynamic behavior of production systems. On the other hand, models can be developed that represent important combinations of operating conditions, each of which represents dynamic behavior under those specific conditions. A set of piecewise linear approximations then can be used to represent non-linear relationships between variables.

Example 2.11 Piecewise Approximation of a Logistic Operating Curve

The relationship between work in progress (WIP) and actual capacity shown in Figure 2.17 is another example of nonlinear behavior. When WIP $w(t)$ is relatively low in a production work system such as that shown in Figure 2.15, production capacity may not be fully utilized and actual capacity can be less than full capacity because of the work content of individual orders and the timing of arriving orders. Conversely, when WIP is relatively high, work is nearly always waiting and the work system is nearly fully utilized. The actual capacity $r_a(t)$ of the work system therefore may depend on both its full capacity r_f and the work in progress $w(t)$.

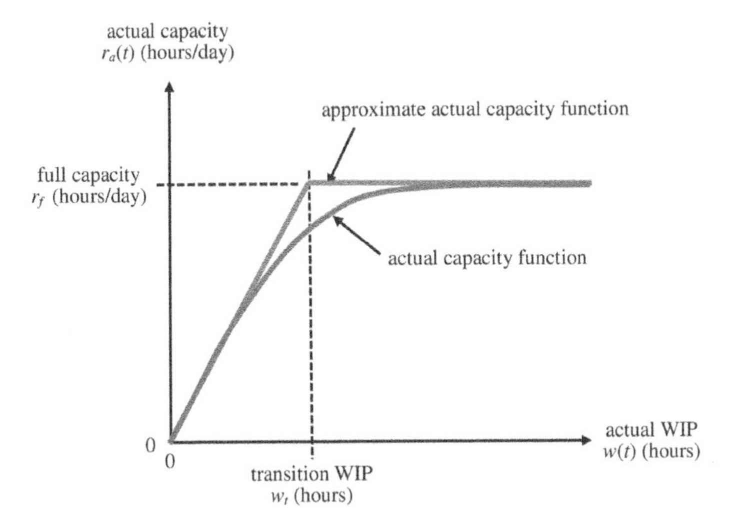

Figure 2.17 Actual production capacity function and a piecewise linear approximation.

As shown in Figure 2.17, the actual capacity function can be approximated in a piecewise manner by two segments, delineated by a WIP transition point w_t hours, where for $w(t) \geq w_t$

$$r_a(t) \approx r_f$$

and for $w(t) < w_t$

$$r_a(t) \approx \frac{r_f}{w_t} w(t)$$

2.5 Summary

The examples that were presented in this chapter illustrate some of the ways that models can be developed for production systems and components. Production systems often have many components, and dynamic models of these components need to be obtained individually using physical analysis or experimental data. Then they can be combined if desired to obtain a model of the dynamic behavior of an entire production system. In Chapter 3, transformation methods will be introduced that allow algebraic equations to be substituted for the linear differential and difference equations that describe the dynamic behavior of components in a production system. These transformations make combining models of components relatively easy and they are compatible with the many analysis and design tools that are implemented in control system engineering software. As also described in Chapter 3, transformed models of production system components can be assembled into block diagrams that graphically represent the input–output relationships between components in production systems, aiding in identifying, understanding, and designing the dynamic behavior of production systems.

3

Transfer Functions and Block Diagrams

Modeling as described in Chapter 2 results in a set of differential, difference, or algebraic equations that describe the dynamic behavior of the components and interactions between the components of a production system. Analysis of the dynamic behavior of a production system and design of decision-making rules that result in favorable dynamic behavior requires combining the models of the production system's components. However, it can be a challenge to combine these equations: there may be a mix of continuous-time and discrete-time models; the structure of the production system may be complicated and this structure will be present in the equations; and while low-order differential or difference equations may describe the dynamic behavior of individual components in a satisfactory manner, combined models may be of significantly higher order. Fortunately, Laplace transforms and Z transforms can be used to convert differential and difference equations, respectively, into algebraic forms that are easily manipulated and readily support mathematical analysis and design tools implemented in control system engineering software.

Transfer functions that represent cause-and-effect relationships between the components can be obtained after transformation of continuous-time and discrete-time models of components of production systems. The structure of these relationships is important and must be well understood in order to proceed to dynamic analysis of the complete production system and design of decision-making. Block diagrams often are used to graphically illustrate the physical and mathematical structure of production systems, and transfer functions are placed within blocks in a block diagram to clearly show how component inputs and outputs are dynamically and mathematically related.

The Laplace and Z transforms are defined in this chapter, and key properties are discussed that allow transfer functions to be readily obtained and manipulated. Although the most fundamental definitions rarely need to be applied in practice by production system engineers, it is necessary to thoroughly understand these theoretical underpinnings of control theory because the practical tools that support analysis and design are built upon them. Production systems can contain both continuous-time and discrete-time components, and an integrated presentation is taken in this chapter in addressing how transfer functions and system models can be obtained; a similar approach often is taken in control system engineering software. Consideration of delay also is integrated throughout this chapter because delay is common in production

Control Theory Applications for Dynamic Production Systems: Time and Frequency Methods for Analysis and Design, First Edition. Neil A. Duffie.

systems. Transfer functions and block diagrams will be used extensively in subsequent chapters in calculation of time and frequency response, evaluation of system stability, and design of decision-making. In this chapter, the focus is on fundamental definitions and illustration of their application using straightforward, practical examples.

3.1 Laplace Transform

The unilateral Laplace transform of a continuous function of time $f(t)$ is defined as

$$\mathcal{L}\{f(t)\} = \int_{0^-}^{\infty} f(t)\mathrm{e}^{-st}dt \tag{3.1}$$

where s is a new, complex variable with real part α and imaginary part β:

$$s = \alpha + \mathrm{j}\beta \tag{3.2}$$

A discontinuity at time $t = 0$ is included when evaluating the integral, hence the notation 0^- in Equation 3.1.

In the following examples, the definition of the Laplace transform in Equation 3.1 is used to obtain the transforms of a unit step function, an exponential function and an exponentially decaying sinusoidal function. These results are listed in Table 3.1 along with the results for several other commonly considered continuous functions of time. These results are well known, and Equation 3.1 therefore rarely needs to be applied in practice in production system engineering.

Table 3.1 Laplace transforms of common continuous functions of time.

	function	$f(t)$ $(f(t) = 0 \text{ for } t < 0)$	$\mathcal{L}\{f(t)\}$
1	unit step	1	$\frac{1}{s}$
2	exponential	$\mathrm{e}^{t/T}$	$\frac{1}{s + 1/T}$
3	sine	$\sin(\omega t)$	$\frac{\omega}{s^2 + \omega^2}$
4	cosine	$\cos(\omega t)$	$\frac{s}{s^2 + \omega^2}$
5	decaying sine	$\mathrm{e}^{-t/T}\sin(\omega t)$	$\frac{\omega}{\left(s + 1/T\right)^2 + \omega^2}$
6	decaying cosine	$\mathrm{e}^{-t/T}\cos(\omega t)$	$\frac{s + 1/T}{\left(s + 1/T\right)^2 + \omega^2}$
7	unit impulse	$\int_{0^-}^{0+} \delta(t)dt = 1$	1

Example 3.1 Laplace Transform of a Unit Step Function

Unit step function $u(t)$ is shown in Figure 3.1. $u(t) = 0$ for time $t < 0$ and $u(t) = 1$ for time $t \geq 0$. The Laplace transform of this function is

$$\mathcal{L}\{u(t)\} = \int_{0^-}^{\infty} u(t)e^{-st}dt \tag{3.3}$$

$$\mathcal{L}\{u(t)\} = \int_{0^-}^{\infty} e^{-st}dt \tag{3.4}$$

$$\mathcal{L}\{u(t)\} = \frac{e^{-s\infty}}{-s} - \frac{e^{-s0^-}}{-s} \tag{3.5}$$

$$\mathcal{L}\{u(t)\} = \frac{1}{s} \tag{3.6}$$

Example 3.2 Laplace Transform of an Exponential Function of Time

An exponential function of time with time constant τ and $f(t) = 0$ for time $t < 0$ is shown in Figure 3.2. The Laplace transform of this function is

$$\mathcal{L}\left\{e^{-t/\tau}\right\} = \int_{0^-}^{\infty} e^{-t/\tau}e^{-st}dt \tag{3.7}$$

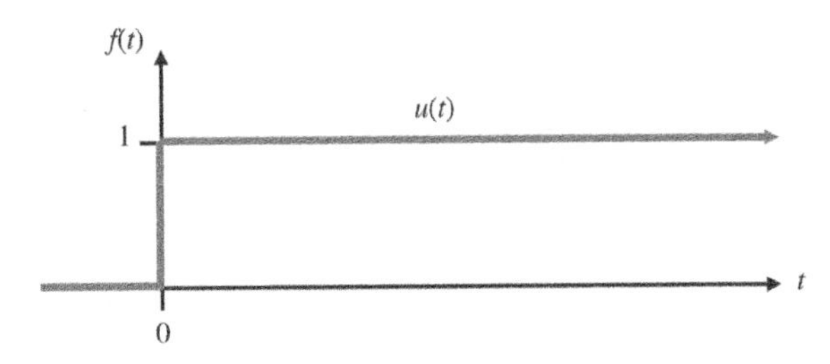

Figure 3.1 Unit step function of time.

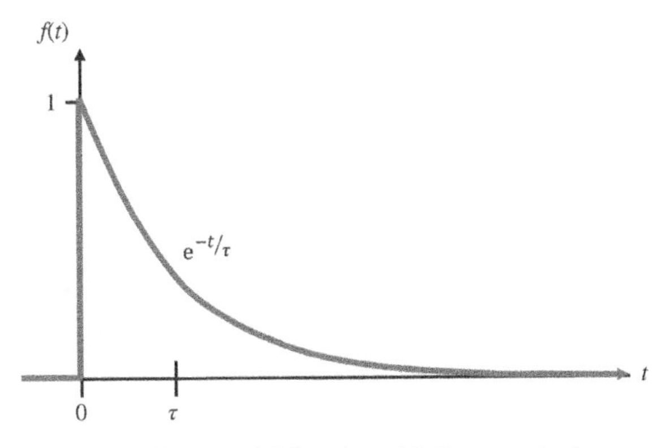

Figure 3.2 Exponential function with time constant τ.

$$\mathcal{L}\left\{e^{-t/\tau}\right\} = \int_{0^-}^{\infty} e^{-\left(s+1/\tau\right)t}\,dt \tag{3.8}$$

$$\mathcal{L}\left\{e^{-t/\tau}\right\} = \frac{e^{-\left(s+1/\tau\right)\infty}}{-\left(s+1/\tau\right)} - \frac{e^{-\left(s+1/\tau\right)0^-}}{-\left(s+1/\tau\right)} \tag{3.9}$$

$$\mathcal{L}\left\{e^{-t/\tau}\right\} = \frac{1}{\left(s+1/\tau\right)} \tag{3.10}$$

Example 3.3 Laplace Transform of a Decaying Sinusoidal Function

A decaying unit amplitude sinusoidal function of time $f(t)$ with time constant τ and $f(t) = 0$ for $t < 0$ is shown in Figure 3.3. The Laplace transform of this function is

$$\mathcal{L}\left\{e^{-t/\tau}\sin(\omega t)\right\} = \int_{0^-}^{\infty} e^{-t/\tau}\sin(\omega t)e^{-st}\,dt \tag{3.11}$$

From Euler's formulas,

$$\mathcal{L}\left\{e^{-t/\tau}\sin(\omega t)\right\} = \int_{0^-}^{\infty}\left(\frac{e^{\left(j\omega-1/\tau\right)t} - e^{-\left(j\omega+1/\tau\right)t}}{2j}\right)e^{-st}\,dt \tag{3.12}$$

$$\mathcal{L}\left\{e^{-t/\tau}\sin(\omega t)\right\} = \frac{1}{2j}\left(\int_{0^-}^{\infty} e^{\left(j\omega-1/\tau\right)t}e^{-st}\,dt - \int_{0^-}^{\infty} e^{-\left(j\omega+1/\tau\right)t}e^{-st}\,dt\right) \tag{3.13}$$

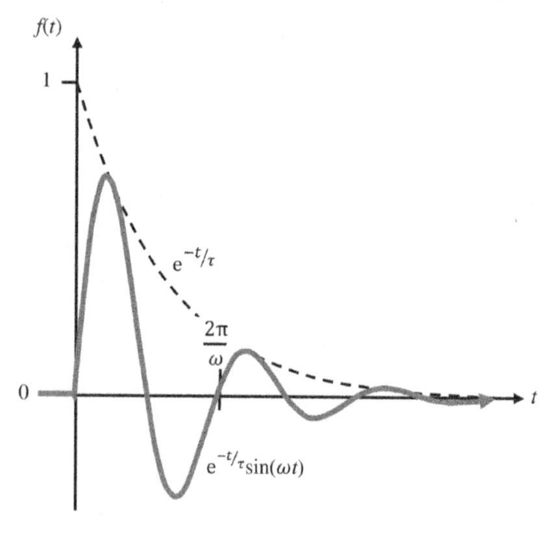

Figure 3.3 Decaying sine function with frequency ω and time constant τ.

Similar to the derivation above for an exponential function,

$$\mathcal{L}\left\{e^{-t/\tau}\sin(\omega t)\right\} = \frac{1}{2j}\left(\frac{1}{s-\left(j\omega - 1/\tau\right)} - \frac{1}{s+\left(j\omega + 1/\tau\right)}\right) \quad (3.14)$$

$$\mathcal{L}\left\{e^{-t/\tau}\sin(\omega t)\right\} = \frac{\omega}{\left(s+1/\tau\right)^2 + \omega^2} \quad (3.15)$$

3.2 Properties of the Laplace Transform

The Laplace transform has useful properties that are taken advantage of in representing differential equations using transfer functions, drawing block diagrams for production systems, analyzing dynamic behavior, and designing decision-making. In the following sections, the definition of the Laplace transform in Equation 3.1 is used to illustrate these properties.

3.2.1 Laplace Transform of a Function of Time Multiplied by a Constant

For constant K,

$$\mathcal{L}\left\{Kf(t)\right\} = \int_{0^-}^{\infty} Kf(t)e^{-st}dt \quad (3.16)$$

$$\mathcal{L}\left\{Kf(t)\right\} = K\int_{0^-}^{\infty} f(t)e^{-st}dt \quad (3.17)$$

$$\mathcal{L}\left\{Kf(t)\right\} = K\mathcal{L}\left\{f(t)\right\} \quad (3.18)$$

3.2.2 Laplace Transform of the Sum of Two Functions of Time

$$\mathcal{L}\left\{f(t)+g(t)\right\} = \int_{0^-}^{\infty} \left(f(t)+g(t)\right)e^{-st}dt \quad (3.19)$$

$$\mathcal{L}\left\{f(t)+g(t)\right\} = \int_{0^-}^{\infty} f(t)e^{-st}dt + \int_{0^-}^{\infty} g(t)e^{-st}dt \quad (3.20)$$

$$\mathcal{L}\left\{f(t)+g(t)\right\} = \mathcal{L}\left\{f(t)\right\} + \mathcal{L}\left\{g(t)\right\} \quad (3.21)$$

3.2.3 Laplace Transform of the First Derivative of a Function of Time

$$\mathcal{L}\left\{\frac{df(t)}{dt}\right\} = \int_{0^-}^{\infty} \frac{df(t)}{dt}e^{-st}dt \quad (3.22)$$

Integrating by parts,

$$\mathcal{L}\left\{\frac{df(t)}{dt}\right\} = f(t)e^{-st}\Big|_{0^-}^{\infty} - \int_{0^-}^{\infty} f(t)\left(-se^{-st}\right)dt \tag{3.23}$$

$$\mathcal{L}\left\{\frac{df(t)}{dt}\right\} = f(\infty)e^{-s\infty} - f\left(0^-\right)e^{-s0^-} + s\int_{0^-}^{\infty} f(t)e^{-st}dt \tag{3.24}$$

$$\mathcal{L}\left\{\frac{df(t)}{dt}\right\} = s\mathcal{L}\left\{f(t)\right\} - f\left(0^-\right) \tag{3.25}$$

3.2.4 Laplace Transform of Higher Derivatives of a Function of Time Function

In a similar manner, the Laplace transform of the i-th derivative of a function of time can be shown to be

$$\mathcal{L}\left\{\frac{df^i(t)}{dt^i}\right\} = s^i\mathcal{L}\left\{f(t)\right\} - s^{i-1}f\left(0^-\right) - \cdots - \frac{df^{i-1}\left(0^-\right)}{dt^{i-1}} \tag{3.26}$$

The initial conditions are often assumed to be zero. In this case,

$$\mathcal{L}\left\{\frac{df^i(t)}{dt^i}\right\} = s^i\mathcal{L}\left\{f(t)\right\} \tag{3.27}$$

3.2.5 Laplace Transform of Function with Time Delay

$$\mathcal{L}\left\{f(t-D)\right\} = \int_{0^-}^{\infty} f(t-D)e^{-st}dt \tag{3.28}$$

If $f(t) = 0$ for $t < 0$ as illustrated in Figure 3.4,

$$\mathcal{L}\left\{f(t-D)\right\} = \int_{D^-}^{\infty} f(t-D)e^{-st}dt \tag{3.29}$$

A change in variable can be introduced where $\tau = t - D$:

$$\mathcal{L}\left\{f(t-D)\right\} = \int_{0^-}^{\infty} f(\tau)e^{-s(\tau+D)}d\tau \tag{3.30}$$

$$\mathcal{L}\left\{f(t-D)\right\} = e^{-Ds}\int_{0^-}^{\infty} f(\tau)e^{-s\tau}d\tau \tag{3.31}$$

$$\mathcal{L}\left\{f(t-D)\right\} = e^{-Ds}\mathcal{L}\left\{f(t)\right\} \tag{3.32}$$

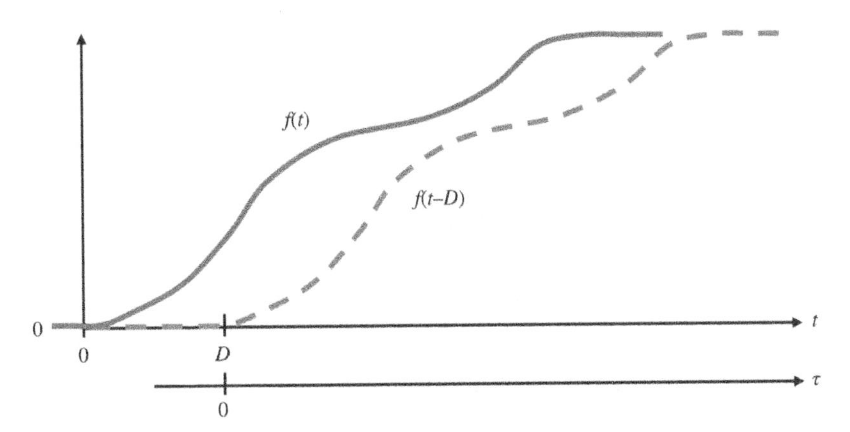

Figure 3.4 Function with time delay D.

3.3 Continuous-Time Transfer Functions

The ratio of the Laplace transform of an output variable of a continuous-time production system or component to the Laplace transform of an input variable, with any other inputs assumed to be zero, is referred to as a continuous-time transfer function. All initial conditions are assumed to be zero.[1] Transfer functions are used extensively in analysis and design of dynamic systems and are used in control system engineering software to represent models of system components and combinations of components. The cause-and-effect physical relationships between components of a production system are important and understanding these input–output relationships is crucial in both obtaining transfer functions and subsequent dynamic analysis and design.

A continuous-time transfer function is obtained by

1) Finding a differential or algebraic equation that models the relationship between output $y(t)$ and input $x(t)$ of a production system or component, with any other inputs assumed to be zero.
2) Obtaining the Laplace transform of the differential or algebraic equation that describes the relationship between output $y(t)$ and input $x(t)$.
3) Assuming initial conditions are zero. The result will be a function of $X(s)$ and $Y(s)$ where the following simplified notation is commonly used to refer to a generic transformed continuous-time function $f(t)$:

$$F(s) = \mathcal{L}\{f(t)\} \tag{3.33}$$

4) Collecting terms of $X(s)$ and $Y(s)$.

1 An important aspect of control system engineering is the ability to model and analyze dynamic relationships in generic ways that are not dependent on specific time-domain input functions and specific initial conditions.

5) Obtaining the ratio $Y(s)/X(s)$.
6) The result will be

$$\frac{Y(s)}{X(s)}=G(s) \tag{3.34}$$

Continuous-time transfer function $G(s)$ has the general form

$$G(s)=\frac{b_m s^m+\cdots+b_2 s^2+b_1 s+b_0}{a_n s^n+\cdots+a_2 s^2+a_1 s+a_0} e^{-Ds} \tag{3.35}$$

where D is the delay between the input and output. Transfer function $G(s)$ represents the dynamic relationship between transformed continuous-time production system or system component output $Y(s)$ and transformed continuous-time input $X(s)$:

$$Y(s)=G(s)X(s) \tag{3.36}$$

Example 3.4 Continuous-Time Transfer Function for WIP in a Production System

It is desired to obtain a continuous-time transfer function that relates work in progress (WIP) $w_w(t)$ hours in the production work system illustrated in Figure 2.1 to the rate of work input to the work system $r_i(t)$ hours/day. In Example 2.1, the following model was found for WIP:

$$\frac{dw_w(t)}{dt}=\frac{dw_d(t)}{dt}+r_i(t)-r_p(t)-r_d(t)$$

where $r_p(t)$ hours/day is the nominal production capacity, $r_d(t)$ hours/day represents capacity disturbances such as worker illness and equipment failures, and $w_d(t)$ hours represents work disturbances such as rush orders and order cancelations.

Applying the Laplace transforms in Equations 3.21 and 3.25,

$$sW_w(s)-w_w(0)=sW_d(s)-w_d(0)+R_i(s)-R_p(s)-R_d(s)$$

The result with all inputs zero except $r_i(t)$ is

$$sW_w(s)-w_w(0)=R_i(s)$$

and with initial conditions zero,

$$sW_w(s)=R_i(s)$$

The continuous-time transfer function that describes the dynamic relationship between WIP and capacity disturbances then is

$$G_w(s)=\frac{W_w(s)}{R_i(s)}=\frac{1}{s}$$

Example 3.5 Continuous-Time Transfer Function for Order Release Rate Decision-Making

It is desired to obtain a continuous-time transfer function that relates order release rate $r_o(t)$ orders/day in the production work system illustrated in Figure 2.3 to backlog deviation $w_e(t)$ orders:

$$w_e(t) = w_b(t) - w_p(t)$$

where $w_b(t)$ orders is the order backlog and $w_p(t)$ orders is the planned order backlog. In Example 2.2, the following decision rule was chosen for adjusting order release rate

$$\frac{dr_o(t)}{dt} = K_1 \frac{dw_e(t)}{dt} + K_2 w_e(t)$$

Applying the Laplace transforms in Equations 3.18, 3.21, and 3.25 with initial conditions zero,

$$sR_o(s) = K_1 sW_e(s) + K_2 W_e(s)$$

The continuous-time transfer function that describes the dynamic relationship between order release rate and backlog deviation, therefore, is

$$G_r(s) = \frac{R_o(s)}{W_e(s)} = \frac{K_1 s + K_2}{s}$$

Example 3.6 Continuous-Time Transfer Function for Two-company Production System with Delays

In the two-company production system illustrated in Figure 2.12, Company A obtains unfinished orders from a supplier, Company B, and then performs the work required to finish them. The relationship found between demand $r_i(t)$ orders/day and Company A order output rate $r_o(t)$ orders/day found in Example 2.7 is

$$r_o(t) = r_i(t - L_B - D_t - L_A)$$

where D days is the delay in shipping and Companies A and B have lead times L_A and L_B days, respectively. Applying the Laplace transform in Equation 3.32,

$$R_o(s) = e^{-(L_B + D_t + L_A)s} R_i(s)$$

The continuous-time transfer function that describes the dynamic relationship between Company A order output rate and demand, therefore, is

$$G_A(s) = \frac{R_o(s)}{R_i(s)} = e^{-(L_B + D_t + L_A)s}$$

Example 3.7 Transfer Function of First-Order Continuous-Time Mixture Heating with Delay

In Example 2.3, a dynamic model was developed for a portion of a production process in which the temperature of a mixture flowing out of a pipe is regulated using a heater and a temperature sensor, both placed at the outlet of the pipe as shown in Figure 2.4. Time constant τ = 49.8 seconds and mixture heating parameter K_h = 0.4°C/V were found using experimental data. As shown in Figure 3.5, regulation of mixture temperature is identical in this example except that the heater is positioned at the pipe inlet and is separated by distance l_p = 60 cm from the mixture temperature sensor at the pipe outlet. Constant mixture flow velocity f_p = 5 cm/second then determines the transportation delay $D_t = l_p/f_p$ = 12 seconds between the heater and the pipe outlet where the temperature sensor is located. This delay must be included in the mixture heating model if it is significant:

$$49.8\frac{d\Delta h(t)}{dt}+\Delta h(t)=0.4v(t-12)$$

Applying the Laplace transforms in Equations 3.18, 3.21, 3.25, and 3.32 with initial conditions zero yields

$$49.8s\Delta H(s)+\Delta H(s)=0.4\mathrm{e}^{-12s}V(s)$$

The continuous-time transfer function for mixture heating is then

$$G(s)=\frac{\Delta H(s)}{V(s)}=\frac{0.4}{49.8s+1}\mathrm{e}^{-12s}$$

Program 3.1 shows how a continuous-time transfer function variable can be created that represents transfer function $G(s)$. This variable then will be available for use in further computations that could include combining it with a transfer function variable that represents the temperature regulation decision-making to obtain a complete dynamic model of mixture temperature regulation.

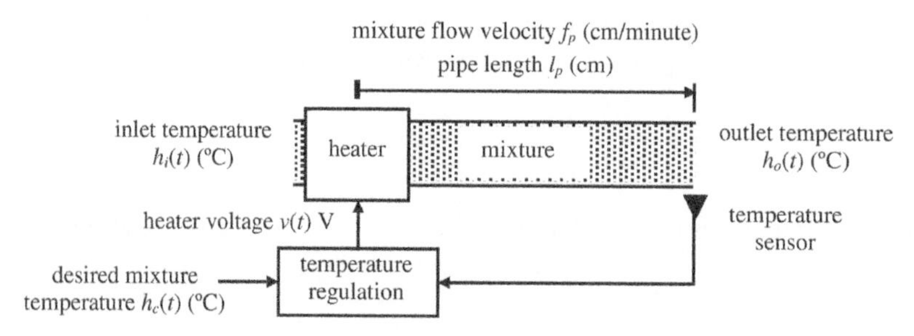

Figure 3.5 Mixture temperature regulation with delay between heater and mixture temperature sensing at the end of pipe.

Program 3.1 Creating a continuous-time transfer function variable with delay

```
tau=49.8;  % mixture heating time constant (seconds)
Kh=0.4;  % mixture heating parameter (°C/V)
lp=60;  % pipe length (cm)
fp=5;  % mixture flow rate (cm/second)

D=lp/fp  % delay (seconds)
```

```
D = 12
```

```
Gs=tf(Kh,[tau, 1]);  % mixture heating transfer function
Gs.OutputDelay=D  % include delay
```

```
Gs =
                        0.4
   exp(-12*s) * ----------
                    49.8 s + 1
Continuous-time transfer function.
```

A continuous-time transfer function variable `Gs` can be created using the `tf` function:

```
Gs=tf(num,den)
```

where `num` and `den` are vectors of numerator and denominator coefficients corresponding to form of the transfer function in Equation 3.35:

```
num=[bm, … b2, b1, b0]
den=[ai, … a2, a1, a0]
```

The numerator and denominator vectors can be included explicitly in the `tf` function:

```
Gs=tf([bm, … b2, b1, b0],[ai, … a2, a1, a0])
```

The default time unit is seconds. The time unit of transfer function variable `Gs` can be changed if desired using

```
Gs.TimeUnit='unit'
```

where `'unit'` can be `'seconds'`, `'minutes'`, `'hours''`, `'days'`, etc. Alternatively, the time unit of transfer function variable `Gs` can be set when it is created using

```
Gs=tf(num,den,'TimeUnit','unit')
```

The default time delay is zero. The output delay in transfer function variable `Gs` can be changed if desired using

```
Gs.OutputDelay=D
```

Alternatively, the output delay D can be included when transfer function variable Gs is created using

```
Gs=tf(num,den,'OutputDelay',D)
```

A continuous-time transfer function variable also can be created algebraically in the form of Equation 3.35 using a transfer function variable that represents the trivial continuous-time transfer function

$$s = \frac{s+0}{1}$$

using a special case of function `tf`:

```
s=tf('s')
```

This is demonstrated in Program 3.2.

Program 3.2 Alternative method for creating a continuous-time transfer function variable

```
tau=49.8;  % mixture heating time constant (seconds)
Kh=0.4;  % mixture heating parameter (°C/V)
lp=60;  % pipe length (cm)
fp=5;  % mixture flow rate (cm/seconds)

D=lp/fp  % delay (seconds)
```

```
D = 12
```

```
s=tf('s')  % define special transfer function variable
```

```
s =
  s
Continuous-time transfer function.
```

```
Gs=exp(-D*s)*Kh/(tau*s+1)  % mixture heating transfer function
```

```
Gs =
                      0.4
    exp(-12*s) * ----------
                  49.8 s + 1
Continuous-time transfer function.
```

```
[num,den]=tfdata(Gs,'v')  % extract numerator and denominator vectors
```

```
num =
         0    0.4000
den =
   49.8000    1.0000
```

```
totaldelay(Gs)  % extract delay
```

```
ans = 12
```

The vectors of numerator and denominator coefficients can be extracted from a transfer function variable using

```
num,den]=tfdata(Gs,'v')
```

and the delay can be extracted using

```
D=totaldelay(Gs)
```

This function sums input delay, output delay and internal delay, which can be specified independently in transfer function variables.

3.4 Z Transform

Consider a sequence such as that shown in Figure 3.6 that consists of values $f(kT)$ separated in time by a constant period T for integer $k = 0,1,2,3,...,\infty$. Sequence $f(kT)$ could be fundamentally discrete, for example, representing a sequence of decisions that are made periodically in a production system. Alternatively, as illustrated in Figure 3.7, successive values of sequence $f(kT)$ could be acquired by sampling continuous function of time $f(t)$. The ideal sampler shown in Figure 3.7 obtains value $f(kT)$ without delay at each instant in time $kT = 0, T, 2T, 3T, ..., \infty$.

The Z transform of sequence $f(kT)$ is defined as

$$\mathcal{Z}\{f(kT)\} = \sum_{k=0}^{\infty} f(kT)z^{-k} \tag{3.37}$$

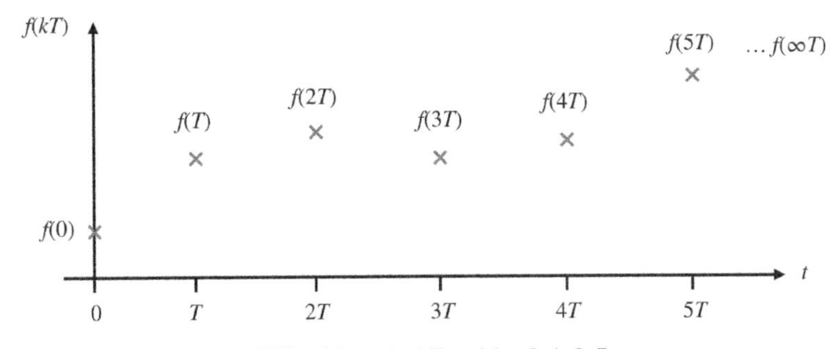

Figure 3.6 Sequence $f(kT)$ with period T and $k = 0, 1, 2, 3, ..., \infty$.

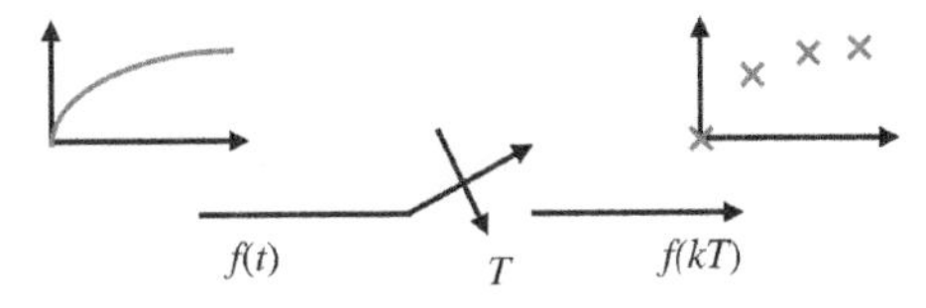

Figure 3.7 Ideal sampler with period T, continuous-time input function $f(t)$, and discrete-time output sequence $f(kT)$.

where z is a new, complex variable that is related to Laplace variable s and has real part α and imaginary part β:

$$z = e^{Ts} \tag{3.38}$$

$$z = \alpha + j\beta \tag{3.39}$$

In the following examples, the definition of the Z transform in Equation 3.37 is used to obtain the transforms of a unit step sequence, an exponential sequence, and an exponentially decaying sinusoidal sequence. These results are listed in Table 3.2 along with the results for several other commonly considered discrete functions of time. These results are well known and rarely need to be derived in practice in production system engineering.

Example 3.8 Z Transform of a Unit Step Sequence

For the unit step sequence shown in Figure 3.8, $u(kT) = 1$ for $k \geq 0$. Then

$$\mathcal{Z}\{u(kT)\} = \sum_{k=0}^{\infty} u(kT) z^{-k} \tag{3.40}$$

Table 3.2 Z transforms of common sequences.

	function	$f(kT)$ ($f(kT) = 0$ for $k < 0$)	$\mathcal{Z}\{f(kT)\}$
1	unit step	1	$\frac{1}{1-z^{-1}}$
2	exponential	$e^{-kT/\tau}$	$\frac{1}{1-e^{-T/\tau}z^{-1}}$
3	sine	$\sin(\omega kT)$	$\frac{\sin(\omega T)z^{-1}}{1-2\cos(\omega T)z^{-1}+z^{-2}}$
4	cosine	$\cos(\omega kT)$	$\frac{1-\cos(\omega T)z^{-1}}{1-2\cos(\omega T)z^{-1}+z^{-2}}$
5	decaying sine	$e^{-kT/\tau}\sin(\omega T)$	$\frac{e^{-T/\tau}\sin(\omega T)z^{-1}}{1-2e^{-T/\tau}\cos(\omega T)z^{-1}+e^{-2T/\tau}z^{-2}}$
6	decaying cosine	$e^{-kT/\tau}\cos(\omega T)$	$\frac{1-e^{-T/\tau}\cos(\omega T)z^{-1}}{1-2e^{-T/\tau}\cos(\omega T)z^{-1}+e^{-2T/\tau}z^{-2}}$
7	unit sample	$f(0) = 1$ for $k = 0$ $f(kT) = 0$ for $k \neq 0$	1

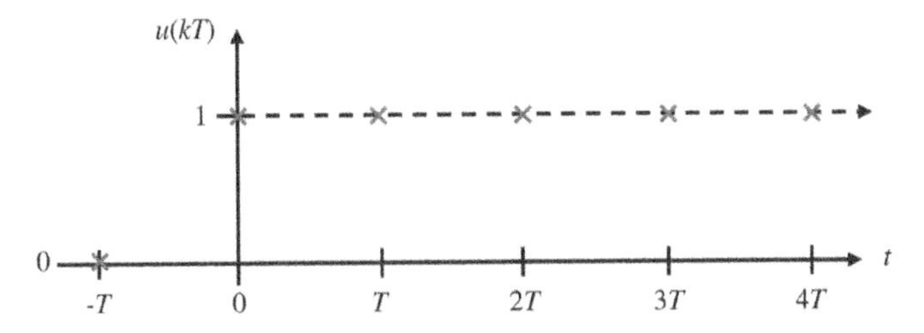

Figure 3.8 Unit step sequence.

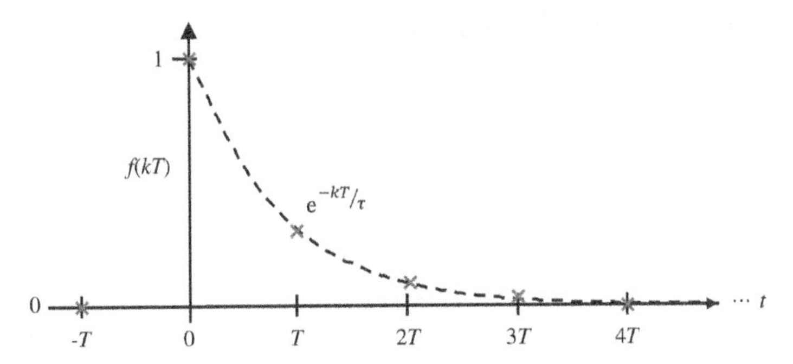

Figure 3.9 Exponential sequence.

The infinite series is

$$\mathcal{Z}\{u(kT)\} = 1 + z^{-2} + z^{-3} + \cdots \tag{3.41}$$

and the closed form of this series is

$$\mathcal{Z}\{u(kT)\} = \frac{1}{1 - z^{-1}} \tag{3.42}$$

Example 3.9 Z Transform of an Exponential Sequence

For the exponential sequence shown in Figure 3.9,

$$f(kT) = e^{-kT/\tau} \tag{3.43}$$

for $k \geq 0$. Then

$$\mathcal{Z}\left\{e^{-kT/\tau}\right\} = 1 + e^{-T/\tau} z^{-1} + e^{-2T/\tau} z^{-2} + \cdots \tag{3.44}$$

The closed form of this series is

$$\mathcal{Z}\left\{e^{-kT/\tau}\right\} = \frac{1}{1 - e^{-T/\tau} z^{-1}} \tag{3.45}$$

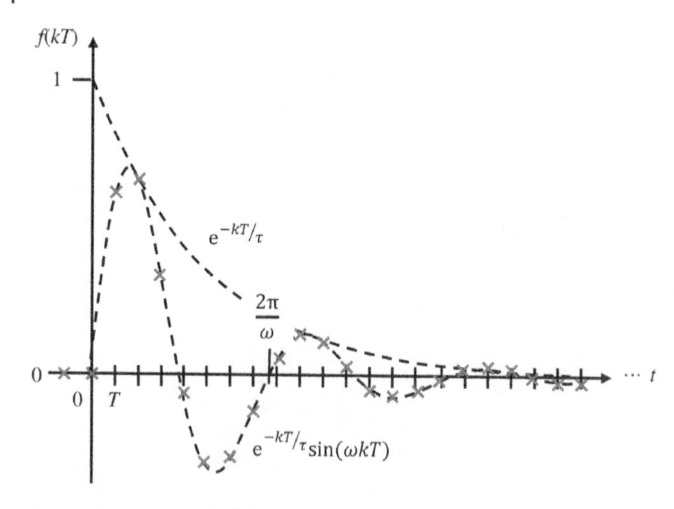

Figure 3.10 Decaying sinusoidal sequence.

Example 3.10 Z Transform of a Decaying Sinusoidal Sequence

For the decaying sinusoidal sequence shown in Figure 3.10,

$$f(kT) = e^{-kT/\tau}\sin(\omega kT) \tag{3.46}$$

for $t \geq 0$. Using Euler's formulas,

$$\mathcal{Z}\left\{e^{-kT/\tau}\sin(\omega kT)\right\} = \mathcal{Z}\left\{e^{-kT/\tau}\left(\frac{e^{j\omega kT} - e^{-j\omega kT}}{2j}\right)\right\} \tag{3.47}$$

From Equation 3.45,

$$\mathcal{Z}\left\{e^{-kT/\tau}\sin(\omega kT)\right\} = \frac{1}{2j}\left(\frac{1}{1 - e^{\left(j\omega - T/\tau\right)T}z^{-1}} - \frac{1}{1 - e^{-\left(j\omega + T/\tau\right)T}z^{-1}}\right) \tag{3.48}$$

Using Euler's formulas again,

$$\mathcal{Z}\left\{e^{-kT/\tau}\sin(\omega kT)\right\} = \left(\frac{e^{-T/\tau}\sin(\omega T)z^{-1}}{1 - 2e^{-T/\tau}\cos(\omega T)z^{-1} + e^{-2T/\tau}z^{-2}}\right) \tag{3.49}$$

3.5 Properties of the Z Transform

Like the Laplace transform, the Z transform has useful properties that are taken advantage of in many aspects of control system engineering. In the following sections, the definition of the Z transform in Equation 3.37 is used to illustrate these properties.

3.5.1 Z Transform of a Sequence Multiplied by a Constant

For constant K,

$$\mathcal{Z}\{Kf(kT)\} = \sum_{k=0}^{\infty} Kf(kT)z^{-k} \tag{3.50}$$

$$\mathcal{Z}\{Kf(kT)\} = K\sum_{k=0}^{\infty} f(kT)z^{-k} \tag{3.51}$$

$$\mathcal{Z}\{Kf(kT)\} = K\mathcal{Z}\{f(kT)\} \tag{3.52}$$

3.5.2 Z Transform of the Sum of Two Sequences

$$\mathcal{Z}\{f(kT)+g(kT)\} = \sum_{k=0}^{\infty}(f(kT)+g(kT))z^{-k} \tag{3.53}$$

$$\mathcal{Z}\{f(kT)+g(kT)\} = \sum_{k=0}^{\infty} f(kT)z^{-k} + \sum_{k=0}^{\infty} g(kT)z^{-k} \tag{3.54}$$

$$\mathcal{Z}\{f(kT)+g(kT)\} = \mathcal{Z}\{f(kT)\} + \mathcal{Z}\{g(kT)\} \tag{3.55}$$

3.5.3 Z Transform of Time Delay dT

A sequence $f(kT)$ is shown in Figure 3.11 along with a corresponding sequence that is delayed by dT, where d is an integer and $d > 0$. The Z transform of the delayed sequence is

$$\mathcal{Z}\{f((k-d)T)\} = f(0)z^{-d} + f(T)z^{-(d+1)} + f(2T)z^{-(d+2)} + \cdots \tag{3.56}$$

$$\mathcal{Z}\{f((k-d)T)\} = z^{-d}\left(f(0) + f(T)z^{-1} + f(2T)z^{-2} + \cdots\right) \tag{3.57}$$

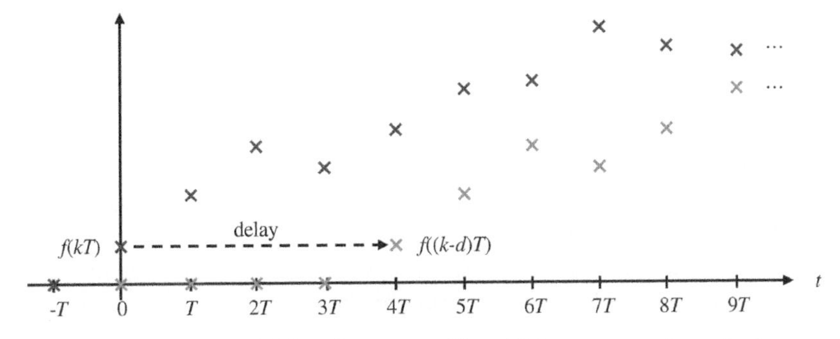

Figure 3.11 Example of delayed sequence $f((k-d)T)$ for the case where $d = 4$.

$$\mathcal{Z}\{f((k-d)T)\} = z^{-d}\sum_{k=0}^{\infty} f(kT)z^{-k} \tag{3.58}$$

$$\mathcal{Z}\{f((k-d)T)\} = z^{-d}\mathcal{Z}\{f(kT)\} \tag{3.59}$$

3.5.4 Z Transform of a Difference Equation

Consider the difference equation

$$\begin{aligned} a_n y((k+n)T) &= b_m x((k+m)T) + \ldots + b_1 x((k+1)T) + b_0 x(kT) \\ &\quad - a_{n-1} y((k+n-1)T) - \ldots - a_1 y((k+1)T) - a_0 y(kT) \end{aligned} \tag{3.60}$$

where $y(kT) = 0$ and $x(kT) = 0$ for integer $k < 0$. Summing with $k = 0, 1, 2, 3, \ldots \infty$ and noting Equation 3.58 yields

$$\begin{aligned} a_n z^n \sum_{k=0}^{\infty} y(kT)z^{-k} &= b_m z^m \sum_{k=0}^{\infty} x(kT)z^{-k} + \ldots + b_1 z \sum_{k=0}^{\infty} x(kT)z^{-k} \\ &\quad + b_0 \sum_{k=0}^{\infty} x(kT)z^{-k} - a_{n-1} z^{n-1} \sum_{k=0}^{\infty} y(kT)z^{-k} \\ &\quad - \ldots - a_1 z \sum_{k=0}^{\infty} y(kT)z^{-k} - a_0 \sum_{k=0}^{\infty} y(kT)z^{-k} \end{aligned} \tag{3.61}$$

Applying the definition of the Z transform in Equation 3.37,

$$\begin{aligned} a_n z^n \mathcal{Z}\{y(kT)\} &= b_m z^m \mathcal{Z}\{x(kT)\} + \ldots + b_1 z \mathcal{Z}\{x(kT)\} + b_0 \mathcal{Z}\{x(kT)\} \\ &\quad - a_{n-1} z^{n-1} \mathcal{Z}\{y(kT)\} - \ldots - a_1 z \mathcal{Z}\{y(kT)\} - a_0 \mathcal{Z}\{y(kT)\} \end{aligned} \tag{3.62}$$

3.6 Discrete-Time Transfer Functions

The ratio of the Z transform of an output variable of a discrete-time production system or component to the Z transform of an input variable, with any other inputs assumed to be zero is referred to as a transfer function. All initial conditions are assumed to be zero. A discrete-time transfer function is obtained by

1) Finding a difference or algebraic equation that models the relationship between output $y(kT)$ and input $x(kT)$ of a production system or component, with any other inputs assumed to be zero.
2) Obtaining the Z transform of the difference or algebraic equation that describes the relationship between output $y(kT)$ and input $x(kT)$.
3) Assuming initial conditions are zero. The result will be a function of $X(z)$ and $Y(z)$ where the following simplified notation is commonly used to refer to a transformed discrete-time function $f(kT)$:

$$F(z) = \mathcal{Z}\{f(kT)\} \tag{3.63}$$

4) Collecting terms of $X(z)$ and $Y(z)$.
5) Obtaining the ratio $Y(z)/X(z)$.
6) The result will be

$$\frac{Y(z)}{X(z)} = G(z) \tag{3.64}$$

Discrete-time transfer function $G(z)$ has the general form

$$G(z) = \frac{b_m z^m + \cdots + b_2 z^2 + b_1 z + b_0}{a_n z^n + \cdots + a_2 z^2 + a_1 z + a_0} \tag{3.65}$$

Transfer function $G(z)$ represents the dynamic relationship between transformed discrete-time production system or system component output $Y(z)$ and transformed discrete-time input $X(z)$:

$$Y(z) = G(z)X(z) \tag{3.66}$$

Example 3.11 Positive and Negative Powers of z

Discrete-time transfer functions can have positive integer and negative integer powers of z. Consider the following transfer function, which is in the form of Equation 3.65 and includes a delay dT days where d is a positive integer:

$$G(z) = \frac{b}{z^{d+1} - az^d}$$

This transfer function is equivalent to

$$G(z) = \frac{b}{z - a} z^{-d}$$

$$G(z) = \frac{bz^{-1}}{1 - az^{-1}} z^{-d}$$

Example 3.12 Discrete-Time Transfer Function Relating WIP to Work Input Rate

It is desired to obtain a discrete-time transfer function that relates work in progress (WIP) $w_w(kT)$ hours in the production work system illustrated in Figure 2.6 to capacity disturbances $r_d(kT)$ hours/day. In Example 2.4, the following discrete-time model was found for WIP when there were no WIP disturbances and the rate of work input was the same as planned capacity:

$$w_w(kT) = w_w((k-1)T) - Tr_d((k-1)T)$$

Capacity disturbances are assumed to be constant during each period T days.

Applying the result in Equation 3.62 yields

$$W_w(z) = z^{-1} W_w(z) - Tz^{-1} R_d(z)$$

The discrete-time transfer function that describes the dynamic relationship between WIP and capacity disturbances then is:

$$G_w(z) = \frac{W_w(z)}{R_d(z)} = \frac{-Tz^{-1}}{1 - z^{-1}}$$

Alternatively,

$$G_w(z) = \frac{W_w(z)}{R_d(z)} = \frac{-T}{z - 1}$$

Example 3.13 Discrete-Time Transfer Function with Delay Relating Permanent Worker Capacity to Demand

In Example 2.8 a discrete-time model was developed that describes adjustment of production capacity as the rate of orders input to a production system fluctuates. As shown in Figure 2.13, a portion of the production capacity is provided by permanent workers; this portion cannot be adjusted quickly, should not be adjusted at high frequencies, and because of logistic issues there is a delay of dT days in implementing permanent worker adjustment decisions where d is a positive integer. The remaining portion is provided by cross-trained workers; this portion can be adjusted immediately. It is desired to obtain the discrete-time transfer functions that relate the capacity provided by permanent and cross-trained workers to order input rate.

Order input rate $r_i(kT)$ is measured weekly, $T = 5$ days, and the portion of production capacity provided by permanent workers $r_p(kT)$ is adjusted weekly. An exponential filter is used to focus adjustments in permanent worker capacity on the relatively low frequency content of fluctuating order input rate $r_i(kT)$. The portion of fluctuating order input that is addressed by permanent worker capacity was found in Example 2.8 to be

$$r_p((k+d)T) = \alpha r_i(kT) + (1-\alpha) r_p((k-1+d)T)$$

where $0 < \alpha \leq 1$.

Applying the Z transforms in Equations 3.59 and 3.62 yields

$$z^d R_p(z) = \alpha R_i(z) + (1-\alpha) z^{(d-1)} R_p(z)$$

The discrete-time transfer function that describes the dynamic relationship between capacity provided by permanent workers and order input rate then is

$$G_p(z) = \frac{R_p(z)}{R_i(z)} = \frac{\alpha}{z^d - (1-\alpha) z^{(d-1)}}$$

Alternatively,

$$G_p(z) = \frac{R_p(z)}{R_i(z)} = \frac{\alpha}{1 - (1-\alpha) z^{-1}} z^{-d}$$

The portion of fluctuating order input that is addressed by cross-trained worker capacity was found in Example 2.8 to be

$$r_c(kT) = r_i(kT) - r_p(kT)$$

Applying the Z transform yields

$$R_c(z) = R_i(z) - R_p(z)$$

and from above,

$$R_c(z) = R_i(z) - \frac{\alpha}{1-(1-\alpha)z^{-1}} z^{-d} R_i(z)$$

The discrete-time transfer function that describes the dynamic relationship between capacity provided by cross-trained workers employees and order input rate then is

$$G_c(z) = \frac{R_c(z)}{R_i(z)} = \frac{1-(1-\alpha)z^{-1} - \alpha z^{-d}}{1-(1-\alpha)z^{-1}}$$

As demonstrated in Program 3.3 for $T = 5$ days, $dT = 10$ days and $\alpha = 0.2$, a discrete-time transfer function variable `Gz` can be created using the `tf` function:

```
Gz=tf(num,den,T)
```

where `T` is the period and `num` and `den` are vectors of the numerator and denominator coefficients, respectively, on the non-negative powers of z in Equation 3.65.

```
num=[bm, … b2, b1, b0]
den=[an, … a2, a1, a0]
```

The default time delay is zero. Output delay dT is changed in discrete-time transfer function variable Gz using

```
Gz=tf(num,den,T,'OutputDelay',d)
```

Program 3.3 Creating a discrete-time transfer function variable

```
T=5;   % period (days)
d=2;   % delay dT (days)
a=0.2;    % weighting factor for exponential filter

% transfer function for permanent workers
Gz=tf([a 0],[1, -(1-a)],T,'TimeUnit','days','OutputDelay',d)
```

```
Gz =
                 0.2z
    z^(-2)  *  -------
                z - 0.8
Sample time: 5 days
Discrete-time transfer function.
```

or

```
Gz.OutputDelay=d
```

where d is a non-negative integer.

Alternatively, the trivial discrete-time transfer function

$$z = \frac{z+0}{1}$$

can be created using a special case of function `tf`:

```
z=tf('z',T)
```

and discrete-time transfer function variables then can be created algebraically in the form of Equation 3.65 as demonstrated in in Program 3.4.

Program 3.4 Alternative method for creating a discrete-time transfer function variable

```
T=5;  % period (days)
d=2;  % delay dT=10 (days)
a=0.2;  % weighting factor for exponential filter

z=tf('z',T,'TimeUnit','days')  % special case
```

```
  z =

    z

  Sample time: 5 days
  Discrete-time transfer function.
```

```
Gz=a/(z^d-(1-a)*z^(d-1))  % transfer function for permanent workers
```

```
  Gz =

         0.2
    -----------
    z^2 - 0.8 z

  Sample time: 5 days
  Discrete-time transfer function.
```

3.7 Block Diagrams

It is important to understand the physical and logistical relationships between components in a productions system because the structure of these relationships contributes to the system's dynamic behavior and significantly influences the design of decision-making components. Block diagrams are often used to graphically illustrate these relationships. Transfer functions are used within blocks to state the modeled cause-and-effect relationships between inputs and outputs. For a continuous-time

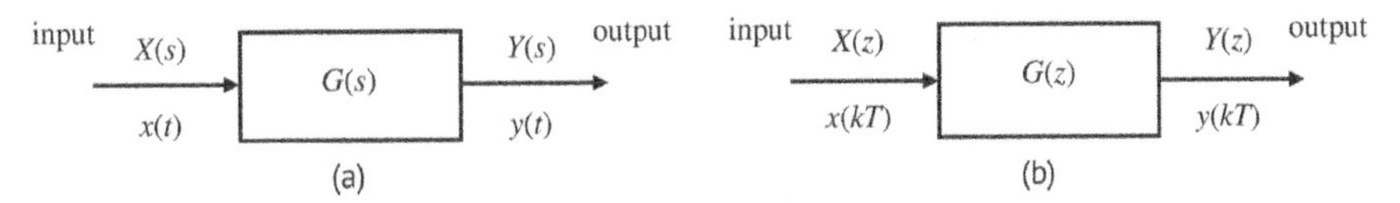

Figure 3.12 Continuous-time and discrete-time blocks. (a) Block with continuous-time transfer function $G(s)=Y(s)/X(s)$. (b) Block with discrete-time transfer function $G(z)=Y(z)/X(z)$.

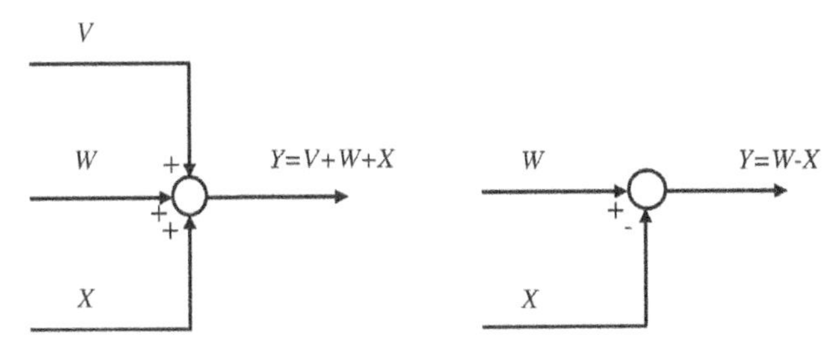

Figure 3.13 Summing and differencing of continuous-time or discrete-time variables in a block diagram.

production system or system component model that has input $x(t)$ and output $y(t)$, a unidirectional block can be drawn as shown in Figure 3.12a. The block indicates that the input results in the output (and not vice versa in general) and it also contains continuous-time transfer function $G(s)$, which represents the dynamic relationship between input variable $x(t)$ and output variable $y(t)$ after the Laplace transform is applied. Similarly, for a discrete-time production system or system component model that has input $x(kT)$ and output $y(kT)$, a unidirectional block can be drawn as shown in Figure 3.12b.

When a production system has multiple components, the blocks that represent the individual components can be connected in a block diagram that shows graphically how the components are structurally and algebraically related. These relationships can be direct, with the output of one block being the input to a subsequent block, or they can involve summing or differencing of two or more system variables as illustrated in Figure 3.13. Furthermore, a block diagram can have external input variables that do not originate in other blocks in the diagram.

Example 3.14 Block Diagram for Work In Progress in a Production Work System

It is desired to use a block diagram to show the relationships between work in progress (WIP) $w_w(t)$ hours in the production work system illustrated in Figure 2.1 and the variables that affect WIP including the rate of work input to the work system $r_i(t)$ hours/day, nominal production capacity $r_p(t)$ hours/day, capacity disturbances $r_d(t)$ hours/day, and work disturbances $w_d(t)$ hours. In Example 2.1, the following model was found for WIP:

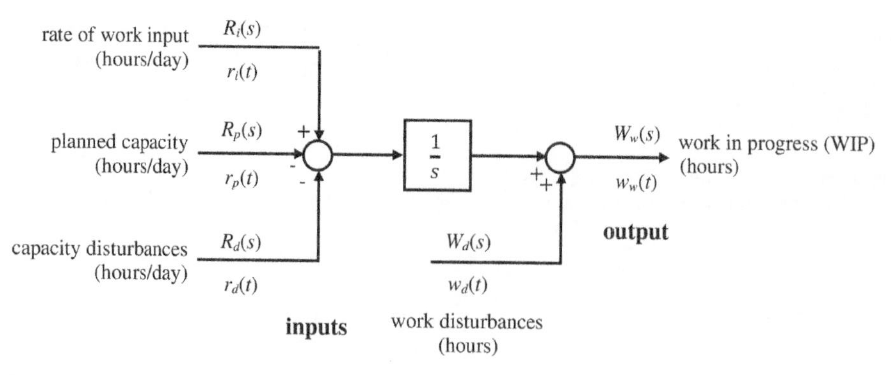

Figure 3.14 Block diagram for WIP in a production work system.

$$\frac{dw_w(t)}{dt} = \frac{dw_d(t)}{dt} + r_i(t) - r_p(t) - r_d(t)$$

Assuming zero initial conditions and applying the Laplace transforms in Equations 3.21 and 3.25,

$$sW_w(s) - w_w(0) = sW_d(s) - w_d(0) + R_i(s) - R_p(s) - R_d(s)$$

$$W_w(s) = W_d(s) + \frac{1}{s}\left(R_i(s) - R_p(s) - R_d(s)\right)$$

The block diagram in Figure 3.14 shows this relationship graphically. WIP $w_w(t)$, the output, is the result of inputs $r_i(t)$, $r_p(t)$, $r_d(t)$ and $w_d(t)$.

Example 3.15 Block Diagram for Adjusting Capacity of Cross-Trained and Permanent Workers

It is desired to use a block diagram to represent the cause and effect relationships in the discrete-time model that was developed in Example 2.8 for making permanent and cross-trained worker capacity adjustment decisions in response to fluctuations in order input rate $r_i(kT)$. As shown in Figure 2.13, the portion of production capacity provided by permanent workers $r_p(kT)$ is focused on lower frequency fluctuations and there is a delay of dT days in implementing permanent worker adjustment decisions where d is a positive integer. The portion of production capacity provided by cross-trained workers $r_c(kT)$ is adjusted immediately.

An exponential filter with weighting parameter $0 < \alpha \leq 1$ and output $r_f(kT)$ is used to focus adjustments in permanent worker capacity on the relatively low frequency content of fluctuating order input rate $r_i(kT)$. The difference and algebraic equations that describe worker capacity decision-making are

$$r_f(kT) = \alpha r_i(kT) + (1-\alpha) r_f((k-1)T)$$

$$r_p\left((k+d)T\right)=r_f\left(kT\right)$$

$$r_c\left(kT\right)=r_i\left(kT\right)-r_p\left(kT\right)$$

Applying the Z transforms in Equations 3.59 and 3.62 yields

$$R_f\left(z\right)=\frac{\alpha}{1-\left(1-\alpha\right)z^{-1}}R_i\left(z\right)$$

$$R_p\left(z\right)=z^{-d}R_f\left(z\right)$$

$$R_c\left(z\right)=R_i\left(z\right)-R_p\left(z\right)$$

The block diagram in Figure 3.15 shows these relationships graphically. The outputs permanent worker capacity $r_p(kT)$ and cross-trained worker capacity $r_c(kT)$, are the result of order input rate $r_i(kT)$.

3.8 Transfer Function Algebra

Transformed models represent algebraic relationships between production system components, regardless of whether they are continuous-time or discrete-time. These relationships can be used to combine models of components in series, in parallel, and in closed-loop topologies; this is equivalent to combining blocks in block diagrams. Common relationships and the results of combining their transfer functions are presented in the following subsections.

3.8.1 Series Relationships

Blocks in series with transfer functions G_1 and G_2 are shown in Figure 3.16a. The transfer functions and the relationships represented are

$$W=G_1X \tag{3.67}$$

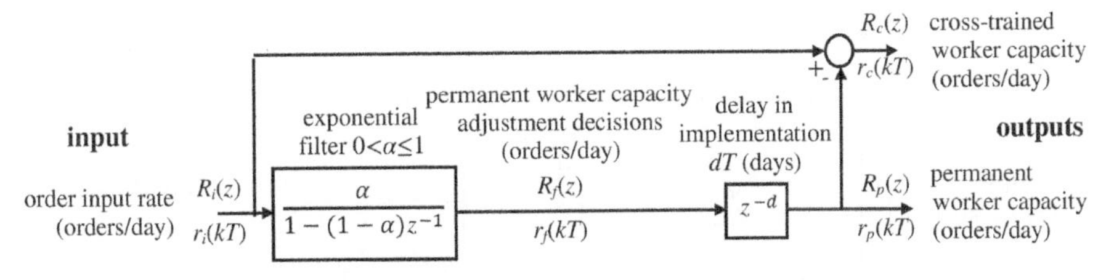

Figure 3.15 Block diagram for adjusting portions of production capacity provided by permanent and cross-trained workers.

$$Y = G_2 W \tag{3.68}$$

$$Y = G_1 G_2 X \tag{3.69}$$

$$\frac{Y}{X} = G_1 G_2 \tag{3.70}$$

The combined transfer function is shown in the block diagram in Figure 3.16b.

Example 3.16 Discrete-Time Transfer Functions in Series

Using the result of Example 3.15, the following discrete-time dynamic model defines the relationship between order input rate $r_i(kT)$ orders/day and the portion of production capacity that is provided by permanent employees $r_p(kT)$ orders/day in a production work system:

$$\frac{R_f(z)}{R_i(z)} = \frac{\alpha}{1-(1-\alpha)z^{-1}}$$

$$\frac{R_p(z)}{R_f(z)} = z^{-d}$$

These transfer functions are in series. Combining these transfer functions yields

$$\frac{R_p(z)}{R_i(z)} = \frac{\alpha}{1-(1-\alpha)z^{-1}} z^{-d}$$

The block diagram in Figure 3.17 shows this combined relationship.

A continuous-time or discrete-time transfer function variable `G` that is the product of two transfer functions in series, represented by variables `G1` and `G2`, can be created using

```
G=series(G1,G2)
```

or transfer function variables can be multiplied:

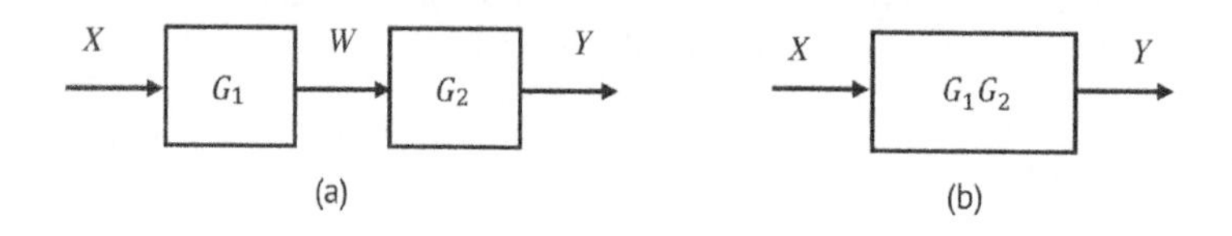

Figure 3.16 Transfer functions and blocks in series.

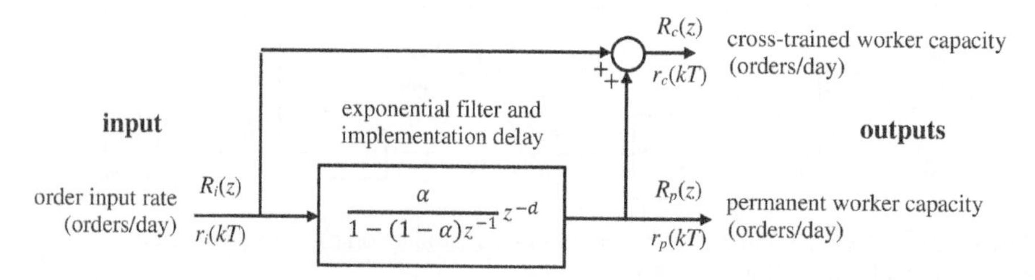

Figure 3.17 Result of combining transfer functions and blocks in series.

```
G=G1*G2
```

This is demonstrated in Program 3.5 in which the above discrete-time transfer functions are combined for capacity adjustment period $T = 5$ days, exponential filter weighting parameter $\alpha = 0.75$ and delay $dT = 10$ days.

Program 3.5 Combining discrete-time transfer functions in series

```
T=5;  % period (days)
d=2;  % delay dT=10 (days)
a=0.2;  % weighting factor for exponential filter

z=tf('z',T,'TimeUnit','days');

Gfz=a/(1-(1-a)*z^-1)  % transfer function for permanent workers
```

```
   Gfz =

        0.2 z
       -------
       z - 0.8

   Sample time: 5 days
   Discrete-time transfer function.
```

```
Gdz=z^-d  % transfer function of delay
```

```
   Gdz =

       1
      ---
      z^2

   Sample time: 5 days
   Discrete-time transfer function.
```

```
Gz=series(Gfz,Gdz)  % filter and delay transfer functions in series
```

```
   Gz =

         0.2 z
     -------------
     z^3 - 0.8 z^2

   Sample time: 5 days
   Discrete-time transfer function.
```

```
Gz=Gfz*Gdz  % filter and delay transfer functions in series
```

```
   Gz =

         0.2 z
     -------------
     z^3 - 0.8 z^2

   Sample time: 5 days
   Discrete-time transfer function.
```

3.8.2 Parallel Relationships

Blocks in parallel with transfer functions G_1 and G_2 are shown in Figure 3.18a. The transfer functions and relationships represented are

$$Y = G_1X + G_2X \tag{3.71}$$

$$Y = (G_1 + G_2)X \tag{3.72}$$

$$\frac{Y}{X} = G_1 + G_2 \tag{3.73}$$

The combined transfer function is shown in the block diagram in Figure 3.18b.

Example 3.17 Continuous-Time Transfer Functions in Parallel

As shown in the block diagram in Figure 3.19a, a continuous-time PID (proportional plus integral plus derivative) decision rule combines the following individual control actions:

Proportional (P) control:

$$m_p(t) = K_p e(t)$$

$$\frac{M_p(s)}{E(s)} = K_p$$

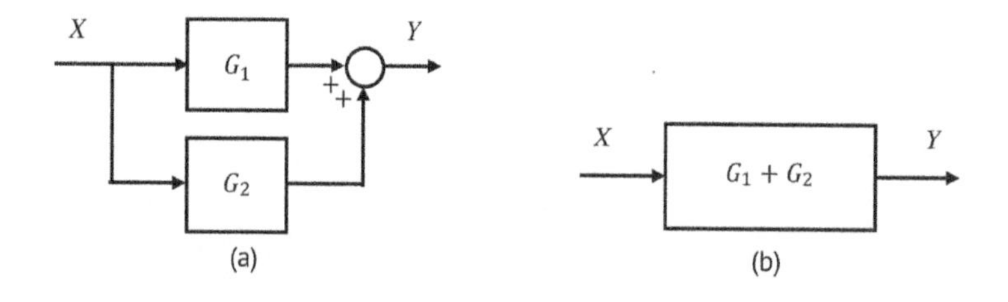

Figure 3.18 Transfer functions and blocks in parallel.

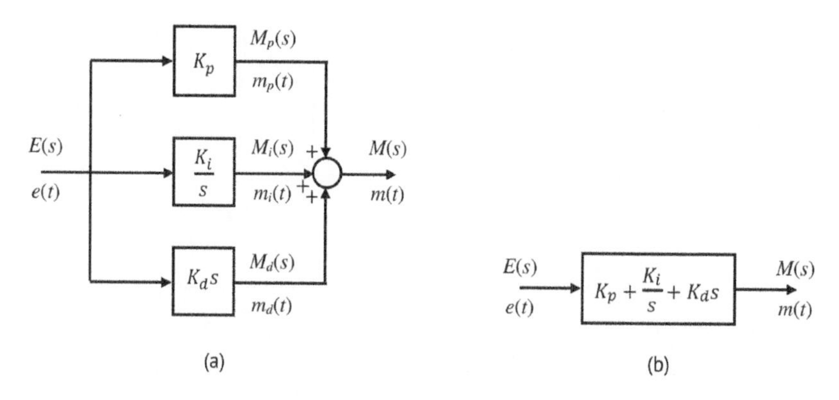

Figure 3.19 PID decision rule block diagrams.

Integral (I) control:

$$m_i(t) = K_i \int_0^t e(t)dt$$

$$\frac{M_i(s)}{E(s)} = \frac{K_i}{s}$$

Derivative (D) control:

$$m_d(t) = K_d \frac{de(t)}{dt}$$

$$\frac{M_d(s)}{E(s)} = K_d s$$

The continuous-time PID decision rule is

$$m(t) = m_p(t) + m_i(t) + m_d(t)$$

Therefore

$$m(t) = K_p e(t) + K_i \int_0^t e(t)dt + K_d \frac{de(t)}{dt}$$

$$M(s) = K_p E(s) + \frac{K_i}{s} E(s) + K_d s E(s)$$

$$M(s) = \left(K_p + \frac{K_i}{s} + K_d s\right) E(s)$$

This result is shown in the single block in Figure 3.19b.

A continuous-time or discrete-time transfer function variable `G` can be created that is the sum of two transfer functions in parallel, represented by variables `G1` and `G2`, using

```
G=parallel(G1,G2)
```

or transfer function variables can be added:

```
G=G1+G2
```

This is demonstrated in Program 3.6 in which the above continuous-time PID transfer functions are combined for $K_p = 8.75$, $K_i = 0.633$, and $K_d = 103$.

Program 3.6 Combining continuous-time transfer functions in parallel

```
Kp=8.75;   % proportional control parameter
Ki=0.633;  % integral control parameter
Kd=103.0;  % derivative control parameter

Gps=Kp  % proportional control transfer function
```

```
Gps = 8.7500
```

```
Gis=Ki*tf(1,[1, 0]) % integral control transfer function
```

```
Gis =
  0.633
  -----
    s
Continuous-time transfer function.
```

```
Gds=Kd*tf([1, 0],1) % derivative control transfer function
```

```
Gds =
  103 s
Continuous-time transfer function.
```

```
Gs=Gps+Gis+Gds % combination of control transfer functions
```

```
Gs =

  103 s^2 + 8.75 s + 0.633
  ------------------------
             s

Continuous-time transfer function.
```

3.8.3 Closed-Loop Relationships

Blocks in a closed loop topology are shown in Figure 3.20a. Transfer function G is in the forward path, transfer function H is in the feedback path and the sign of the feedback is negative. The transfer function that relates output Y to input X is

$$W = X - HY \tag{3.74}$$

$$Y = GW \tag{3.75}$$

$$Y = G(X - HY) \tag{3.76}$$

$$(1+GH)Y = GX \tag{3.77}$$

$$\frac{Y}{X} = \frac{G}{1+GH} \tag{3.78}$$

This result is shown in the block diagram in Figure 3.20b. Similarly, if the sign of the feedback is positive as shown in Figure 3.20c, then the result is

$$\frac{Y}{X} = \frac{G}{1-GH} \tag{3.79}$$

This result is shown in the block diagram in Figure 3.20d.

Other relationships may be of interest. For example, the transfer function that relates output X to input W in Figure 3.20a can be found from Equations 3.74 and 3.75:

$$W = X - GHW \tag{3.80}$$

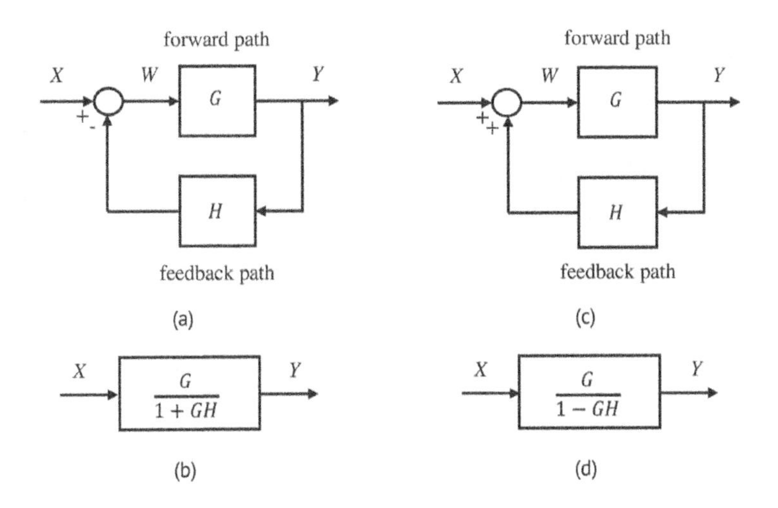

Figure 3.20 Transfer functions in a closed-loop topology: negative feedback (a) and (b); positive feedback (c) and (d).

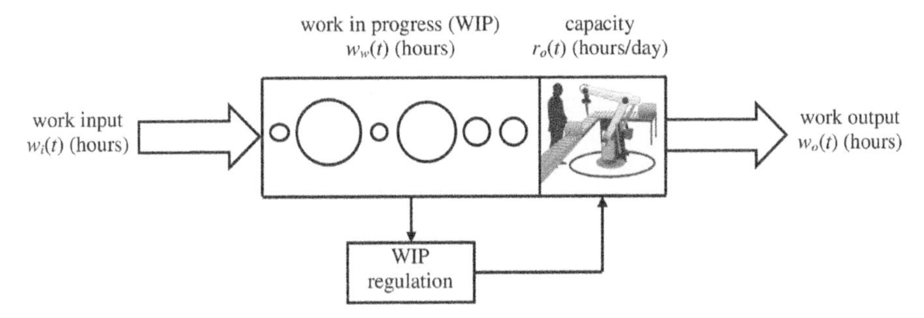

Figure 3.21 Production system with capacity adjustment.

$$(1+GH)W = X \tag{3.81}$$

$$\frac{W}{X} = \frac{1}{1+GH} \tag{3.82}$$

Example 3.18 Closed-Loop Transfer Function for Continuous-Time Capacity Adjustment

It is desired to obtain a continuous-time model for the production work system illustrated in Figure 3.21. The following proportional capacity adjustment rule is used for production capacity $r_o(t)$ hours/day:

$$r_o(t) = K_p w_w(t)$$

where $w_w(t)$ hours is the WIP and positive K_p days^{-1} is an adjustable decision-making parameter. Production capacity is increased as WIP accumulates, and this tends to reduce and eventually eliminate further accumulation of WIP. If there is no WIP, production capacity is zero.

The WIP is

$$w_w(t) = w_i(t) - w_o(t)$$

where $w_i(t)$ hours is the work input and $w_o(t)$ hours is the work output. The relationship between work output hours and production capacity is

$$w_o(t) = w_o(0) + \int_0^t r_o(t)\,dt$$

or

$$\frac{dw_o(t)}{dt} = r_o(t)$$

Assuming zero initial conditions, the Laplace transforms of these equations are

$$R_o(s) = K_p W_w(s)$$

$$W_w(s) = W_i(s) - W_o(s)$$

$$W_o(s) = \frac{1}{s} R_o(s)$$

The closed-loop topology of capacity adjustment as represented by these equations is shown in the block diagram in Figure 3.22.

The transfer function that relates work output to work input can be obtained using Equations 3.70 and 3.78:

$$\frac{W_o(s)}{W_i(s)} = \frac{K_p \frac{1}{s}}{1 + K_p \frac{1}{s}}$$

$$\frac{W_o(s)}{W_i(s)} = \frac{K_p}{s + K_p}$$

while the transfer function that relates WIP to work input can be obtained using Equations 3.70 and 3.82:

$$\frac{W_w(s)}{W_i(s)} = \frac{1}{1 + K_p \frac{1}{s}}$$

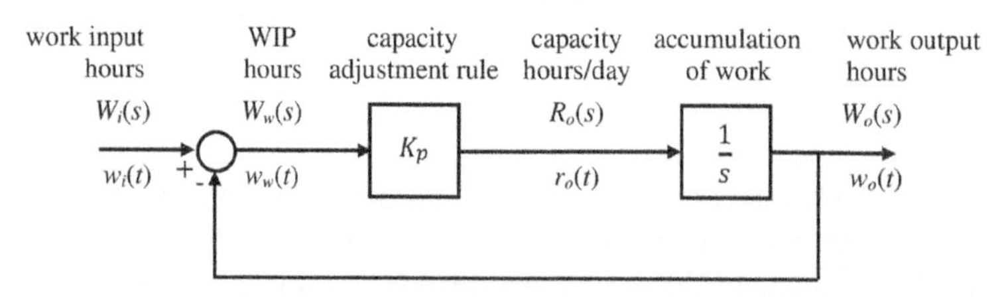

Figure 3.22 Block diagram for production system with capacity adjustment.

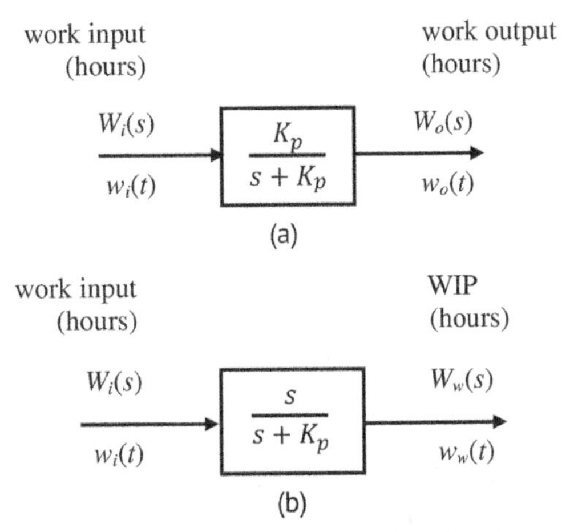

Figure 3.23 Closed-loop transfer functions for capacity adjustment. (a) Dynamic relationship between work output and work input. (b) Dynamic relationship between WIP and work input.

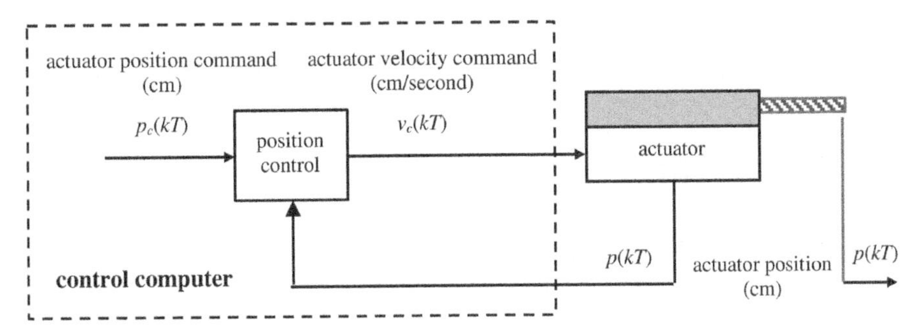

Figure 3.24 Computer-controlled actuator with desired actuator velocity input $v_c(kT)$ and position output $p(kT)$.

$$\frac{W_w(s)}{W_i(s)} = \frac{s}{s + K_p}$$

These results are shown in Figures 3.23a and 3.23b, respectively.

Example 3.19 Closed-Loop Transfer Function for Discrete-Time Control of Actuator Position

For the actuator shown in Figure 3.24, the following discrete-time transfer function represents the dynamic relationship between desired actuator velocity $v_c(kT)$ cm/second and actuator position $p(kT)$ cm:

$$\frac{P(z)}{V_c(z)} = \frac{b_1 z + b_0}{z^2 + a_1 z + a_0}$$

where T seconds is the period between adjustments in desired actuator velocity calculated and output by the control computer and

$$b_1 = T - \tau\left(1 - e^{-T/\tau}\right)$$

$$b_0 = -\left(Te^{-T/\tau} - \tau\left(1 - e^{-T/\tau}\right)\right)$$

$$a_1 = -\left(1 + e^{-T/\tau}\right)$$

$$a_0 = e^{-T/\tau}$$

A position control with command input $p_c(kT)$ cm also is shown in Figure 3.24, and a closed-loop block diagram for the actuator with proportional control is shown in Figure 3.25a where the position control transfer function is

$$\frac{V_c(z)}{E(z)} = K_p$$

The discrete-time transfer function of the closed-loop system can be found using Equation 3.78:

$$\frac{P(z)}{P_c(z)} = \frac{K_p\left(\dfrac{b_1 z + b_0}{z^2 + a_1 z + a_0}\right)}{1 + K_p\left(\dfrac{b_1 z + b_0}{z^2 + a_1 z + a_0}\right)}$$

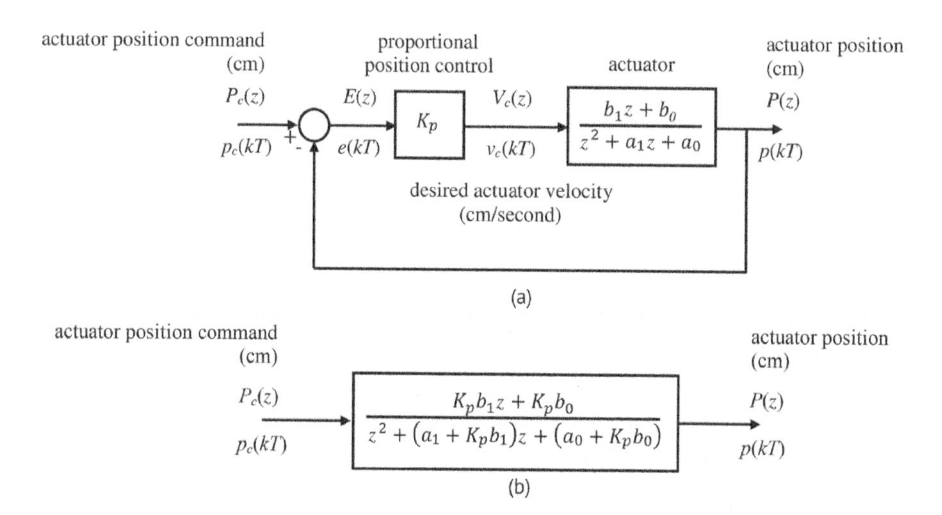

Figure 3.25 Block diagrams for discrete-time proportional control of actuator position. (a) Closed-loop control topology. (b) Closed-loop transfer function.

$$\frac{P(z)}{P_c(z)} = \frac{K_p b_1 z + K_p b_0}{z^2 + (a_1 + K_p b_1)z + (a_0 + K_p b_0)}$$

This result is shown in the block diagram in Figure 3.25b.

A continuous-time or discrete-time transfer function variable F can be created for a closed-loop topology with transfer function variable G in the forward path, H in the feedback path, and negative feedback using

```
F=feedback(G,H)
```

or

```
F=G/(1+G*H)
```

This is demonstrated in Program 3.7 in which the above closed-loop transfer function is calculated for proportional control parameter $K_p = 0.25$ seconds^{-1}, period $T = 0.005$ seconds and time constant $\tau = 0.05$ seconds. In this case, there are no components in the feedback path and $H(z) = 1$ when using `feedback`.

Program 3.7 Calculation of closed-loop transfer function

```
T=0.005;  % period (seconds)
tau=0.05;  % actuator time constant (seconds)
Kp=2.5;  %proportional control parameter (seconds^-1)

b1=T-tau*(1-exp(-T/tau));  % numerator and denominator coefficients
b0=-(T*exp(-T/tau)-tau*(1-exp(-T/tau)));
a1=-(1+exp(-T/tau));
a0=exp(-T/tau);

Gpz=tf([b1, b0],[1, a1, a0],T)  % actuator position transfer function
```

```
   Gpz =

     0.0002419 z + 0.0002339
     -----------------------
     z^2 - 1.905 z + 0.9048

   Sample time: 0.005 seconds
   Discrete-time transfer function.
```

```
Gclz=feedback(Kp*Gpz,1)  % closed-loop transfer function
```

```
   Gclz =

     0.0006047 z + 0.0005849
     -----------------------
     z^2 - 1.904 z + 0.9054

   Sample time: 0.005 seconds
   Discrete-time transfer function.
```

3.8.4 Transfer Functions of Production Systems with Multiple Inputs and Outputs

Most models of production systems have more than one output or more than one input. Rush orders and equipment failures are examples of inputs at different locations in the topology of a production system, and hence they may affect the dynamic behavior of the system differently. Similarly, backlog and production capacity are at different locations in the topology of a production system and may be outputs of interest for different reasons. Design of decision-making is often focused on achieving favorable combinations of input–output behavior.

A single transfer function only represents the dynamic cause-and-effect relationship between one input and one output; however, in a production system with M outputs and N inputs, $M \times N$ transfer functions represent the relationships between individual inputs and outputs. For a given output of interest, a transfer function for each of the inputs can be obtained by assuming that all other inputs are zero. Also, as illustrated in the following examples, these transfer functions can be included in a single equation that represents how the output is affected by contributions of all of the inputs.

Example 3.20 Closed-Loop Continuous-Time Transfer Functions for Backlog Regulation

The continuous-time model found in Example 2.2 for the production system with backlog regulation shown in Figure 2.3 has multiple inputs and outputs. Accumulation of orders was modeled using

$$\frac{dw_b(t)}{dt} = \frac{dw_d(t)}{dt} + r_i(t) - r_o(t)$$

where $w_b(t)$ orders is the order backlog, $w_d(t)$ orders represents disturbances such as rush orders and order cancellations, $r_i(t)$ orders/day is the order input rate and $r_o(t)$ orders/day is the order release rate. The decision rule chosen for backlog regulation was

$$\frac{dr_o(t)}{dt} = K_1 \frac{d\left(w_b(t) - w_p(t)\right)}{dt} + K_2\left(w_b(t) - w_p(t)\right)$$

where $w_p(t)$ orders is the planned order backlog. K_1 days^{-1} and K_2 days^{-2} are decision rule parameters that are designed to obtain favorable closed-loop backlog regulation dynamic behavior.

Assuming zero initial conditions, the Laplace transforms are

$$sW_b(s) = sW_d(s) + R_i(s) - R_o(s)$$

$$sR_o(s) = K_1 s\left(W_b(s) - W_p(s)\right) + K_2\left(W_b(s) - W_p(s)\right)$$

and

$$W_b(s) = W_d(s) + \frac{1}{s}\left(R_i(s) - R_o(s)\right)$$

$$R_o(s) = \left(K_1 + \frac{K_2}{s}\right)\left(W_b(s) - W_p(s)\right)$$

These relationships are shown in the block diagram in Figure 3.26.

In this model there are three inputs: planned backlog, order input rate, and order disturbances. Combining the above equations yields the relationships between these inputs and the two outputs, which are backlog and order release rate:

$$W_b(s) = \frac{\left(K_1 + \frac{K_2}{s}\right)\frac{1}{s}}{1 + \left(K_1 + \frac{K_2}{s}\right)\frac{1}{s}} W_p(s) + \frac{\frac{1}{s}}{1 + \left(K_1 + \frac{K_2}{s}\right)\frac{1}{s}} R_i(s) + \frac{1}{1 + \left(K_1 + \frac{K_2}{s}\right)\frac{1}{s}} W_d(s)$$

$$R_o(s) = -\frac{K_1 + \frac{K_2}{s}}{1 + \left(K_1 + \frac{K_2}{s}\right)\frac{1}{s}} W_p(s) + \frac{\left(K_1 + \frac{K_2}{s}\right)\frac{1}{s}}{1 + \left(K_1 + \frac{K_2}{s}\right)\frac{1}{s}} R_i(s) + \frac{K_1 + \frac{K_2}{s}}{1 + \left(K_1 + \frac{K_2}{s}\right)\frac{1}{s}} W_d(s)$$

or

$$W_b(s) = \frac{(K_1 s + K_2) W_p(s) + s R_i(s) + s^2 W_d(s)}{s^2 + K_1 s + K_2}$$

$$R_o(s) = \frac{-\left(K_1 s^2 + K_2 s\right) W_p(s) + (K_1 s + K_2) R_i(s) + \left(K_1 s^2 + K_2 s\right) W_d(s)}{s^2 + K_1 s + K_2}$$

The individual transfer function that relates each output to each input is present in the above equations. These transfer functions also are shown in the block diagrams in Figure 3.27. It should be noted that while the numerators of the six transfer functions differ, their denominators are identical. This property of closed-loop topologies is advantageous because, as described in subsequent chapters, the denominator is related to fundamental dynamic properties of the modeled production system.

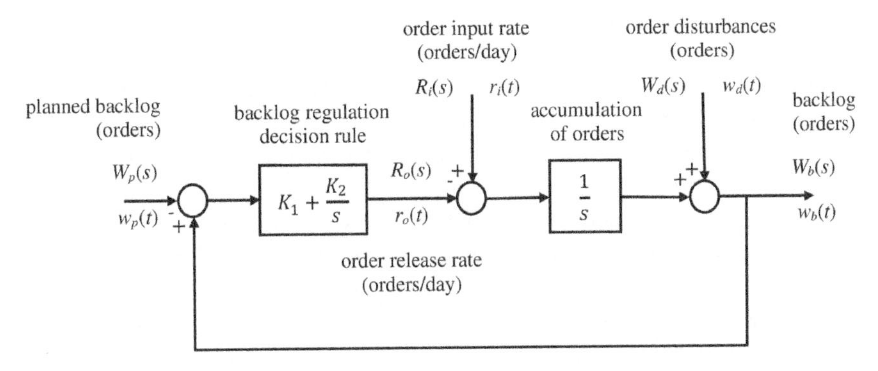

Figure 3.26 Block diagram for a production system with closed-loop backlog regulation.

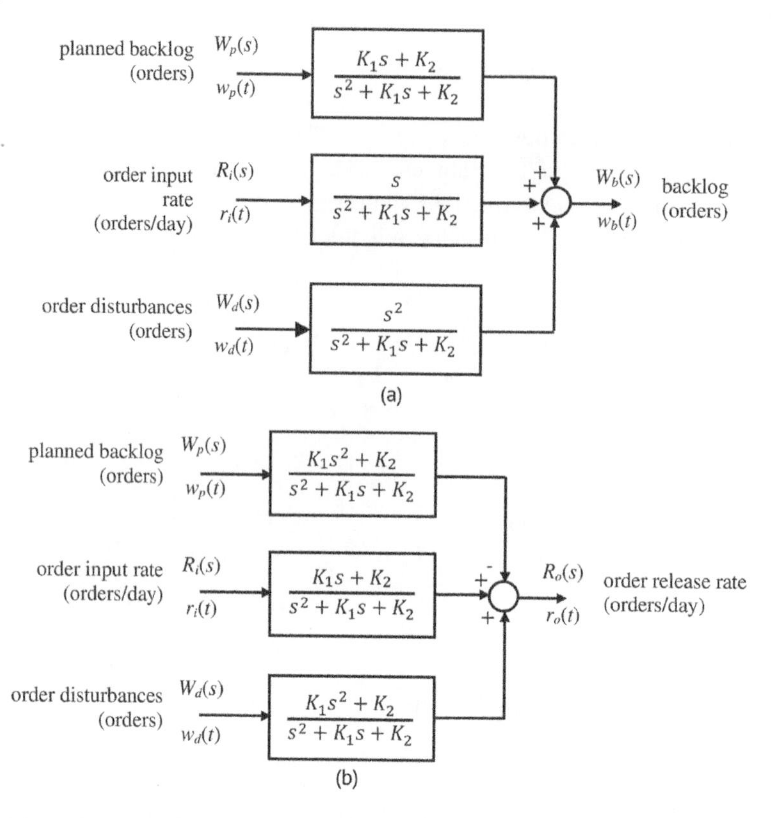

Figure 3.27 Block diagrams showing backlog regulation transfer functions. (a) Dynamic relationships for backlog. (b) Dynamic relationships for order release rate.

Example 3.21 Closed-Loop Discrete-Time Transfer Functions for Pressing Operation Control

A portion of a production system in which a pressing operation is used to process material is illustrated in Figure 3.28. In this pressing operation both actuator position and applied force need to be controlled. Desired force to be applied as a function of time in pressing is one input and the thickness of the material to be pressed is the other input. The material thickness is variable but is unknown in the pressing operation; it is neither sensed nor available in production data. The actual force applied in pressing is of interest, as is actuator position. A model of the dynamic relationships between these outputs and the force and material thickness inputs is needed that can guide design of decision-making in the force and actuator position controls and help predict and verify the operation of the press.

The force control, implemented in a control computer and not yet designed, has transfer function $G_{cf}(z)$ and adjusts actuator position command $p_c(kT)$ cm based on the error $e_f(z)$ N between force command $f_c(kT)$ N and pressing force $f(kT)$ N measured using a force sensor. The position control, also implemented in the control computer and not yet designed, has transfer function $G_{cp}(z)$ and adjusts desired actuator velocity $v_c(kT)$ cm/second based on the error $e_p(kT)$ cm between position command $p_c(kT)$ cm and actual actuator position $p(kT)$ cm.

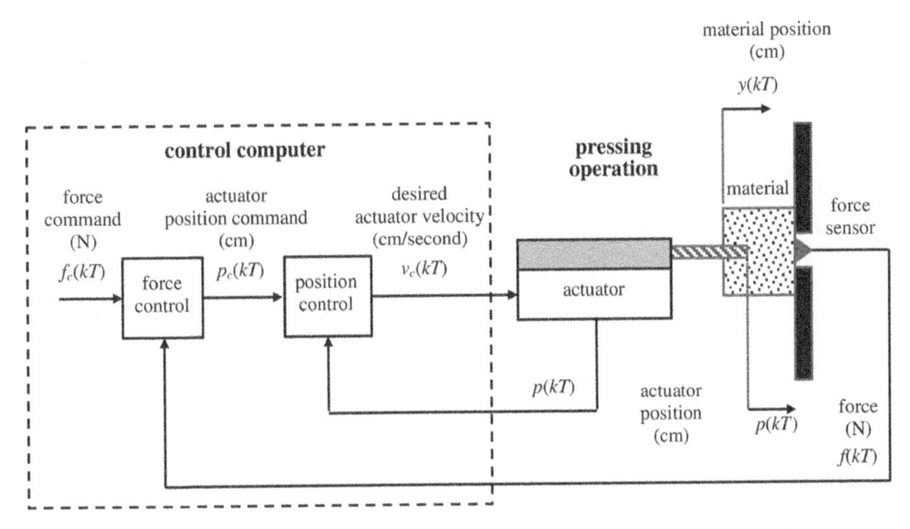

Figure 3.28 Pressing operation with computer control of force and actuator position.

A block diagram for this production process is shown in Figure 3.29a that visually emphasizes force command $f_c kT)$ as an important input; however, the equivalent block diagram in Figure 3.29b visually emphasizes material position $y(kT)$ as an important input. Identical algebraic relationships are represented by these block diagrams, and these relationships can be used to find the transfer functions that describe input–output relationships in a dynamic model of this production process.

The following discrete-time transfer function for the actuator was used in Example 3.19:

$$G_a(z) = \frac{P(z)}{V_c(z)} = \frac{\left(T - \tau\left(1 - e^{-T/\tau}\right)\right)z - \left(Te^{-T/\tau} - \tau\left(1 - e^{-T/\tau}\right)\right)}{z^2 - \left(1 + e^{-T/\tau}\right)z + e^{-T/\tau}}$$

The inner position control loop has the following transfer function, which can be found using Equations 3.70 and 3.78:

$$G_p(z) = \frac{P(z)}{P_c(z)} = \frac{G_{cp}(z)G_a(z)}{1 + G_{cp}(z)G_a(z)}$$

This allows the block diagrams in Figure 3.29 to be simplified as shown in Figure 3.30 to focus on the outer force control loop.

The relationship between pressing force and actuator position can be modelled using

$$f(kT) = K_f\left(p(kT) - y(kT)\right)$$

where K_f N/cm is the stiffness of the material being pressed. The equations for force control are then

$$F(z) = K_f\left(P(z) - Y(z)\right)$$

$$P(z) = \left(F_c(z) - F(z)\right)G_{cf}(z)G_p(z)$$

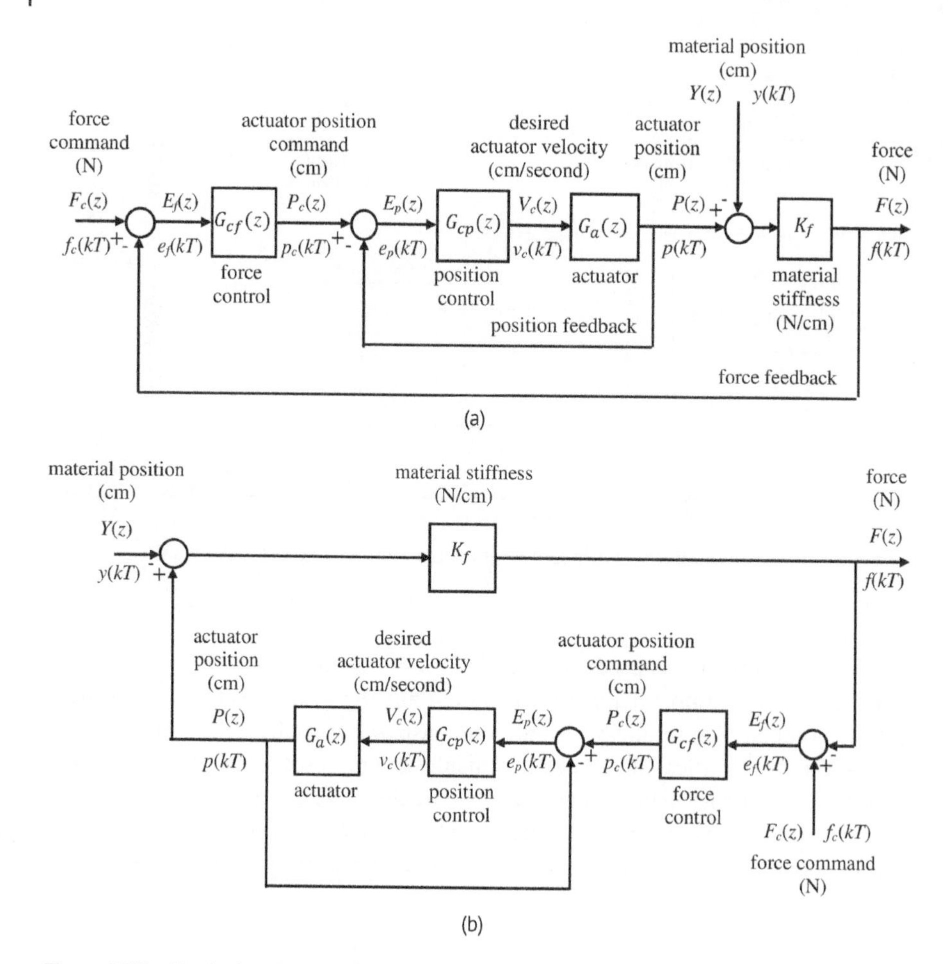

Figure 3.29 Equivalent block diagrams for pressing process control.

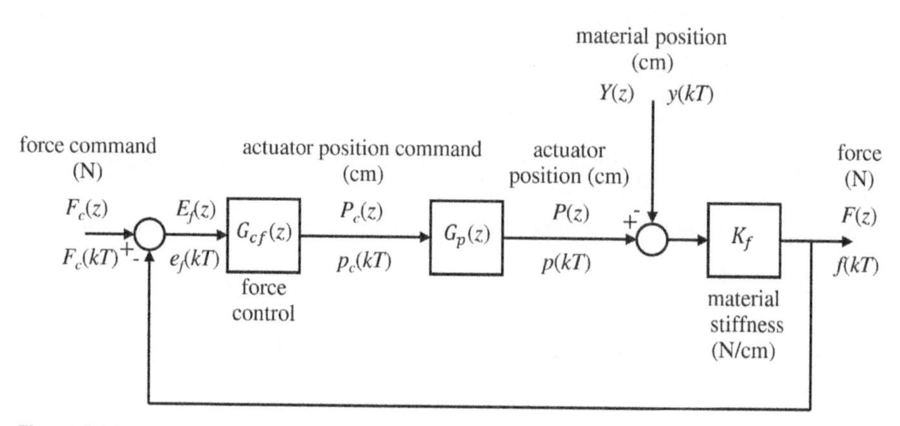

Figure 3.30 Simplified block diagram for pressing process control.

Combining these yields the following input–output relationships

$$F(z) = \frac{K_f G_{cf}(z) G_p(z)}{1 + K_f G_{cf}(z) G_p(z)} F_c(z) - \frac{K_f}{1 + K_f G_{cf}(z) G_p(z)} Y(z)$$

$$P(z) = \frac{G_{cf}(z) G_p(z)}{1 + K_f G_{cf}(z) G_p(z)} F_c(z) - \frac{K_f G_{cf}(z) G_p(z)}{1 + K_f G_{cf}(z) G_p(z)} Y(z)$$

The individual transfer functions that model the dynamic relationship between each output and each input in the above equations are shown in the block diagrams in Figure 3.31. These relationships will facilitate the design of pressing force control and actuator position control as represented by transfer functions $G_{cf}(z)$ and $G_{cp}(z)$, respectively, using methods such as those described in Chapter 6.

3.8.5 Matrices of Transfer Functions

For both continuous-time or discrete-time models, the transfer functions that relate outputs $X_1, X_2, \ldots, X_m, \ldots, X_M$ to inputs $U_1, U_2, \ldots, U_n, \ldots, U_N$ can be represented in matrix form:

$$\begin{bmatrix} X_1 \\ X_2 \\ \vdots \\ X_m \\ \vdots \\ X_M \end{bmatrix} = \begin{bmatrix} G_{11} & G_{12} & \cdots & G_{1n} & \cdots & G_{1N} \\ G_{21} & G_{22} & \cdots & G_{2n} & \cdots & G_{2N} \\ \vdots & \vdots & \ddots & \vdots & \vdots & \vdots \\ G_{m1} & G_{m2} & \cdots & G_{mn} & \cdots & G_{mN} \\ \vdots & \vdots & \vdots & \vdots & \ddots & \vdots \\ G_{M1} & G_{M2} & \cdots & G_{Mn} & \cdots & G_{MN} \end{bmatrix} \begin{bmatrix} U_1 \\ U_2 \\ \vdots \\ U_n \\ \vdots \\ U_N \end{bmatrix} \tag{3.83}$$

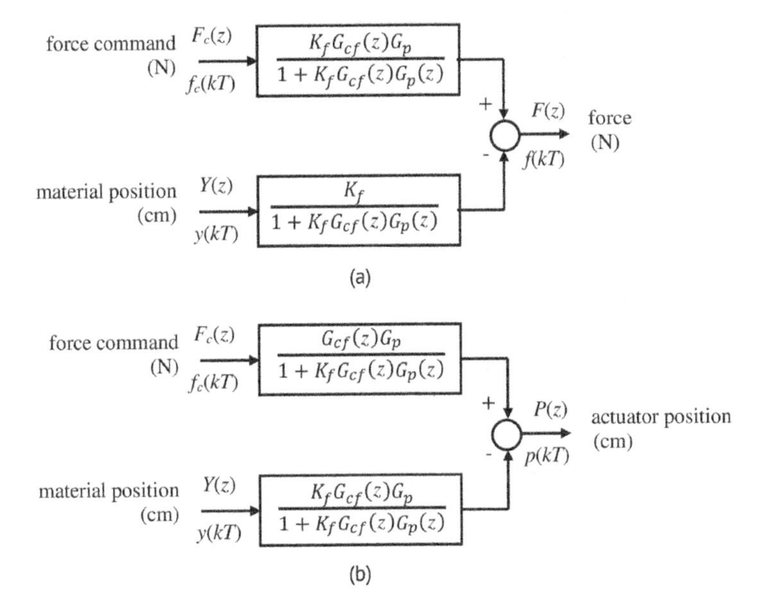

Figure 3.31 Block diagrams for closed-loop pressing system with two inputs.

where G_{mn} is the transfer function that relates output x_m to input u_n. Alternatively,

$$\boldsymbol{X} = \boldsymbol{G}\boldsymbol{U} \tag{3.84}$$

where $\boldsymbol{X}$ is the vector of outputs, $\boldsymbol{U}$ is the vector of inputs and $\boldsymbol{G}$ is the matrix of transfer functions.

If the model for decision-making in a production system has the form shown in Figure 3.32 where $\boldsymbol{R}$ is a vector of P inputs, $\boldsymbol{U}$ is a vector of N decision-making outputs (also N production system components inputs), $\boldsymbol{X}$ is the vector of M production system outputs, $\boldsymbol{G}_c$ is $N{\times}P$ matrix of controller transfer functions, $\boldsymbol{G}_p$ is the $M{\times}N$ matrix of production system component transfer functions, and $\boldsymbol{H}$ is the $P{\times}N$ matrix of feedback transfer functions, then

$$\boldsymbol{U} = \boldsymbol{G}_c\left(\boldsymbol{R} - \boldsymbol{H}\boldsymbol{X}\right) \tag{3.85}$$

$$\boldsymbol{X} = \boldsymbol{G}_p\boldsymbol{G}_c\left(\boldsymbol{R} - \boldsymbol{H}\boldsymbol{X}\right) \tag{3.86}$$

$$\left(\mathbf{I} + \boldsymbol{G}_p\boldsymbol{G}_c\boldsymbol{H}\right)\boldsymbol{X} = \boldsymbol{G}_p\boldsymbol{G}_c\boldsymbol{R} \tag{3.87}$$

$$\boldsymbol{X} = \left(\mathbf{I} + \boldsymbol{G}_p\boldsymbol{G}_c\boldsymbol{H}\right)^{-1}\boldsymbol{G}_p\boldsymbol{G}_c\boldsymbol{R} \tag{3.88}$$

where $\mathbf{I}$ is the $M{\times}M$ identity matrix. The matrix of transfer functions $\boldsymbol{G}$ that describes the dynamic behavior of the closed-loop production system with input vector $\boldsymbol{R}$ and output vector $\boldsymbol{X}$ then is

$$\boldsymbol{G} = \left(\mathbf{I} + \boldsymbol{G}_p\boldsymbol{G}_c\boldsymbol{H}\right)^{-1}\boldsymbol{G}_p\boldsymbol{G}_c \tag{3.89}$$

Example 3.22 Matrix of Continuous-Time Transfer Functions Modeling Warehouse with Two Inputs and One Output

Consider Product X that flows into and out of a warehouse as illustrated in Figure 3.33. An approximate continuous-time model for inventory $x(t)$ items of Product X in the warehouse is

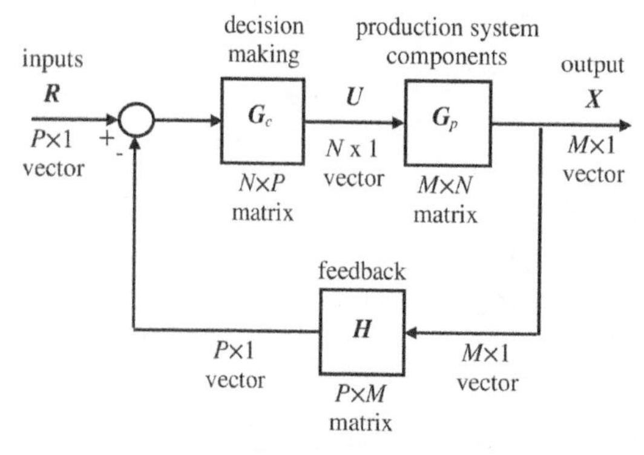

Figure 3.32 Closed-loop production system model with matrices of transfer functions.

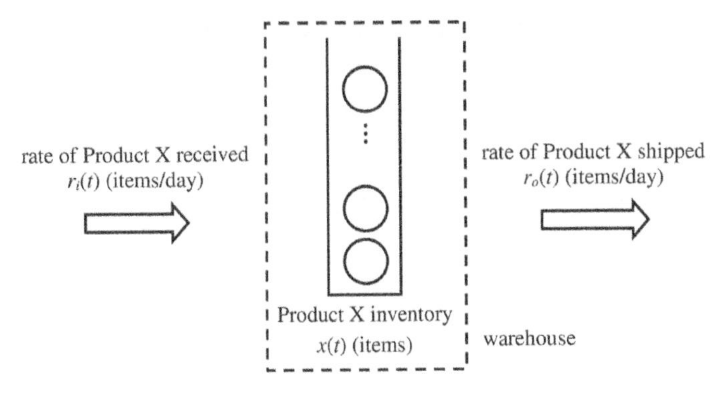

Figure 3.33 Warehouse with Product X input rate $r_i(t)$, output rate $r_o(t)$ and inventory $x(t)$.

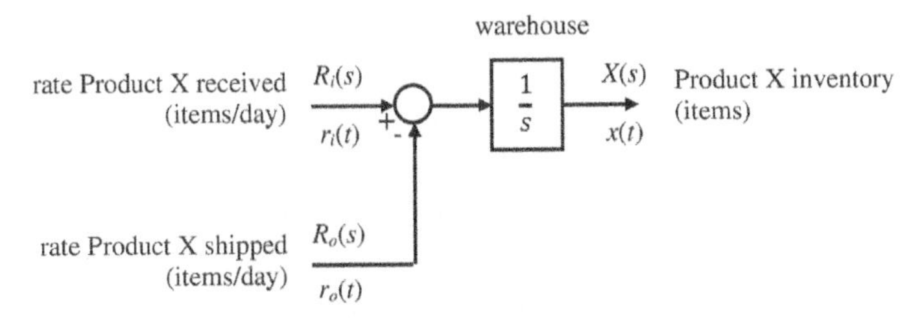

Figure 3.34 Block diagram for Product X inventory in warehouse.

$$\frac{dx(t)}{dt} = r_i(t) - r_o(t)$$

where $r_i(t)$ items/day and $r_o(t)$ items/day are the rates at which Product X is received into and shipped from the warehouse, respectively. The Laplace transform is

$$X(s) = \frac{1}{s}\left(R_i(s) - R_o(s)\right)$$

This model is represented by the block diagram in Figure 3.34. In matrix form, the model becomes

$$\left[X(s)\right] = \begin{bmatrix} \frac{1}{s} & -\frac{1}{s} \end{bmatrix} \begin{bmatrix} R_i(s) \\ R_o(s) \end{bmatrix}$$

or

$$\left[X(s)\right] = \mathbf{G}(s) \begin{bmatrix} R_i(s) \\ R_o(s) \end{bmatrix}$$

where there is one output $X(s)$ and two inputs $R_i(s)$ and $R_o(s)$. The transfer function matrix is

$$\mathbf{G}(s) = \begin{bmatrix} \frac{1}{s} & -\frac{1}{s} \end{bmatrix}$$

Program 3.8 illustrates how a matrix of transfer function variables can be created.

Program 3.8 Creation of continuous-time transfer function matrix

```
s=tf('s','TimeUnit','days');

Gps=[1/s, -1/s]  % transfer function matrix [X(s)/Ri(s) X(s)/Ro(s)]
```

```
Gps =

  From input 1 to output:
  1
  -
  s
  From input 2 to output:
  -1
  --
   s
Continuous-time transfer function.
```

Example 3.23 Matrix of Discrete-Time Transfer Functions Modeling Closed-Loop Control of a Pressing Operation

The following dynamic input–output relationships can be obtained from those found in Example 3.21 for closed-loop control of a pressing operation:

$$P(z) = G_p(z)P_c(z)$$

$$P_c(z) = G_{cf}(z)\left(F_c(z) - F(z)\right)$$

$$F(z) = K_f G_p(z)P_c(z) - K_f Y(z)$$

In Figure 3.35, these equations are expressed in the matrix form shown in Figure 3.32. The matrix of discrete-time closed-loop transfer functions that relates the force command and material position inputs to the force and actuator position outputs can be found using Equation 3.88:

$$\begin{bmatrix} F(z) \\ P(z) \end{bmatrix} = \left(\begin{bmatrix} 1 & 0 \\ 0 & 1 \end{bmatrix} + \begin{bmatrix} K_f G_p(z) & -K_f \\ G_p(z) & 0 \end{bmatrix} \begin{bmatrix} G_{cf}(z) & 0 \\ 0 & 1 \end{bmatrix} \begin{bmatrix} 1 & 0 \\ 0 & 0 \end{bmatrix} \right)^{-1} \times$$
$$\begin{bmatrix} K_f G_p(z) & -K_f \\ G_p(z) & 0 \end{bmatrix} \begin{bmatrix} G_{cf}(z) & 0 \\ 0 & 1 \end{bmatrix} \begin{bmatrix} F_c(z) \\ Y(z) \end{bmatrix}$$

$$\begin{bmatrix} F(z) \\ P(z) \end{bmatrix} = \begin{bmatrix} \dfrac{K_f G_{cf}(z) G_p(z)}{1 + K_f G_{cf}(z) G_p(z)} & \dfrac{-K_f}{1 + K_f G_{cf}(z) G_p(z)} \\ \dfrac{G_{cf}(z) G_p(z)}{1 + K_f G_{cf}(z) G_p(z)} & \dfrac{-K_f G_{cf}(z) G_p(z)}{1 + K_f G_{cf}(z) G_p(z)} \end{bmatrix} \begin{bmatrix} F_c(z) \\ Y(z) \end{bmatrix}$$

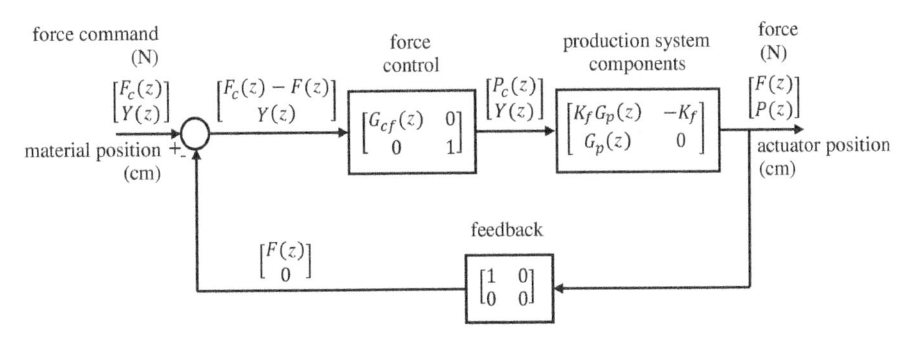

Figure 3.35 Block diagram for closed-loop control of pressing operation in matrix form.

3.8.6 Factors of Transfer Function Numerator and Denominator

It often is useful to factor the numerator and denominator polynomials of a continuous-time or discrete-time transfer function. Analysis of the roots of the denominator polynomial, which can be obtained from its factors, yields important information about the fundamental dynamic behavior of the system or system component represented by the transfer function; this will be discussed in detail in subsequent chapters.

The result of factoring the continuous-time transfer function in Equation 3.35 is

$$G(s) = K_s \frac{(s-c_1)(s-c_2)\ldots(s-c_m)}{(s-r_1)(s-r_2)\ldots(s-r_n)} e^{-sD} \tag{3.90}$$

where roots r_1, r_2, ..., r_n and c_1, c_2, ..., c_m can be real or complex.[2] K_s is a scalar and is often referred to as the gain of the transfer function. Similarly, as the result of factoring the discrete-time transfer function in Equation 3.65 is

$$G(z) = K_z \frac{(z-c_1)(z-c_2)\ldots(z-c_m)}{(z-r_1)(z-r_2)\ldots(z-r_n)} \tag{3.91}$$

Example 3.24 Factored Continuous-Time Transfer Function

Transfer function

$$G(s) = \frac{b_1 s + b_0}{s^2 + a_1 s + a_0}$$

can be factored yielding

$$G(s) = K_s \frac{(s-c_1)}{(s-r_1)(s-r_2)}$$

where

$$K_s = b_1$$

2 The roots r_1, r_2, ..., r_n of the denominator polynomial often are referred to as the poles of transfer function G, and roots c_1, c_2, ..., c_m of the numerator polynomial often are referred to as the zeros of transfer function G.

$$c_1 = -\frac{b_0}{b_1}$$

$$r_{1,2} = \frac{-a_1 \pm \sqrt{a_1^2 - 4a_0}}{2}$$

Example 3.25 Factored Discrete-Time Transfer Function

The following transfer function was used in Example 3.19 to represent the dynamic behavior of the linear actuator shown in Figure 3.24:

$$\frac{P(z)}{V_c(z)} = \frac{\left(T - \tau\left(1 - e^{-T/\tau}\right)\right)z - \left(Te^{-T/\tau} - \tau\left(1 - e^{-T/\tau}\right)\right)}{z^2 - \left(1 + e^{-T/\tau}\right)z + e^{-T/\tau}}$$

This transfer function can be factored into

$$G(z) = K_z \frac{(z - c_1)}{(z - r_1)(z - r_2)}$$

where

$$K_z = T - \tau\left(1 - e^{-T/\tau}\right)$$

$$c_1 = \frac{Te^{-T/\tau} - \tau\left(1 - e^{-T/\tau}\right)}{T - \tau\left(1 - e^{-T/\tau}\right)}$$

$$r_1 = 1$$

$$r_2 = e^{-T/\tau}$$

A continuous-time or discrete-time transfer function variable `G` can be displayed in factored form using

```
zpk(G)
```

Furthermore,

```
[Z,P,K]=zpkdata(G,'v')
```

returns the roots of the numerator polynomial in vector `Z`, the roots of the denominator polynomial in vector `P` and `K` corresponds to either K_s in Equation 3.90 or K_z in Equation 3.91. This is demonstrated in Program 3.9 in which the above discrete-time transfer function is factored when $T = 0.005$ seconds and $\tau = 0.05$ seconds.

3.8.7 Canceling Common Factors in a Transfer Function

The results obtained when combining or manipulating transfer function variables sometimes will be of higher order than expected because there are common factors in

Program 3.9 Factoring a discrete-time transfer function

```
T=0.005;  % period (seconds)
tau=0.05;  % time constant (seconds)

b1=T-tau*(1-exp(-T/tau));  % numerator and denominator coefficients
b0=-(T*exp(-T/tau)-tau*(1-exp(-T/tau)));
a1=-(1+exp(-T/tau));
a0=exp(-T/tau);

Gpz=tf([b1, b0],[1, a1, a0],T)  % actuator transfer function
```

```
  Gpz =
 
      0.0002419 z + 0.0002339
      -----------------------
      z^2 - 1.905 z + 0.9048
 
  Sample time: 0.005 seconds
  Discrete-time transfer function.
```

```
zpk(Gpz)  % factors of actuator transfer function
```

```
  ans =
 
      0.00024187 (z+0.9672)
      ---------------------
        (z-1) (z-0.9048)
 
  Sample time: 0.005 seconds
  Discrete-time zero/pole/gain model.
```

```
[C,R,Kz]=zpkdata(Gpz,'v')  % gain and roots of numerator and denominator
```

```
  C = -0.9672
  R =
      1.0000
      0.9048
  Kz = 2.4187e-04
```

the numerator and denominator. Canceling these common factors simplifies the result, which is beneficial in subsequent analysis and controller design.

Example 3.26 Common Factors in a Continuous-Time Transfer Function

Figure 2.4 shows the mixture heating and mixture temperature regulation components that form a portion of a production process. A continuous-time model for the change in mixture temperature $\Delta h(t)$ °C that results from voltage $v(t)$ input to the heater was found in Example 2.3:

$$\tau_h \frac{d\Delta h(t)}{dt} + \Delta h(t) = K_h v(t)$$

where τ_h seconds is the mixture heating time constant and K_h °C/V is the mixture heating parameter. The mixture heating transfer function is

$$\frac{\Delta H(s)}{V(s)} = \frac{K_h}{\tau_h s + 1}$$

Furthermore, the outlet temperature $h_o(t)$ °C is a function of inlet temperature $h_i(t)$ °C and temperature change $\Delta h(t)$ °C due to heating:

$$H_o(s) = H_i(s) + \Delta H(s)$$

The temperature error is

$$e(t) = h_c(t) - h_o(t)$$

where $h_c(t)$ °C is the desired mixture temperature at the outlet. The corresponding transform is

$$E(s) = H_c(s) - H_o(s)$$

The temperature error is used in the heater voltage adjustment rule

$$\frac{dv(t)}{dt} = K_c\left(\tau_c \frac{de(t)}{dt} + e(t)\right)$$

where K_c and τ_c are temperature regulation parameters for which values must be chosen. The corresponding transfer function is

$$\frac{V(s)}{E(s)} = \frac{K_c(\tau_c s + 1)}{s}$$

A block diagram for mixture temperature regulation is shown in Figure 3.36.

For $\tau_h = 49.8$ seconds and $K_h = 0.4$°C/V, choosing $\tau_c = \tau_h$ and $K_c = 0.06$ (V/second)/°C results in a common factor $(s + 0.02008)$ in the numerator and denominator of the closed-loop temperature regulation transfer function $H(s)/H_c(s)$ calculated using Program 3.10. Such common factors often arise when using control system engineering software, and cancellations often are not performed automatically.

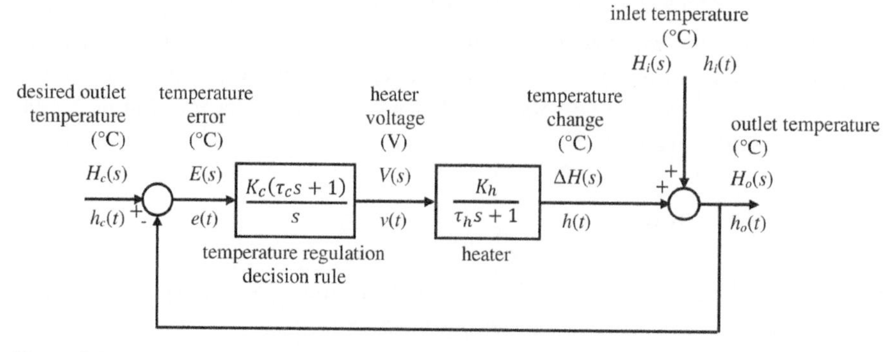

Figure 3.36 Block diagram for mixture temperature regulation.

A transfer function variable G can be simplified by canceling common factors in the numerator and denominator using

```
G=mineral(G)
```

Cancellation of the common factor in the temperature regulation transfer function $H(s)/H_r(s)$ using `minreal` is demonstrated in Program 3.10. The result is a simpler 1st-order transfer function.

Program 3.10 Canceling common factors in a continuous-time transfer function

```
Kh=0.4;  % mixture heating parameter (°C/V)
tauh=49.8;  % mixture heating time constant (seconds)

Kc=0.06;  % temperature regulation parameter ((V/second)/°C)
tauc=tauh;  % temperature regulation parameter (seconds)

Ghs=tf(Kh, [tauc, 1])  % mixture heating transfer function
```

```
  Ghs =

          0.4
    ------------
     49.8 s + 1

  Continuous-time transfer function.
```

```
Gcs=Kc*tf([tauh, 1],[1, 0]);  % regulation rule transfer function

Gcls=feedback(Gcs*Ghs,1)  % closed-loop transfer function
```

```
  Gcls =

            1.195 s + 0.024
    ---------------------------
    49.8 s^2 + 2.195 s + 0.024

  Continuous-time transfer function.
```

```
zpk(Gcls)  % factors
```

```
  ans =

       0.024 (s+0.02008)
    ----------------------
    (s+0.024) (s+0.02008)

  Continuous-time zero/pole/gain model.
```

```
Gcls=minreal(Gcls)  % cancel common factors
```

```
  Gcls =

        0.024
    ------------
      s + 0.024

  Continuous-time transfer function.
```

3.8.8 Padé Approximation of Continuous-Time Delay

Some of the analyses described in Chapter 4 can require replacement of non-zero time delay D in a continuous-time transfer function in the form of Equation 3.35 with an approximating transfer function that does not contain the term e^{-Ds}.

An equivalent polynomial expression for a continuous time delay is

$$e^{-Ds} = 1 - Ds + \frac{D^2}{2}s^2 - \frac{D^3}{6}s^3 + \frac{D^4}{24}s^4 + \cdots \tag{3.92}$$

But this infinite series generally is not helpful. A better option is to replace e^{-Ds} with a Padé approximation.

A 1st-order Padé approximation of delay is

$$e^{-Ds} \approx \frac{-\frac{D}{2}s + 1}{\frac{D}{2}s + 1} \tag{3.93}$$

and a second-order Padé approximation of delay D is

$$e^{-Ds} \approx \frac{\frac{D^2}{12}s^2 - \frac{D}{2}s + 1}{\frac{D^2}{12}s^2 + \frac{D}{2}s + 1} \tag{3.94}$$

Third-order and higher-order Padé approximations also can be applied, but a 2nd-order approximation often is adequate.

Example 3.27 Substitution of Padé Approximation for Continuous-Time Delay

In Example 2.7 the following continuous-time model was developed for the two-company production system with lead time and transportation delays shown in Figure 2.12:

$$r_o(t) = r_i(t - L_B - D_t - L_A)$$

where the input rate to Company B is demand $r_i(t)$ orders/day and the output rate from Company A is $r_o(t)$ orders/day, Companies A and B have lead times L_A and L_B days, respectively, and the transportation time between an order leaving Company B and arriving at Company A is D days. The corresponding transfer function is

$$\frac{R_o(s)}{R_i(s)} = e^{-(L_B + D_t + L_A)s}$$

Substituting the 2nd-order Padé approximation in Equation 3.94,

$$\frac{R_o(s)}{R_i(s)} \approx \frac{\frac{(L_B + D_t + L_A)^2}{12}s^2 - \frac{(L_B + D_t + L_A)}{2}s + 1}{\frac{(L_B + D_t + L_A)^2}{12}s^2 + \frac{(L_B + D_t + L_A)}{2}s + 1}$$

As demonstrated in Program 3.11, function pade replaces the delay in a continuous-time transfer function variable Gs with an Nth-order Padé approximation:

```
Gs=pade(Gs,N)
```

pade sets delay in the resulting transfer function variable to zero.

In Program 3.11, a second-order Padé approximation is used to eliminate the time delay term in the two-company transfer function variable that represents the relationship between order output rate and order input rate.

3.8.9 Absorption of Discrete Time Delay

It has been shown in Example 3.11 that when time delay is present, discrete-time transfer functions often are in the form

$$G(z) = \frac{b_m z^m + \cdots + b_2 z^2 + b_1 z + b_0}{a_n z^n + \cdots + a_2 z^2 + a_1 z + a_0} z^{-d} \tag{3.95}$$

Delay parameter d is a non-negative integer and signifies time delay dT, which is an integer multiple of period T. Because the powers of z in discrete-time transfer functions are integers, delay d can be absorbed into the denominator if desired:

$$G(z) = \frac{b_m z^m + \cdots + b_2 z^2 + b_1 z + b_0}{a_n z^{d+n} + \cdots + a_2 z^{d+2} + a_1 z^{d+1} + a_0 z^d} \tag{3.96}$$

This form, without explicit delay, can be advantageous in some computations performed using control system engineering software.

Program 3.11 Padé approximation of continuous-time delay

```
LA=7;  % Company A lead time delay (days)
LB=11;  % Company B lead time delay (days)
Dt=4;  % transportation delay (days)

% 2-company delay transfer function
Gs=tf(1,1,'TimeUnit','days','OutputDelay',LA+Dt+LB)
```

```
  Gs =
     exp(-22*s) * (1)
  Continuous-time transfer function.
```

```
Gs=pade(Gs,2)  % 2nd-order pade approximation
```

```
  Gs =
     s^2 - 0.2727 s + 0.02479
     -------------------------
     s^2 + 0.2727 s + 0.02479
  Continuous-time transfer function.
```

Example 3.28 Absorption of Delay in a Discrete-Time Transfer Function

In Example 3.16 the following transfer function was obtained that describes the relationship between the portion of production capacity provided by permanent workers $r_p(kT)$ orders/day and production system order input rate $r_i(kT)$ orders/day:

$$\frac{R_p(z)}{R_i(z)} = \frac{\alpha}{1-(1-\alpha)z^{-1}} z^{-d}$$

α is an exponential filter weighting parameter for focusing adjustments in the portion of production capacity provided by permanent workers on lower frequency fluctuations in order input rate, and there is a delay dT in implementing these adjustments. The block diagram in Figure 3.15 contains this relationship.

If transfer function variable Gz explicitly includes delay, it can be absorbed into the denominator polynomial using

```
Gz=absorbDelay(Gz)
```

and the explicit delay then is zero. This is demonstrated in Program 3.12 where delay d is absorbed into the denominator of the above transfer function in a case where the capacity adjustment period is $T = 5$ days, exponential filter weighting parameter $\alpha = 0.75$ and delay in implementing capacity adjustments $dT = 10$ days ($d = 2$ adjustment periods).

Program 3.12 Absorption of delay in a discrete-time transfer function

```
T=5;   % adjustment period (days)
d=2;   % dT=10 delay in implementation (days)
a=0.75;   % exponential filter parameter

% transfer function of exponential filter and time delay in
implementation
Gdz=tf([a, 0],[1, -(1-a)],T,'OutputDelay',d,'TimeUnit','days')

  Gdz =

                    0.75 z
        z^(-2) * ------------
                  z - 0.25

  Sample time: 5 days
  Discrete-time transfer function.

Gdz.OutputDelay  % transfer function with explicit delay

  ans = 2

Gdz=absorbDelay(Gdz)   % absorb output delay into denominator

  Gdz =

           0.75 z
     ------------------
      z^3 - 0.25 z^2

  Sample time: 5 days
  Discrete-time transfer function.
```

3.9 Production Systems with Continuous-Time and Discrete-Time Components

Production systems often have continuous-time and discrete-time inputs, outputs and components. Components that are associated with physical variables often are continuous-time in nature, while components that are associated with decision-making often are discrete-time in nature. Continuous-time and discrete-time models of production system components need to be combined for the purpose of understanding the dynamic behavior of a production system and designing its decision rules, and continuous-time component models must be converted to discrete-time models before transfer function algebra and associated analyses can be applied.

When a continuous-time portion of the topology of a production system transitions to a discrete-time portion, a sampling or measurement component is present in the production system. A model of sampling already has been introduced in Section 3.4. The output of an ideal sampler is a discrete-time variable and the input to a sampler can be either a continuous-time system input or the output of a continuous-time system component. On the other hand, when a discrete-time portion of the topology of a production system transitions to a continuous-time portion, a holding or implementation component is present in the model of the production system. The output of a hold is a continuous-time variable and the input to a hold can be either a discrete-time system input or the output of a discrete-time system component.

3.9.1 Transfer Function of a Zero-Order Hold (ZOH)

The holding function illustrated in Figure 3.37a is referred to as a zero-order hold (ZOH). The hold can take various forms in production systems. One example of a ZOH is implementation of production capacity adjustment decisions that are made periodically and hence are discrete-time. After implementation of the most recent discrete-time decision, continuous-time production capacity remains constant until implementation of the next decision. Another example of a ZOH is a control computer's digital-to-analog converter that produces a continuous-time voltage in production equipment that is proportional to the current value of a discrete-time variable computed in software. The voltage is held constant, and only changes when the value of the discrete-time variable is recomputed.

Discrete-time ZOH input $x(kT)$ has values $x(0)$, $x(T)$, $x(2T)$, ... separated in time by period T, whereas continuous-time output $h(t) = x(kT)$ for each period $kT \leq t < (k+1)T$. The relationship is described by

$$h(t)=\sum_{k=0}^{\infty}x(kT)\big(u(t-kT)-u(t-(k+1)T)\big) \tag{3.97}$$

where $u(t) = 0$ for $t < 0$ and $u(t) = 1$ for $t \geq 0$. The corresponding Laplace transform is found using Equations 3.6 and 3.32:

$$H(s)=\sum_{k=0}^{\infty}x(kT)\left(\frac{e^{-Ts}}{s}-\frac{e^{-(k+1)Ts}}{s}\right) \tag{3.98}$$

$$H(s)=\left(\frac{1-e^{-Ts}}{s}\right)\sum_{k=0}^{\infty}x(kT)e^{-kTs} \tag{3.99}$$

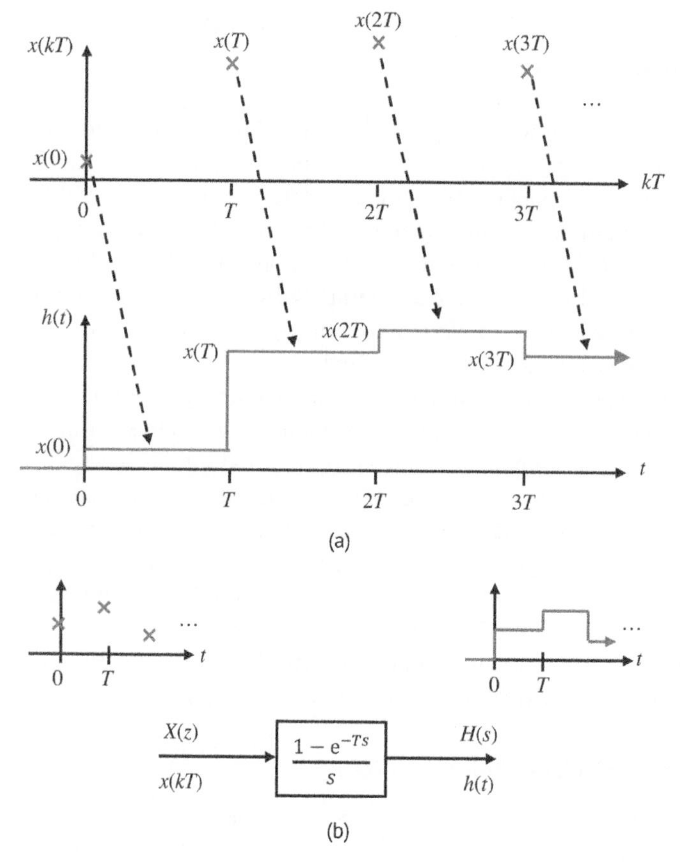

Figure 3.37 Zero-order hold with input $x(kT)$ and output $h(t)$. (a) Example of input $x(kT)$ and output $h(t)$ of zero-order hold (ZOH). (b) Zero-order hold (ZOH) block.

Applying the definitions of the Z transform in Equations 3.37 and 3.38,

$$H(s) = \left(\frac{1 - e^{-Ts}}{s}\right) X(z) \tag{3.100}$$

A zero-order hold (ZOH) block and an example of its discrete-time input and continuous-time output are shown in Figure 3.37b.

3.9.2 Discrete-Time Transfer Function Representing Continuous-Time Components Preceded by a Hold and Followed by a Sampler

A transfer function $G(s)$ that models a continuous-time component or collection of continuous-time components of a production system is shown in Figure 3.38a, preceded by a hold and followed by a sampler. The input to the hold is discrete-time variable $x(kT)$ and the output of the hold is continuous-time variable $h(t)$. Transfer function

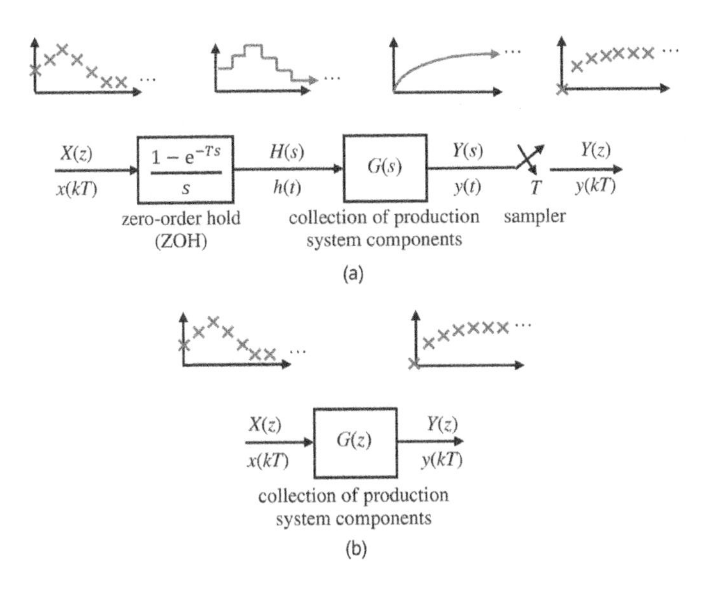

Figure 3.38 Discrete-time transfer function $G(z)$ representing combination of hold, collection production system components represented by transfer function $G(s)$, and sampler. (a) Example of discrete-time input to hold, continuous-time output of hold, continuous-time output of collection of production system components, and discrete-time output of sampler. (b) Corresponding example of relationship between discrete-time input to hold and discrete-time output of sampler.

$G(s)$ models the dynamic relationship between input $h(t)$ of the collection of continuous-time components and output $y(t)$ of the collection of components, which also is the input to the sampler. The output of the sampler is discrete-time variable $y(kT)$. The corresponding discrete-time transfer function $G(z)$ that models the relationship between discrete-time output $y(kT)$ and discrete-time input $x(kT)$ is shown in Figure 3.38b. The dynamic behavior represented by $G(z)$ is the result of the combined dynamic behavior of the hold, collection of continuous components represented by $G(s)$, and the sampler.

Example 3.29 Discrete-Time Model of WIP Reported after Delay

A model of a production system has the topography shown in the block diagram in Figure 3.39a. This model contains both continuous-time and discrete-time variables and components, and it is desired to obtain a discrete-time model that can be used in analysis of the dynamic behavior of the system. The continuous-time model of accumulation of work in the production system is

$$w_w(t) = w_d(t-D) + w_w(-D) + \int_0^t \left(r_i(t-D) - r_o(t-D)\right)dt$$

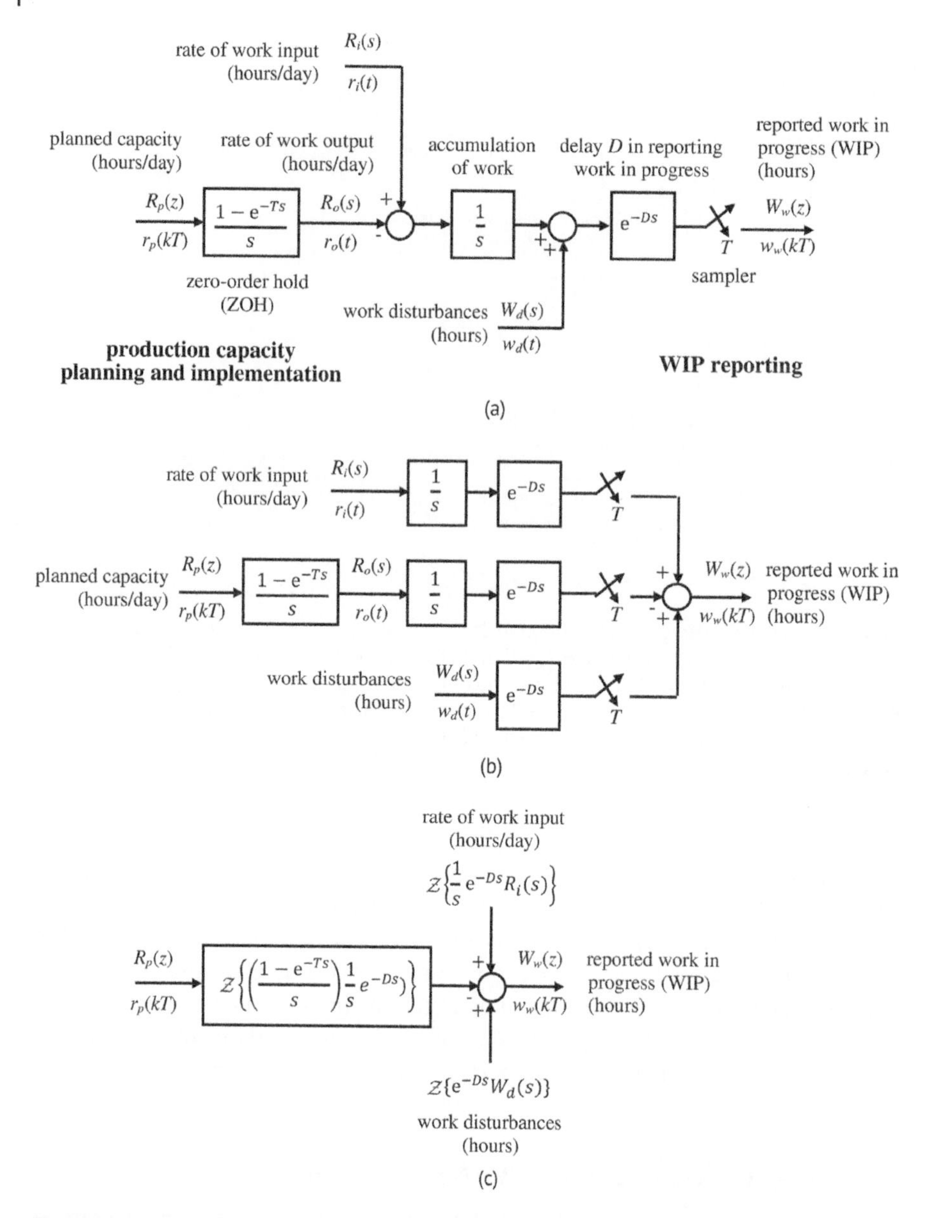

Figure 3.39 Block diagram with discrete-time planned capacity input, continuous-time work input rate and work disturbance inputs, and discrete-time reported WIP output. (a) Block diagram for WIP reported with delay D days after measurement. (b) Equivalent block diagram with continuous-time and discrete-time variables. (c) Equivalent block diagram with only discrete-time variables.

Planned capacity $r_p(kT)$ orders/day is changed with period T days. During each period $kT \leq t < (k+1)T$ the rate of work done $r_o(t) = r_p(kT)$ hours/day is constant. The rate of work input is $r_i(t)$ hours/day. The work in progress (WIP) that has accumulated in the production system depends on $r_i(t)$, $r_o(t)$, and work disturbances $w_d(t)$ hours. WIP is measured, but there is a delay of $D \geq 0$ days in reporting measured WIP.

The block diagram in Figure 3.39a can be redrawn as shown in Figure 3.39b to clearly show how each of the inputs affects reported WIP. The discrete-time relationship between these inputs and reported WIP is

$$W_w(z) = \mathcal{Z}\left\{\frac{1}{s}e^{-Ds}R_i(s)\right\} - G(z)R_p(z) + \mathcal{Z}\left\{e^{-Ds}W_d(s)\right\}$$

as illustrated in Figure 3.39c where the discrete-time transfer function that represents the combination of the ZOH, accumulation of WIP, delay in reporting WIP, and sampling is

$$G(z) = \mathcal{Z}\left\{\left(\frac{1-e^{-Ts}}{s}\right)\frac{1}{s}e^{-Ds}\right\}$$

which depends on period T and delay D. The delay can be separated into

$$D = dT + D'$$

where $0 \leq D' < T$ and d is an integer:[3]

$$d = \text{floor}\left(\frac{D}{T}\right)$$

This Z transform is well known:[4]

$$G(z) = \frac{T\left(1 - \frac{D'}{T}\right)z^{-1} + D'z^{-2}}{1 - z^{-1}}z^{-d}$$

The Z transforms that are functions of given continuous-time inputs $r_i(t)$ and $w_d(t)$ can be obtained if necessary using the definition in Equation 3.37:

$$\mathcal{Z}\left\{\frac{1}{s}e^{-Ds}R_i(s)\right\} = \sum_{k=0}^{\infty}\left(r_i(-D) + \int_0^{kT} r_i(t-D)\,dt\right)z^{-k}$$

$$\mathcal{Z}\left\{e^{-Ds}W_d(s)\right\} = \sum_{k=0}^{\infty} w_d(kT - D)z^{-k}$$

Program 3.13 demonstrates how discrete-time transfer function $G(z)$ can be obtained, given continuous-time transfer function $G(s)$, WIP sampling period T and the presence

---

3 floor(x) is the largest integer not greater than x.

4 This is often referred to as a modified Z transform because it incorporates delay D that can be a non-integer multiple of period T. Tables of Z and modified Z transforms can be found in many control theory publications, and control system engineering software provides functions for numerically calculating these transforms.

Program 3.13 Discrete-time transfer function representing ZOH, accumulation of work, delay and WIP sampler

```
T=5;   % period T (days)
D=1;   % delay in reporting WIP measurements (days)

% continuous-time transfer function for accumulation of work and delay
Gs=tf(1,[1, 0],'TimeUnit','days','OutputDelay',D)
```

```
Gs =

                  1
    exp(-1*s)  *  -
                  s

Continuous-time transfer function.
```

```
% tranfer function for ZOH, accumulation of WIP, delay, and sampler
Gz=c2d(Gs,T)
```

```
Gz =

               4 z + 1
  z^(-1)  *  ----------
                z - 1

Sample time: 5 days
Discrete-time transfer function.
```

of the zero-order hold. Discrete-time transfer function variable `Gz` can be obtained from continuous-time transfer function variable `Gs` for sample period `T` using

```
Gz=c2d(Gs,T,'zoh')
```

where `'zoh'` is the default type of hold and can be omitted. The transfer functions of all of the continuous-time components between the hold and the WIP sampler must be combined into continuous-time transfer function variable `Gs`.

The result obtained in Program 3.13 is for the case where sample period $T = 5$ days and delay $D = 1$ day.

Example 3.30 Discrete-Time Model of Mixture Temperature Regulation

In Example 2.3, a dynamic model was developed for a portion of a production process in which the temperature of a mixture flowing out of a pipe is regulated using a heater and a temperature sensor placed at the end of the pipe as shown in Figure 2.4. Time constant $\tau = 49.8$ seconds and heater parameter $K_h = 0.4$ °C/V were found using experimental data, and the following continuous-time dynamic model of mixture heating was developed:

$$\tau \frac{d\Delta h(t)}{dt} + \Delta h(t) = K_h v(t)$$

where $\Delta h(t)$ °C is the change in temperature of the mixture due to heating and $v(t)$ is the voltage applied to the heater. The mixture inlet temperature is $h_i(t)$ °C and the mixture outlet temperature $h_o(t)$ °C after heating is

$$h_o(t) = h_i(t) + \Delta h(t)$$

and the Laplace transforms of this heater model are

$$\Delta H(s) = \frac{K_h}{\tau s + 1} V(s)$$

$$H_o(s) = H_i(s) + \Delta H(s)$$

In this example, mixture temperature regulation is implemented using a control computer. The topology of temperature regulation is shown in the block diagram in Figure 3.40, which includes a zero-order hold (ZOH), an amplifier to supply the heater voltage, and a sampler for the output of the mixture temperature sensor. The discrete-time temperature regulation decision rule chosen, which is implemented in the control computer, is

$$v_c(kT) = v_c((k-1)T) + K_1(h_c(kT) - h_o(kT)) + K_2(h_c((k-1)T) - h_o((k-1)T))$$

where $h_c(kT)$ °C is the desired mixture outlet temperature, $v_c(kT)$ V is the calculated heater voltage manipulation, and K_1 V/°C and K_2 V/°C are temperature regulation parameters that need to be designed to obtained favorable dynamic behavior. The Z transform for temperature regulation decision-making then is

$$V_c(z) = \frac{K_1 + K_2 z^{-1}}{1 - z^{-1}} (H_c(z) - H_o(z))$$

The block diagram in Figure 3.41a more clearly shows the discrete-time nature of closed-loop mixture temperature regulation, and leads to the discrete-time model for heating, which combines the ZOH, amplifier, continuous-time model of mixture heating and temperature sampling shown in Figure 3.41b:

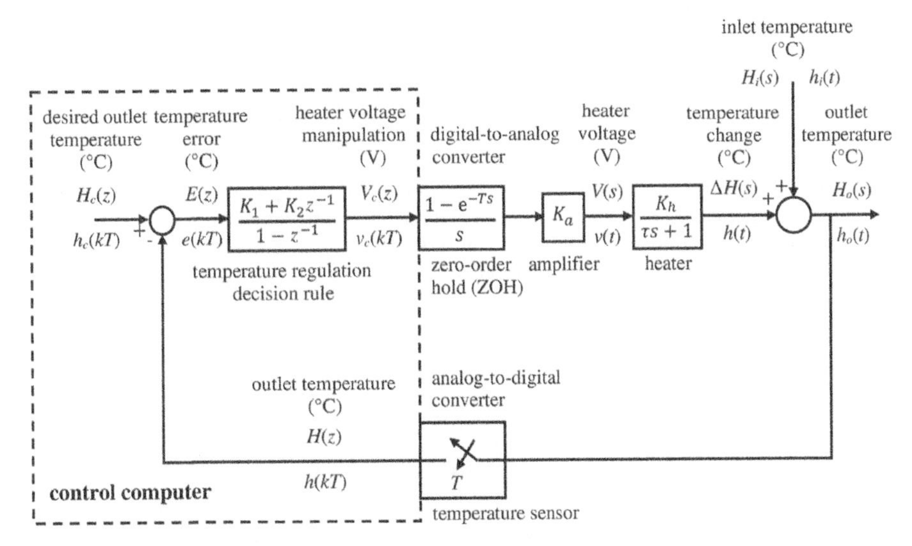

Figure 3.40 Block diagram for computer mixture temperature regulation with discrete-time and continuous-time variables and transfer functions.

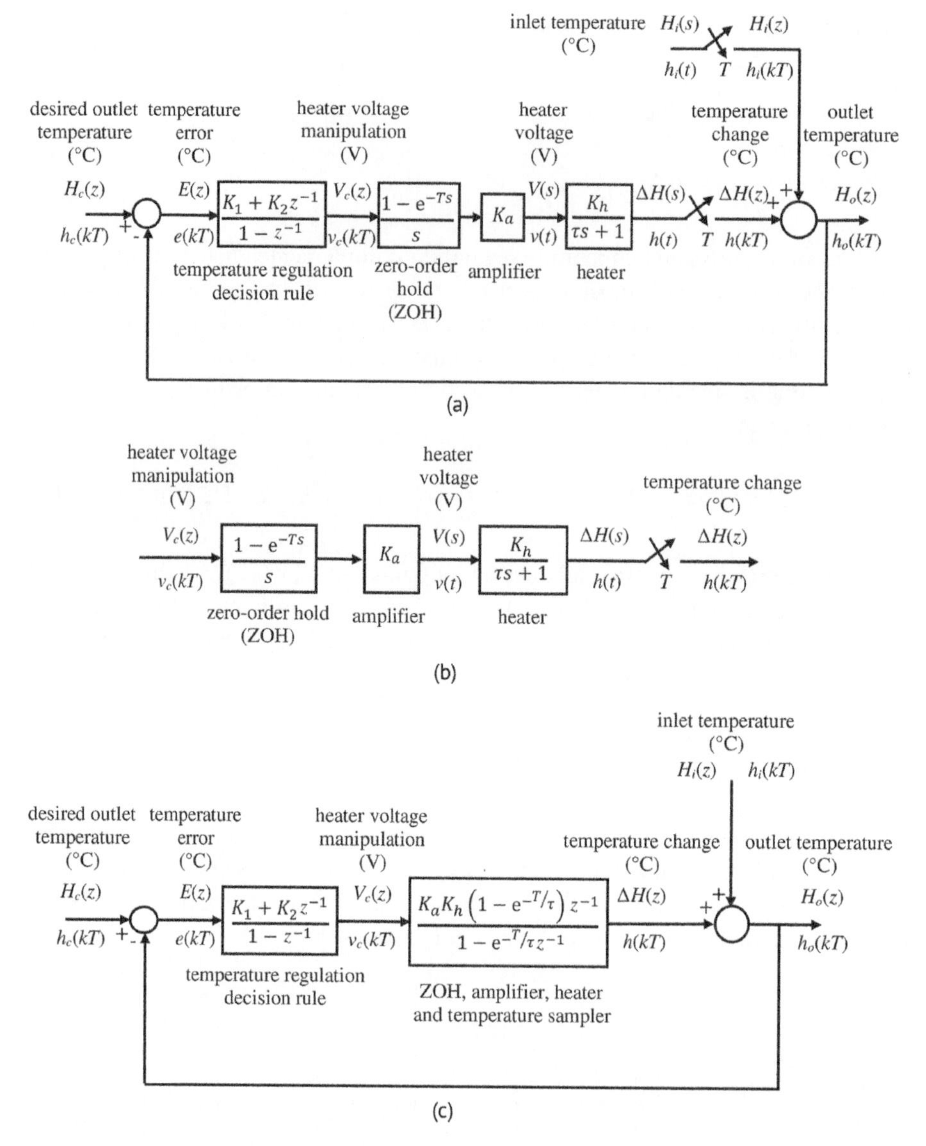

Figure 3.41 Replacement of zero-order hold, continuous-time model of heating and sampler with discrete-time model of heating. (a) Block diagram with relocated sampling. (b) Blocks for obtaining discrete-time mixture heating transfer function. (c) Block diagram with discrete-time model of mixture temperature regulation.

$$\Delta H(z) = \mathcal{Z}\left\{\left(\frac{1 - e^{-Ts}}{s}\right) K_a \frac{K_h}{\tau s + 1}\right\} V_c(z)$$

where K_a V/V is the amplifier parameter.

The continuous-time model for mixture heating can be used to obtain a new model in a discrete-time form for temperature regulation period T. At time T for constant heater voltage manipulation $v_c(0)$ V during the interval $0 \le t < T$,

$$\Delta h(T) = e^{-T/\tau} \Delta h(0) + K_a K_h \left(1 - e^{-T/\tau}\right) v_c(0)$$

For constant heater voltage manipulation $v_c(kT)$ V during the interval $kT \le t < (k+1)T$:

$$\Delta h\big((k+1)T\big) = e^{-T/\tau} \Delta h(kT) + K_a K_h \left(1 - e^{-T/\tau}\right) v_c(kT)$$

and as shown in Figure 3.41c, the corresponding discrete-time transfer function has the well-known form

$$\frac{\Delta H(z)}{V_c(z)} = \frac{K_a K_h \left(1 - e^{-T/\tau}\right) z^{-1}}{1 - e^{-T/\tau} z^{-1}}$$

Program 3.14 can be used to obtain a discrete-time transfer function variable for mixture heating without using the above derivation for the components in Figure 3.41b when $T = 5$ seconds, $\tau = 49.8$ seconds, $K_a = 20$ V/V, and $K_h = 0.4$°C/V.

Program 3.14 Discrete-time transfer function representing ZOH, amplifier, mixture heating, and temperature measurement

```
T=5;  % period (seconds)
tau=49.8;  % mixture heating time constant (seconds)
Kh=0.4;  % mixture heating parameter (°C/V)
Ka=20;  % amplifier parameter (V/V)

Ghs=tf(Kh,[tau, 1])  % continuous-time mixture heating transfer function
```

```
Ghs =

       0.4
  ----------
  49.8 s + 1

Continuous-time transfer function.
```

```
Gz=c2d(Ka*Ghs,T,'zoh')  % discrete-time heating transfer function
```

```
Gz =

    0.7642
  ----------
  z - 0.9045

Sample time: 5 seconds
Discrete-time transfer function.
```

3.10 Potential Problems in Numerical Computations Using Transfer Functions

The results of numerical computation of transfer functions using control system engineering software sometimes do not precisely match the theoretical results that are expected from symbolic manipulations. Results can be obtained that are of higher order than expected, coefficients that are expected to be zero may have small values instead, factors in the numerator and denominator are expected to cancel but do not, and values of roots may differ somewhat from what is expected. Many computational problems are related to challenges of numerical methods for handling polynomials. The coefficients in transfer function polynomials often have a relatively large range of values, and combining transfer function models sometimes leads to ill-conditioned computations. As discussed in subsequent chapters, fundamental dynamic properties are directly related to the roots of denominator polynomials, but the roots found by numerical computations can be unexpectedly sensitive to the values and precision of coefficients. Multiple equal roots and higher orders tend to make computational problems worse, as does transfer function algebra involving addition or subtraction.[5]

When problems are encountered in numerical computations using transfer functions in control system engineering software, they often can be reduced by one or a combination of the following approaches:

- Convert transfer function models to state-space models[6] before combining them.
- Avoid numerically adding or subtracting transfer functions.
- Ensure that known common denominator polynomials remain numerically identical, for example, by numerically adding or subtracting only the numerator polynomials.
- Cancel common factors early in numerical computations.
- Obtain theoretical, symbolic[7] solutions and substitute numerical values in later steps.

Detecting, solving and avoiding numerical problems in the use of control system engineering software can be challenging, and more than a cursory consideration is beyond the scope here. The following simple example illustrates some of the alternative computational approaches that can be used and some of the unexpected results that can be obtained.

---

5 There are barriers of attainable accuracy in standard root-finding methods used in control engineering software. Alternative methods exist that can find multiple roots with non-trivial multiplicities with significantly improved accuracy. The reader is referred to the Bibliography and many other publications on mathematical software for more information on this topic.

6 Definition of system and component models in state-space form, the format and implications of the matrices used in state-space models, modeling of internal delays, and concepts such as observability and controllability are not discussed here. Nevertheless, the state-space form is recognized as an alternative mathematical representation of transfer functions that has advantages in the numerical computations performed by control system engineering software. The interested reader is referred to the Bibliography and many other publications in which the state-space approach to modeling, analysis, and design of control systems is discussed.

7 Control system engineering software often supports symbolic definitions and manipulations of equations, including finding symbolic expressions for roots of polynomials and simplification of symbolic equations. The application of this functionality in control system analysis and design is not discussed here, and the reader is referred to control system engineering software documentation to review supported symbolic tools.

Example 3.31 Numerical Inaccuracies in Addition of Transfer Functions

A production network has five identical work systems in parallel as shown in the block diagram in Figure 3.42. Twenty percent of the input rate to the production network goes to each work system. The work output rate of the production work systems can be changed but not immediately, and a time constant τ days approximately describes the first-order lagging discrete-time relationship between production work system input rate and output rate:

$$r_o(kT) = e^{-T/\tau} r_o((k-1)T) + \left(1 - e^{-T/\tau}\right) r_i((k-1)T)$$

$$G(z) = \frac{R_o(z)}{R_i(z)} = \frac{\left(1 - e^{-T/\tau}\right) z^{-1}}{1 - e^{-T/\tau} z^{-1}}$$

where T is the sampling period, daily for example ($T = 1$ day). The production network capacity transfer function is simply

$$\frac{R_a(z)}{R_i(z)} = G(z) + G(z) + G(z) + G(z) + G(z) = 5G(z)$$

Several methods are used in Program 3.15 to compute this transfer function, and the results obtained illustrate both the challenges of numerical computation of transfer function algebra and how performing computations with state-space variables and avoiding addition may improve numerical results.

Transfer function variable G can be converted to a state-space variable using

```
Gss=ss(G)
```

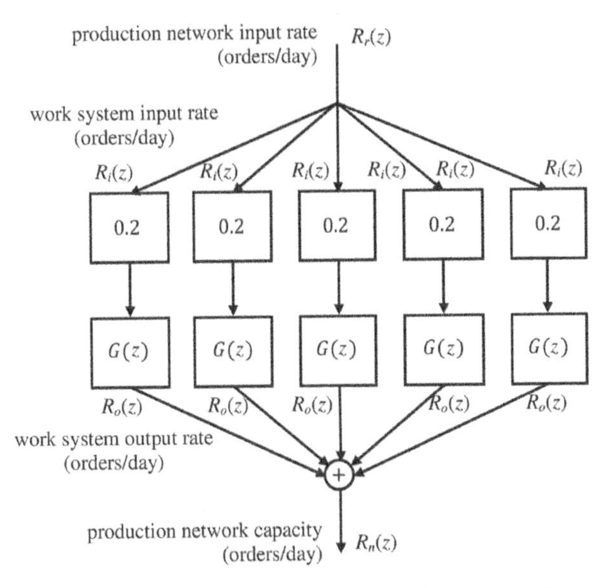

Figure 3.42 Production network with five identical work systems in parallel.

Program 3.15 Approaches for summing transfer function variables

```
T=1;  % period (days)
tau=10;  % time constant (days)
G=tf(1-exp(-T/tau),[1, -exp(-T/tau)],T,'TimeUnit','days')
```

```
G =

     0.09516
  --------------
   z - 0.9048

Sample time: 1 days
Discrete-time transfer function.
```

```
G5=5*G  % five transfer functions in parallel, expected result
```

```
G5 =

      0.4758
  --------------
    z - 0.9048

Sample time: 1 days
Discrete-time transfer function.
```

```
Gsum=G+G+G+G+G;
zpk(Gsum)  % unexpected result
```

```
ans =

  0.47581 (z-0.9048)^4
  ---------------------
       (z-0.9048)^5

Sample time: 1 days
Discrete-time zero/pole/gain model.
```

```
minreal(Gsum,10^-3)  % relax cancelation tolerance, unexpected result
```

```
ans =

     0.4758
  ----------
  z - 0.9035

Sample time: 1 days
Discrete-time transfer function.
```

```
Gss=ss(G)+G+G+G+G;  % force computations to state-space
tf(minreal(Gss))  % expected result
```

```
4 states removed.
ans =

     0.4758
  -----------
  z - 0.9048

Sample time: 1 days
Discrete-time transfer function.
```

```
[num,den]=tfdata(G,'v');  % obtain numerator and denominator polynomials
Gnum=tf(num+num+num+num+num,den,T);  % sum numerator polynomials
Gnum.TimeUnit='days'  % expected result
```

```
Gnum =

        0.4758
     ----------
     z - 0.9048

Sample time: 1 days
Discrete-time transfer function.
```

where G is a transfer function variable and Gss is a corresponding state-space variable. The state-space representation takes precedence in numerical computations, and when one variable in an operation such as addition or multiplication is a state-space variable, the other variable is converted to state-space and the result is a state space variable. State-space variables can be converted to transfer function variables using

```
G=tf(Gss)
```

In Program 3.15, the work system transfer function $G(z)$ is calculated for $T = 1$ days and $\tau = 10$ days. Next, the production network transfer function is calculated by multiplying: $5G(z)$; as expected, the result has the same denominator. Next, the work system transfer function is summed: $G(z) + G(z) + G(z) + G(z) + G(z)$. The result has extra, identical factors in the numerator and denominator, and they do not cancel because of issues such as root finding precision. Knowledge that the transfer functions being summed are identical has not been taken advantage of in the numerical computations. Unfortunately, `minreal` with its default cancellation tolerance does not remove what should be common factors in the numerator and denominator polynomials; the inaccuracies in root finding with repeated roots are greater than the default `minreal` tolerance of 10^{-6}. Relaxing this tolerance to 10^{-3} results in cancellation of common factors, but the denominator of the result differs from the expected result and what would have been obtained by symbolic manipulations. Even though this is a simple system, the difference is numerically significant.

Two additional approaches for summing the transfer function are also used in Program 3.15. First, the summation is done using state-space variables. In this case, application of `minreal` with the default tolerance cancels common factors and the expected result is obtained. Second, the numerator and denominator polynomials are obtained from work system transfer function $G(z)$, and the numerator polynomials then are summed to obtain the numerator polynomial of the production network transfer function for which the denominator polynomial is known to be the same as the denominator polynomial of $G(z)$. In this case, the expected result also is obtained.

3.11 Summary

The Laplace and Z transforms were defined in this chapter, and it was shown how transfer functions based on these transforms can be obtained and manipulated. Transfer functions represent input–output relationships models of components in production systems, as well as input–output relationships in combined models of production systems. Block diagrams were shown to be useful for graphically illustrating the topography of these relationships; this is important because these relationships must be well understood in order to perform the dynamic analyses of complete production systems and design of decision-making that will be described in subsequent chapters.

Production systems often have a mix of continuous-time and discrete-time components, and it was shown how a discrete-time transfer function can be obtained for continuous-time components that are preceded by a zero-order hold and followed by a sampler, allowing a discrete-time transfer function to be obtained that represents the dynamic behavior of the production system. A variety of other aspects of obtaining and manipulating transfer functions also were addressed that will be important in subsequent chapters.

4

Fundamental Dynamic Characteristics and Time Response

Fundamental dynamic characteristics of production systems and components can be obtained directly from the transfer functions that describe their behavior. These characteristics are important and fundamental because, like transfer functions, they do not depend on particular inputs; rather, they describe how quickly the outputs of a system or component tend to respond to inputs in general; for example, whether responses tend to be oscillatory and whether the system or component is stable or not stable. If stable, outputs tend to decay to a steady state with time; if not stable, outputs tend to continue to grow with time and not reach a steady state. Important fundamental dynamic characteristics are defined in this chapter, and their influence on time response is illustrated. This is useful because production engineers are familiar with time response and are comfortable with models and simulations that predict how production systems and their components respond to various inputs. In subsequent chapters, it will be shown how fundamental dynamic characteristics influence frequency response and how they can be used to design decision making for production systems.

Time constants, damping ratios, natural frequencies and stability are fundamental characteristics of dynamic behavior. These characteristics are defined in this chapter for continuous-time and discrete-time models of production systems and components, and they are obtained directly from transfer functions without requiring simulation or calculation of time response. This is done by decomposing transfer functions into simpler elements, the characteristics of which are valuable in assessing whether dynamic behavior is favorable and appropriate. Production systems can have unstable components or can become unstable when numerical values of decision-making parameters are chosen incorrectly. Criteria for stability of continuous-time and discrete-time systems and components are derived in this chapter, and examples are given that illustrate how the elements of transfer functions are investigated for the purpose of assessing stability.

The concepts of transient and steady-state time response are discussed along with basic methods for calculating time response. It often is useful to consider the response to step input functions, and specific characteristics of step response such as final value, settling time and overshoot often can anticipated using fundamental dynamic

Control Theory Applications for Dynamic Production Systems: Time and Frequency Methods for Analysis and Design, First Edition. Neil A. Duffie.

characteristics.[1] However, response to more complicated functions of time also may be of interest when it is known that a production system will be subjected to specific inputs when in operation. Responses to both simple and general input functions of time are readily calculated using control system engineering software as illustrated in examples presented.

4.1 Obtaining Fundamental Dynamic Characteristics from Transfer Functions

Fundamental dynamic characteristics of a production system or component can be obtained from the roots of the characteristic equation associated with the continuous-time or discrete-time transfer function that represents its dynamic behavior. These fundamental characteristics include time constants, damping ratios, natural frequencies and whether the system or component is stable.

4.1.1 Characteristic Equation

A characteristic equation is obtained from the denominator of a transfer function. It was noted in Section 3.8.6 that continuous-time transfer function,

$$G(s)=\frac{Y(s)}{X(s)}=\frac{b_m s^m+\cdots+b_2 s^2+b_1 s+b_0}{a_n s^n+\cdots+a_2 s^2+a_1 s+a_0}\mathrm{e}^{-Ds} \tag{4.1}$$

can be factored into the form

$$G(s)=\frac{Y(s)}{X(s)}=K_s\frac{(s-c_1)(s-c_2)\cdots(s-c_m)}{(s-r_1)(s-r_2)\cdots(s-r_n)}\mathrm{e}^{-Ds} \tag{4.2}$$

The characteristic equation associated with this continuous-time transfer function is

$$a_n s^n+\cdots+a_2 s^2+a_1 s+a_0=0 \tag{4.3}$$

the roots of which appear in the factors of the denominator of the transfer function:

$$(s-r_1)(s-r_2)\cdots(s-r_i)\cdots(s-r_n)=0 \tag{4.4}$$

Roots $r_1, r_2, \ldots, r_i, \ldots, r_n$ can be real or complex and have the general form

$$r_i=\alpha_i+\mathrm{j}\beta_i \tag{4.5}$$

Similarly, the discrete-time transfer function

$$G(z)=\frac{Y(z)}{X(z)}=\frac{b_m z^m+\cdots+b_2 z^2+b_1 z+b_0}{a_n z^n+\cdots+a_2 z^2+a_1 z+a_0} \tag{4.6}$$

1 Response to sinusoidal inputs is also of significant interest, and this topic is treated separately in Chapter 5: Frequency Response.

can be factored into the form

$$G(z)=\frac{Y(z)}{X(z)}=K_z\frac{(z-c_1)(z-c_2)\cdots(z-c_m)}{(z-r_1)(z-r_2)\cdots(z-r_n)} \tag{4.7}$$

The characteristic equation associated with this discrete-time transfer function is

$$a_n z^n+\cdots+a_2z^2+a_1z+a_0=C \tag{4.8}$$

the roots of which appear in the factors of the denominator of the transfer function:

$$(z-r_1)(z-r_2)\cdots(z-r_i)\cdots(z-r_n)=0 \tag{4.9}$$

Again, roots $r_1, r_2, ..., r_i, ..., r_n$ can be real or complex.

4.1.2 Fundamental Continuous-Time Dynamic Characteristics

The roots $r_1, r_2, ..., r_i, ..., r_n$ of a characteristic equation associated with the transfer function of a continuous-time production system or system component convey important information about the fundamental dynamic characteristics of that system or component. The contribution of each factor of the denominator of the transfer function, for real roots, and each pair of factors, for complex conjugate pairs of roots, can be examined by expanding Equation 4.2 into partial fractions. For the case when there are no repeated roots and $m < n$,[2]

$$G(s)=\frac{Y(s)}{X(s)}=\left(\frac{q_1}{s-r_1}+\frac{q_2}{s-r_2}+\ldots+\frac{q_i}{s-r_i}+\ldots+\frac{q_n}{s-r_n}\right)e^{-Ds} \tag{4.10}$$

where coefficient q_i is real when root r_i is real, and coefficients q_i and q_{i+1} are complex conjugate pairs when roots r_i and r_{i+1} are complex conjugate pairs. This equation shows that dynamic behavior can be viewed as a sum of contributing behaviors associated with the roots of the characteristic equation.

The individual first-order transfer function associated with real root $r_i = \alpha_i + j0$ is

$$\frac{Y_i(s)}{X(s)}=\frac{q_i}{s-\alpha_i}=\frac{K_i}{\tau_i s+1} \tag{4.11}$$

where the constant of proportionality is

$$K_i=-\frac{q_i}{\alpha_i} \tag{4.12}$$

and the time constant is

$$\tau_i=-\frac{1}{\alpha_i} \tag{4.13}$$

2 The reader is referred to the Bibliography and many other control theory publications for treatment of partial fraction expansion when there are repeated roots or direct terms. The results of the continuous-time and discrete-time analyses that follow are similar for these cases, including stability criteria.

Note that $\tau_i = \infty$ for the common case $\alpha_i = 0$. Response to a unit step input is helpful in understanding the significance of the time constant, and this is illustrated in Example 4.1.

The individual transfer function associated with each complex conjugate pair of roots $r_i = \alpha_i + j\beta_i$ and $r_{i+1} = \alpha_i - j\beta_i$ is

$$\frac{Y_i(s)}{X(s)} = \frac{q_i}{s - r_i} + \frac{q_{i+1}}{s - r_{i+1}} = \frac{K_i(\upsilon_i s + 1)}{\frac{1}{\omega_{n_i}^2} s^2 + \frac{2\zeta_i}{\omega_{n_i}} s + 1} \quad (4.14)$$

where q_i and q_{i+1} are complex conjugates:

$$q_i = \sigma_i + j\gamma_i \quad (4.15)$$

$$q_{i+1} = \sigma_i - j\gamma_i \quad (4.16)$$

The damping ratio associated with the complex conjugate pair of roots is

$$\zeta_i = -\cos(\angle r_i) = \frac{-\alpha_i}{\sqrt{\alpha_i^2 + \beta_i^2}} \quad (4.17)$$

the natural frequency is

$$\omega_{n_i} = |r_i| = \sqrt{\alpha_i^2 + \beta_i^2} \quad (4.18)$$

the constant of proportionality is

$$K_i = -2(\alpha_i \gamma_i + \beta_i \sigma_i) \quad (4.19)$$

and the numerator time constant is

$$\upsilon_i = \frac{-2\sigma_i}{K_i} \quad (4.20)$$

Furthermore, the time constant of exponential decay associated with the pair of complex conjugate roots is

$$\tau_i = \frac{1}{\zeta_i \omega_{n_i}} = -\frac{1}{\alpha_i} \quad (4.21)$$

Response to a unit step input is helpful in understanding the significance of the damping ratio and natural frequency, and this is illustrated in Example 4.2.

Time constants, damping ratios and natural frequencies are dynamic characteristics that are obtained without considering response to specific inputs; instead, they are derived directly from the roots of characteristic equations without solving differential equations or inspecting responses to specific inputs. Therefore, they are fundamental dynamic characteristics that enhance understanding of how a production system or component tends to respond dynamically regardless of input. This is important because the specific nature of inputs and their fluctuation in the future operation of production systems often is unknown. Nevertheless, it is useful to illustrate how dynamic behavior is related to these characteristics, and response to step inputs will be used for this purpose in the next examples.

Example 4.1 Effect of Time Constant on Step Response of a First-Order Continuous-Time Production System or Component

For the first-order continuous-time transfer function in Equation 4.11 and the common case of delay $D = 0$, the corresponding differential equation is

$$\tau_i \frac{dy(t)}{dt} + y_i(t) = K_i x(t)$$

which has the following well-known solution when $x(t)$ is the unit step input shown in Figure 3.1 and the initial condition is $y_i(0) = 0$:

$$y_i(t) = K_i\left(1 - \mathrm{e}^{-t/\tau_i}\right)$$

This response shown in Figure 4.1a for $\tau_i > 0$. The response is exponential in nature as expected, has no overshoot, is 63.2% complete at time $t = \tau_i$, and is 98.2% complete at time $t = 4\tau_i$. Other measures such as final value and settling time are discussed Section 4.2.2

The response for $\tau_i < 0$ is shown in Figure 4.1b. This response grows without bound towards $-\infty$ as time t increases towards $+\infty$.

Example 4.2 Effect of Damping Ratio and Natural Frequency on Step Response of a Second-Order Continuous-Time Production System or Component

For the second-order continuous-time transfer function in Equation 4.14 and the common case of delay $D = 0$ and numerator time constant $\upsilon_i = 0$, the corresponding differential equation is

$$\frac{1}{\omega_{n_i}^2}\frac{d^2 y_i(t)}{dt^2} + \frac{2\zeta_i}{\omega_{n_i}}\frac{dy_i(t)}{dt} + y_i(t) = K_i x(t)$$

which has the following well-known solution for damping ratio $-1 < \zeta_i < 1$ when $x(t)$ is the unit step input shown in Figure 3.1 and the initial conditions are $y_i(0) = 0$ and $dy_i(0)/dt = 0$:

$$y_i(t) = K_i\left(1 - \mathrm{e}^{-\zeta_i \omega_{n_i} t}\left(\cos\left(\sqrt{1-\zeta_i^2}\,\omega_{n_i} t\right) + \frac{\zeta_i}{\sqrt{1-\zeta_i^2}}\sin\left(\sqrt{1-\zeta_i^2}\,\omega_{n_i} t\right)\right)\right)$$

This response shown in Figure 4.2 for various damping ratios $0 < \zeta_i < 1$; the response is oscillatory and often is referred to as underdamped. The amplitude of oscillation decays exponentially with time constant $\tau_i = 1/\zeta_i\omega_{ni}$; this is illustrated for the case of $\zeta_i = 0.1$.

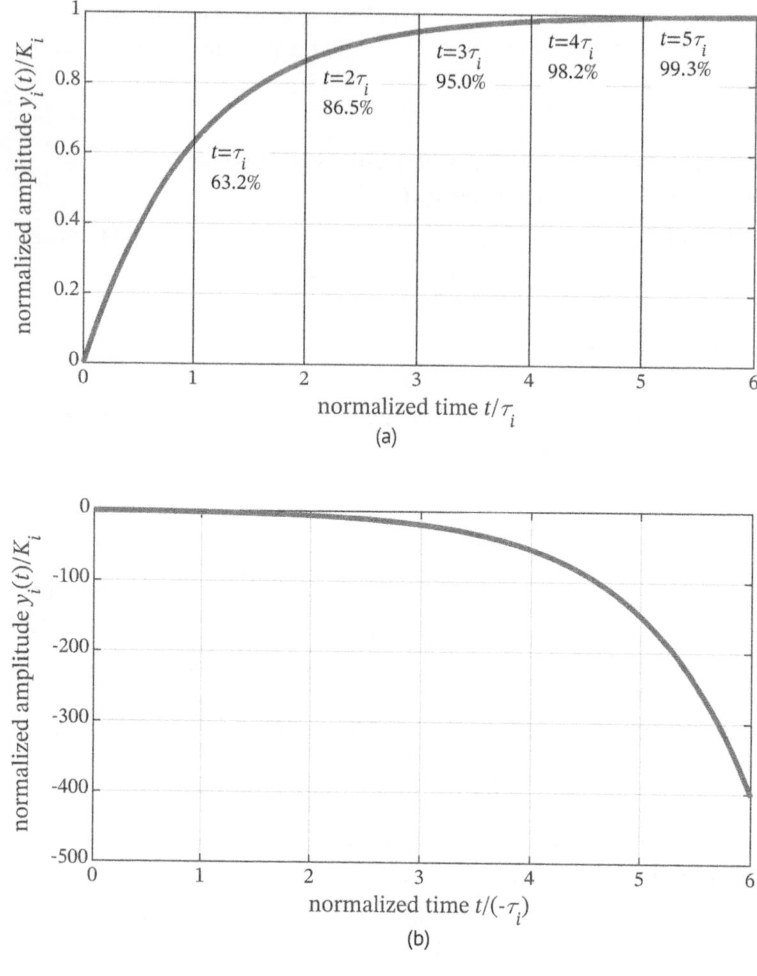

Figure 4.1 First-order continuous-time unit step response with time constant τ_i. (a) $\tau_i > 0$. (b) $\tau_i < 0$.

4.1.3 Continuous-Time Stability Criterion

Stability is an important characteristic for production systems. From a practical perspective, the dynamic behavior of an unstable production system is likely to be unreliable, with system outputs diverging unacceptably even when not driven by system inputs. It is well known that production systems can be unstable; an example is the bullwhip effect.[3] A production system can be stable even though it has unstable components, and properly designed decision rules can make a closed-loop production system stable even though one or more of its components is unstable. Conversely, improperly designed decision rules can make a closed-loop production system unstable even though all of its components, including decision making components, are stable.

3 The bullwhip effect refers to increasing fluctuations in inventory moving up the supply chain in response to fluctuations in demand.

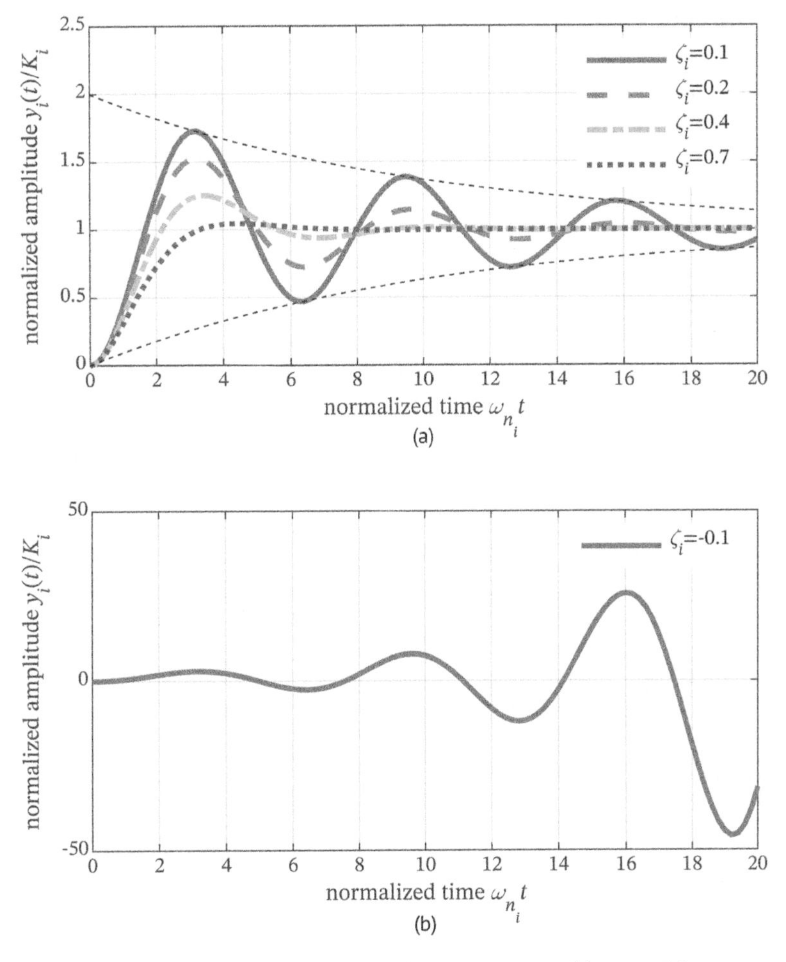

Figure 4.2 Examples of second-order unit step response with natural frequency ω_{ni}, gain K_i and various damping ratios ζ_i. (a) $\zeta_i > 0$; also shown is the envelope of exponential decay for $\zeta_i = 0.1$. (b) $\zeta_i < 0$.

Bounded-input, bounded output stability is one point of view for defining stability. If any bounded input results in an output that is not bounded, then the system is not stable. In practice, inputs to production systems are bounded; for example, they may be constant, change from one level to another or fluctuate between some upper and lower bounds. Unbounded outputs tend to grow, in theory, towards infinity with time[4] or oscillate with ever increasing amplitude, while in practice they may grow towards physical limits.

An alternative, conservative point of view is to define a production system or component as stable if, as time approaches infinity, its outputs to decay from initial

4 In some circumstances this behavior may be acceptable. An example is when production rate is the input and the work completed is the output. In this case a constant (bounded) production rate results in a continually increasing (unbounded) accumulation of completed work.

conditions to zero when all inputs are zero. This is often referred to as internal or asymptotic stability. Internal stability is a fundamental dynamic characteristic of a system or component and can be assessed by examining the roots of its characteristic equation which, as described in Section 4.1.1, is obtained from the denominator of transfer function of the system or component. Specific inputs and the numerator of the transfer function are not considered.

The differential equation associated with Equation 4.11 is

$$\frac{dy_i(t)}{dt} - \alpha_i y_i(t) = q_i x(t) \tag{4.22}$$

which has the following well-known solution for input $x(t) = 0$ and initial condition $y_i(0)$:

$$y_i(t) = y_i(0)e^{\alpha_i t} \tag{4.23}$$

The portion of the response of output $y(t)$ associated with real root $r_{i,} = \alpha_i + j0$ decays towards zero with increasing t when $\alpha_i < 0$; conversely, the portion of response $y(t)$ associated with real root $r_{i,} = \alpha_i + j0$ approaches infinity with increasing t when $\alpha_i > 0$.

The differential equation associated with Equation 4.14 is

$$\frac{d^2y_i(t)}{dt^2} - 2\alpha_i\frac{dy_i(t)}{dt} + \left(\alpha_i^2 + \beta_i^2\right)y_i(t) = 2\left(\alpha_i\sigma_i + \beta_i\gamma_i\right)x(t) \tag{4.24}$$

which has the following well-known solution for input $x(t) = 0$ and initial conditions $y_i(0)$ and $dy_i(0)/dt$:

$$y_i(t) = e^{\alpha_i t}\left(y_i(0)\cos(\beta_i t) + \frac{1}{\beta_i}\left(\frac{dy_i(0)}{dt} - \alpha_i y_i(0)\right)\sin(\beta_i t)\right)$$

The portion of the response of output $y(t)$ associated with a pair of complex conjugate roots $r_{i,} = \alpha_i + j\beta_i$ and $r_{i+1} = \alpha_i - j\beta_i$ decays towards zero with increasing t when $\alpha_i < 0$ and oscillation amplitude approaches infinity with increasing t when $\alpha_i > 0$. Hence, a production system or component is not stable if any portion of the model in the form of Equation 4.10 does not tend to decay towards zero with increasing t when input $x(t) = 0$. Regardless of whether roots of the characteristic equation are real or complex, the continuous-time criterion for system or component stability is

$$\alpha_i < 0 \text{ for all roots } r_i \tag{4.25}$$

Referring to Equations 4.13, 4.17 and 4.21, the production system or component is not stable if any time constant $\tau_i \leq 0$ or any damping ratio $\zeta_i \leq 0$ for all roots r_i.

Example 4.3 Relationship Between Proportional Decision-Rule Parameter and Stability of a Continuous-Time First-Order Production System

K_p days^{-1} is an adjustable decision-making parameter in the continuous-time model of the closed-loop production system described in Example 3.18, and it is of interest to determine the relationship between values chosen for K_p and the stability of the closed-loop system. The block diagram in Figure 3.22 shows the topology of relationships in the production system, in which the following capacity adjustment rule is used to adjust production capacity $r_o(t)$ hours/day in proportion to the work in progress (WIP) $w_w(t)$ hours:

$$r_o(t) = K_p w_w(t)$$

The following transfer function was obtained that relates work output $w_o(t)$ hours to work input $w_i(t)$ hours:

$$\frac{W_o(s)}{W_i(s)} = \frac{K_p}{s + K_p}$$

The characteristic equation is

$$s + K_p = 0$$

for which the root is $r = -K_p$. Applying the stability criterion in Equation 4.25, this production system is stable for all $K_p > 0$. There are no complex conjugate roots and the system therefore is not fundamentally oscillatory. From Equation 4.13, time constant is $\tau = 1/K_p$. Therefore, larger K_p results in smaller τ and a production system that responds more quickly to fluctuations in work input.

Example 4.4 Relationship Between Proportional Decision-Rule Parameter and Stability of a Continuous-Time Second-Order Production System

The following input-output relationships were found in Example 3.20 for the production system with backlog regulation shown in Figure 2.3:

$$W_b(s) = \frac{\left(K_1 + \frac{K_2}{s}\right)\frac{1}{s}}{1 + \left(K_1 + \frac{K_2}{s}\right)\frac{1}{s}} W_p(s) + \frac{\frac{1}{s}}{1 + \left(K_1 + \frac{K_2}{s}\right)\frac{1}{s}} R_i(s) + \frac{1}{1 + \left(K_1 + \frac{K_2}{s}\right)\frac{1}{s}} W_d(s)$$

$$R_o(s) = -\frac{K_1 + \frac{K_2}{s}}{1 + \left(K_1 + \frac{K_2}{s}\right)\frac{1}{s}} W_p(s) + \frac{\left(K_1 + \frac{K_2}{s}\right)\frac{1}{s}}{1 + \left(K_1 + \frac{K_2}{s}\right)\frac{1}{s}} R_i(s) + \frac{K_1 + \frac{K_2}{s}}{1 + \left(K_1 + \frac{K_2}{s}\right)\frac{1}{s}} W_d(s)$$

where $w_b(t)$ orders is the order backlog, $r_o(t)$ orders/day is the order release rate, $w_p(t)$ orders is the planned order backlog, $r_i(t)$ orders/day is the order input rate, and $w_d(t)$ orders represents disturbances such as rush orders and order cancellations. K_1 days^{-1} and K_2 days^{-2} are decision rule parameters that are designed to obtain favorable closed-loop backlog regulation dynamic behavior.

The decision rule for adjusting order release rate based on the difference between planned and actual backlog has proportional and integral components, K_1 and K_2, respectively:

$$r_o(t) = K_1(w_b(t) - w_p(t)) + K_2 \int_0^t (w_b(t) - w_p(t))dt$$

The topology of this example of backlog regulation is shown in the block diagram in Figure 3.26.

All of the transfer functions have the same characteristic equation:

$$s^2 + K_1 s + K_2 = 0$$

The roots of this characteristic equation are

$$r_{1,2} = \frac{-K_1 \pm \sqrt{K_1^2 - 4K_2}}{2}$$

The roots are complex conjugates and responses tend to be oscillatory when

$$K_1^2 < 4K_2$$

In this case, the production system is stable for $K_1 > 0$ because the real part of the roots is negative, satisfying the criterion in Equation 4.25. The damping ratio from Equation 4.17 with $K_1 > 0$ is

$$\zeta = \frac{K_1}{2\sqrt{K_2}}$$

and the natural frequency ω_n days^{-1} from Equation 4.18 with $K_2>0$ is

$$\omega_n = \sqrt{K_2}$$

Therefore, K_1 and K_2 can be chosen to obtain a desired damping ratio and natural frequency.

On the other hand, there are two real roots when

$$K_1^2 \geq 4K_2$$

In this case, the roots are negative, and the production system is stable when $K_1 > 0$ and $K_2 > 0$. From Equation 4.13, the time constants associated with these roots are

$$\tau_1 = \frac{2}{K_1 - \sqrt{K_1^2 - 4K_2}}$$

$$\tau_2 = \frac{2}{K_1 + \sqrt{K_1^2 - 4K_2}}$$

and K_1 and K_2 can be chosen to obtain desired time constants.

Example 4.5 Stability of a Production System with Continuous-Time Capacity Adjustment with Delay

The addition of delay of $D > 0$ days in implementing capacity adjustments in the production system described in Example 4.3 can significantly affect its stability. The new decision rule for adjusting production capacity $r_o(t)$ hours/day is

$$r_o(t) = K_p w_w(t - D)$$

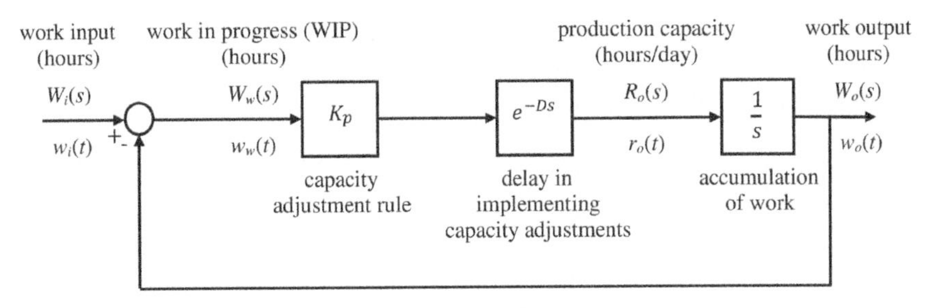

Figure 4.3 Production system with delay D days in implementing capacity adjustments.

where $w_w(t)$ hours is the work in progress (WIP) and K_p days^{-1} is an adjustable decision-making parameter. The new block diagram in Figure 4.3 includes the delay, and it is desired to investigate the effect of choice of decision-making parameter K_p on the stability of the production system.

The transfer function that represents the relationship between work output $w_o(t)$ and work input $w_i(t)$ is

$$\frac{W_o(s)}{W_i(s)} = \frac{K_p e^{-Ds}}{s + K_p e^{-Ds}}$$

and the characteristic equation is

$$s + K_p e^{-Ds} = 0$$

Because of the internal delay D, direct analyses of roots cannot be carried out to determine fundamental dynamic characteristics such as damping ratios and stability. However, a Padé approximation of the delay can be substituted as described in Section 3.8.8. If a second-order Padé approximation of delay D is substituted into the transfer function, the result is

$$\frac{W_o(s)}{W_i(s)} \approx \frac{K_p \left(\dfrac{\frac{D^2}{12}s^2 - \frac{D}{2}s + 1}{\frac{D^2}{12}s^2 + \frac{D}{2}s + 1} \right)}{s + K_p \left(\dfrac{\frac{D^2}{12}s^2 - \frac{D}{2}s + 1}{\frac{D^2}{12}s^2 + \frac{D}{2}s + 1} \right)}$$

The following approximating characteristic equation then can be used to investigate the effect of choice of decision-making parameter K_p on fundamental dynamic characteristics:

$$\frac{D^2}{12}s^3 + \left(\frac{D}{2} + K_p \frac{D^2}{12} \right) s^2 + \left(1 - K_p \frac{D}{2} \right) s + K_p \approx 0$$

It is convenient to use control system engineering software to calculate roots and characteristics such as time constants, damping ratios and natural frequencies. As demonstrated in Program 4.1, the roots of the characteristic equation of a transfer function variable G along with the associated natural frequencies, damping ratios, and time constants can be calculated using

```
damp(G)
```

The natural frequencies, damping ratios and roots can be obtained for the purpose of further analysis using

```
[Wn,zeta,P]=damp(G)
```

where Wn is a vector of natural frequencies, zeta is a vector of damping ratios and P is the vector of roots of the characteristic equation.

For stable real roots $r_i = \alpha_i + j0$ with $\alpha_i < 0$, function damp returns damping ratio $\zeta_i = 1$ and natural frequency $\omega_{ni} = |\alpha_i|$, while for unstable real roots with $\alpha_i > 0$, damp returns damping ratio ζ_i = -1 and natural frequency $\omega_{ni} = |\alpha_i|$. For real roots $r_i = 0$, damp returns damping ratio ζ_i = -1, natural frequency $\omega_{ni} = 0$ and $\tau_i = \infty$.

Program 4.1 illustrates the results for three choices of K_p days^{-1} when delay $D = 3$ days. Inclusion of the second-order Padé approximation of the delay results in three roots of the approximating characteristic equation that have associated approximate dynamic characteristics. For $K_p = 0.1$, the results of damp are three negative real roots; this predicts that choice of $K_p = 0.1$ will result in stable behavior that tends to be non-oscillatory. The longest approximate time constant is 6.1 days.

For $K_p = 0.2$, the results of damp include a pair of complex conjugate roots with negative real parts; this predicts that choice of $K_p = 0.2$ will result in stable behavior that tends to be oscillatory. The associated approximate damping ratio is 0.57 and the associated approximate time constant is 4.5 days.

For $K_p = 1$, the results of damp include a pair of complex conjugate roots with positive real roots; this predicts that choice of $K_p = 1$ will result in unstable behavior that tends to be oscillatory. The associated approximate damping ratio is -0.25 and the associated approximate time constant is -6.2 days. Oscillations tend to grow in amplitude with time in this case, regardless of the nature of the inputs.

Program 4.1 Calculation of fundamental dynamic characteristics of production system with continuous-time capacity adjustment and delay

```
D=3;  % delay (days)
Gs=tf(1,[1, 0],'TimeUnit','days','OutputDelay',D)  % work and delay
```

```
  Gs =

                  1
    exp(-3*s) * -
                  s

  Continuous-time transfer function.
```

```
Gs=pade(Gs,2)  % replace delay with Pade approximation
```

```
  Gs =

          s^2 - 2 s + 1.333
     --------------------------
        s^3 + 2 s^2 + 1.333 s

  Continuous-time transfer function.
```

```
Kp=0.1;  % parameter resulting in stable real roots
damp(minreal(Kp*Gs/(1+Kp*Gs)))  % characteristics of Wo(s)/Wi(s)
```

```
      Pole              Damping          Frequency        Time Constant
                                         (rad/days)          (days)
   -1.63e-01           1.00e+00          1.63e-01           6.13e+00
   -6.21e-01           1.00e+00          6.21e-01           1.61e+00
   -1.32e+00           1.00e+00          1.32e+00           7.60e-01
```

```
Kp=0.2;  % parameter resulting in stable complex roots
damp(minreal(Kp*Gs/(1+Kp*Gs)))  % characteristics of Wo(s)/Wi(s)
```

```
          Pole                   Damping       Frequency      Time Constant
                                               (rad/days)        (days)
   -2.23e-01 + 3.20e-01i        5.71e-01       3.90e-01        4.49e+00
   -2.23e-01 - 3.20e-01i        5.71e-01       3.90e-01        4.49e+00
   -1.75e+00                    1.00e+00       1.75e+00        5.70e-01
```

```
Kp=1;  % parameter resulting in unstable complex roots
damp(minreal(Kp*Gs/(1+Kp*Gs)))  % characteristics of Wo(s)/Wi(s)
```

```
          Pole                   Damping       Frequency      Time Constant
                                               (rad/days)        (days)
    1.61e-01 + 6.13e-01i       -2.54e-01       6.34e-01       -6.22e+00
    1.61e-01 - 6.13e-01i       -2.54e-01       6.34e-01       -6.22e+00
   -3.32e+00                    1.00e+00       3.32e+00        3.01e-01
```

4.1.4 Fundamental Discrete-Time Dynamic Characteristics

As with continuous-time production systems or components, the roots r_1, r_2, ..., r_i, ..., r_n of a characteristic equation of a transfer function of a discrete-time system or system component convey important information about its fundamental dynamic characteristics. The contribution of each factor of the denominator of a discrete-time transfer function can be examined by expanding Equation 4.7 into partial fractions. For the case when there are no repeated roots and $m < n$,

$$G(z)=\frac{Y(z)}{X(z)}=\left(\frac{q_1}{z-r_1}+\frac{q_2}{z-r_2}+\cdots+\frac{q_i}{z-r_i}+\cdots+\frac{q_n}{z-r_n}\right)z^{-d} \tag{4.26}$$

where coefficient q_i is real when root r_i is real, and q_i and q_{i+1} are complex conjugate pairs when roots r_i and r_{i+1} are complex conjugate pairs. The dynamic behavior of a discrete-time system can be viewed as a sum of contributing behaviors associated with the roots of the characteristic equation. Discrete-time time constants, damping ratios and natural frequencies that are equivalent to those defined for continuous-time systems can be obtained by relating a discrete time root r_z to an equivalent continuous-time root r_s using the definition of z in Equation 3.38:

$$r_z = e^{Tr_s} \tag{4.27}$$

or

$$r_s = \frac{\ln(r_z)}{T} \tag{4.28}$$

The first-order discrete-time transfer function associated with real root $r_i = \alpha_i + j0$ is

$$\frac{Y_i(z)}{X(z)} = \frac{q_i}{z - \alpha_i} = \frac{K_i(1 - \alpha_i)}{z - \alpha_i} \tag{4.29}$$

and the time constant of exponential decay associated with this real root $r_i = \alpha_i + j0$ is

$$\tau_i = -\frac{T}{\ln|\alpha_i|} \tag{4.30}$$

and for $\alpha_i > 0$, the relationship between real root $r_i = \alpha_i + j0$ and the corresponding continuous time constant of exponential decay τ_i is

$$\alpha_i = e^{-T/\tau_i} \tag{4.31}$$

The response to a unit step input sequence is helpful in understanding the significance of the time constant, and this is illustrated in Example 4.6.

The second-order discrete-time transfer function associated with a complex conjugate pair of roots $r_i = \alpha_i + j\beta_i$ and $r_{i+1} = \alpha_i - j\beta_i$ is

$$\frac{Y_i(z)}{X(z)} = \frac{q_i}{z - r_i} + \frac{q_{i+1}}{z - r_{i+1}} = \frac{p_{1_i} z + p_{0_i}}{z^2 - 2e^{-\zeta_i \omega_{n_i} T} \cos\left(\sqrt{1 - \zeta_i^2}\, \omega_{n_i} T\right) z + e^{-2\zeta_i \omega_{n_i} T}} \tag{4.32}$$

The damping ratio associated with this complex conjugate pair of roots is

$$\zeta_i = -\cos(\angle \ln(r_i)) \tag{4.33}$$

and the natural frequency is

$$\omega_{n_i} = \left|\frac{\ln(r_i)}{T}\right| \tag{4.34}$$

Furthermore, the time constant of exponential decay associated with the pair of complex conjugate roots is

$$\tau_i = \frac{1}{\zeta_i \omega_{n_i}} = -\frac{T}{\ln|r_i|} \tag{4.35}$$

Examples of the response to a unit step sequence input shown are helpful in understanding the significance of the damping ratio and natural frequency, and this is illustrated in Example 4.7.

As is the case with continuous-time production systems or components, these time constants, damping ratios and natural frequencies are fundamental dynamic characteristics that are obtained without considering response to specific inputs and are derived directly from the roots of the characteristic equation. Their fundamental nature significantly facilitates analysis of the dynamic behavior and guiding design of decision rules. Nevertheless, it is useful to illustrate how dynamic behavior is related to these characteristics, and response to step inputs will be used for this purpose in the next examples.

Example 4.6 Effect of Time Constant on Step Response of a First-Order Discrete-Time Production System or Component

For the first-order continuous-time transfer function in Equation 4.29 with delay $dT = 0$, the corresponding difference equation is

$$y_i(kT) = \alpha_i y_i((k-1)T) + K_i(1-\alpha_i)x((k-1)T)$$

which has the following well-known solution for $k \geq 0$ when $x(kT)$ is the unit step sequence input shown in Figure 3.8 and the initial condition is $y_i(0) = 0$:

$$y_i(kT) = K_i\left(1-\alpha_i^k\right)$$

Response sequences for $0 < \alpha_i < 1$ with $T = 0.8\tau_i$ and $T = 0.1\tau_i$ are shown in Figure 4.4a where time constant τ_i is obtained using Equation 4.30. In this case, $\tau_i > 0$. As expected, the response sequence is exponential in nature, converges to a final value and has no overshoot. Other measures such as settling time and rise time are of interest, and these are discussed in Section 4.2.2.

The response sequence for $\alpha_i > 1$ is shown in Figure 4.4b. In this case, $\tau_i < 0$. This response sequence is exponential in nature but grows without bound towards $-\infty$ as time kT increases towards $+\infty$.

This response sequence for $-1 < \alpha_i < 0$ is shown in Figure 4.4c. In this case, $\tau_i > 0$. The response sequence has the same exponential envelope as the response sequence in Figure 4.4a, but values in the response sequence alternate with respect to the final value at successive times kT.

Example 4.7 Effect of Damping Ratio and Natural Frequency on Step Response of a Second-Order Discrete-Time Production System or Component

For the second-order continuous-time transfer function in Equation 4.32, the corresponding difference equation is

$$y_i(kT) = 2e^{-\zeta_i\omega_{n_i}T}\cos\left(\sqrt{1-\zeta_i^2}\,\omega_{n_i}T\right)y_i((k-1)T) - e^{-2\zeta_i\omega_{n_i}T}y_i((k-2)T)$$
$$+ p_{1_i}x((k-1)T) + p_{0_i}x((k-2)T)$$

For the common case

$$p_{1_i} = K_i\left(1-e^{-\zeta_i\omega_{n_i}T}\left(\cos\left(\sqrt{1-\zeta_i^2}\,\omega_{n_i}T\right)+\frac{\zeta_i}{\sqrt{1-\zeta_i^2}}\sin\left(\sqrt{1-\zeta_i^2}\,\omega_{n_i}T\right)\right)\right)$$

$$p_{0_i} = K_i e^{-\zeta_i\omega_{n_i}T}\left(e^{-\zeta_i\omega_{n_i}T} - \cos\left(\sqrt{1-\zeta_i^2}\,\omega_{n_i}T\right)+\frac{\zeta_i}{\sqrt{1-\zeta_i^2}}\sin\left(\sqrt{1-\zeta_i^2}\,\omega_{n_i}T\right)\right)$$

the well-known solution for $k \geq 0$ when $x(kT)$ is the unit step sequence input shown in Figure 3.8, with initial conditions $y_i(0) = 0$ and $y_i(-T) = 0$, is similar to the continuous-time solution in Example 4.2:

$$y_i(kT) = K_i\left(1-\left(e^{-\zeta_i\omega_{n_i}T}\right)^k\left(\cos\left(\sqrt{1-\zeta_i^2}\,\omega_{n_i}kT\right)+\frac{\zeta_i}{\sqrt{1-\zeta_i^2}}\sin\left(\sqrt{1-\zeta_i^2}\,\omega_{n_i}kT\right)\right)\right)$$

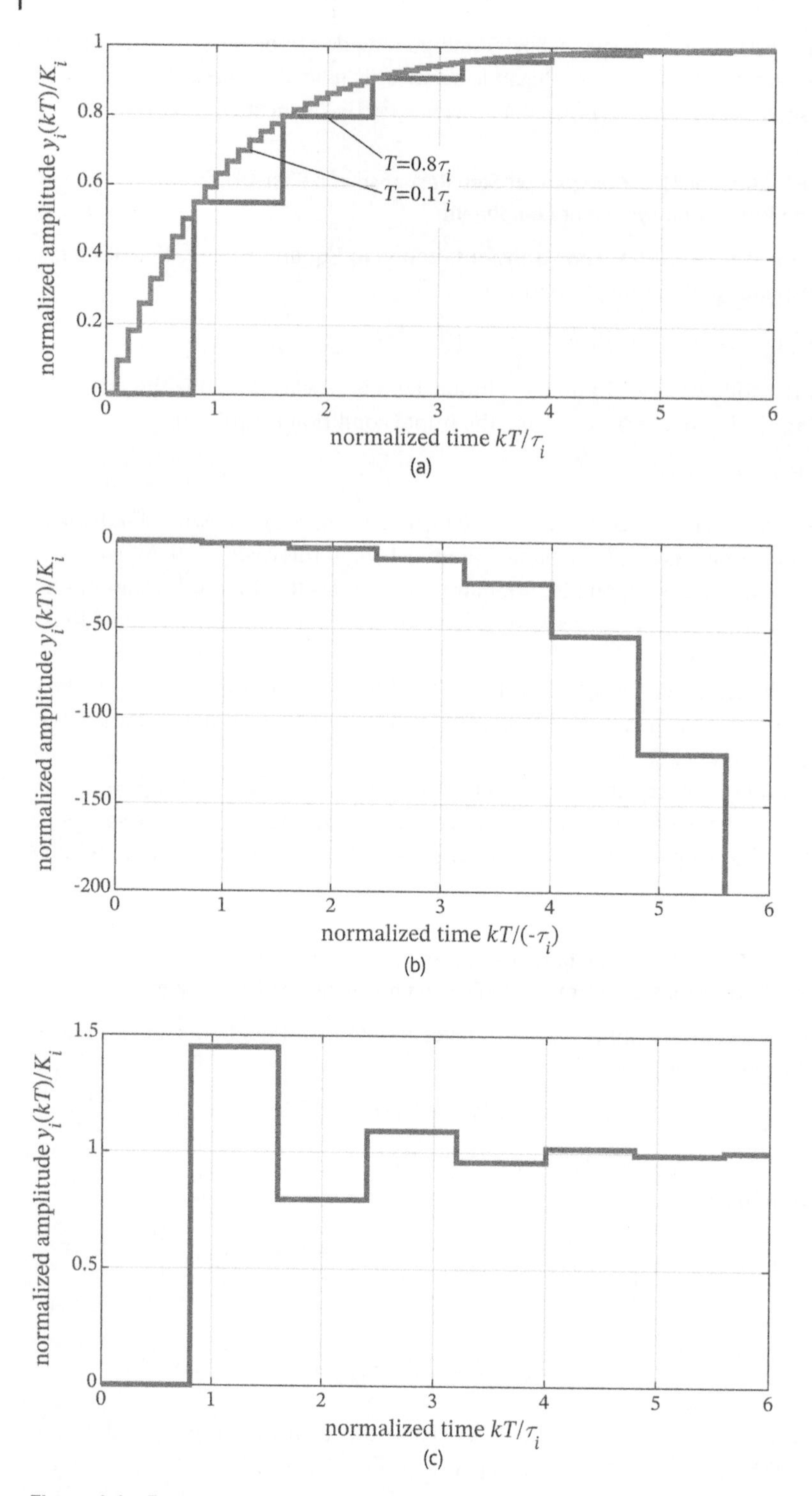

Figure 4.4 Examples of first-order discrete-time unit step response. (a) $0 < \alpha_i < 1, \tau_i > 0$. (b) $\alpha_i > 1, \tau_i < 0, T = 0.8\tau_i$. (c) $-1 < \alpha_i < 0, \tau_i > 0, T = 0.8\tau_i$.

Response sequences are shown in Figure 4.5 for $T = \tau_i$ where time constant τ_i is obtained using Equation 4.35. The response sequence for damping ratio $\zeta_i = 0.1$ is shown in Figure 4.5a. Oscillations in the response sequence decay exponentially, and the response sequence reaches a final value.

The response sequence for $\zeta_i = -0.1$ is shown in Figure 4.5b. The amplitude of the oscillations in this response sequence grows without bound as time kT increases.

4.1.5 Discrete-Time Stability Criterion

The stability of a production system or component represented by a discrete-time model is assessed by examining the roots of its characteristic equation obtained from the denominator of its discrete-time transfer function as described in Section 4.1.1. As is the case with continuous-time systems, internal stability is a fundamental characteristic of dynamic behavior; specific inputs and the numerator of the transfer function are not considered.

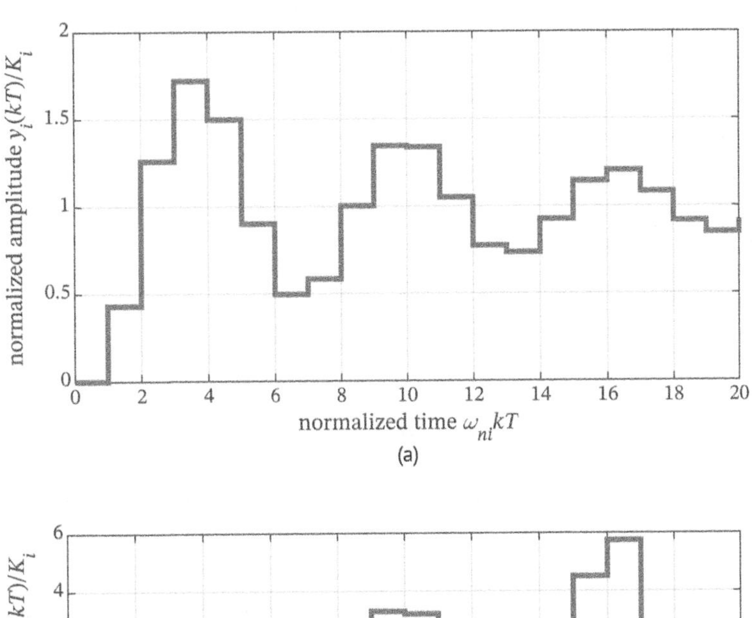

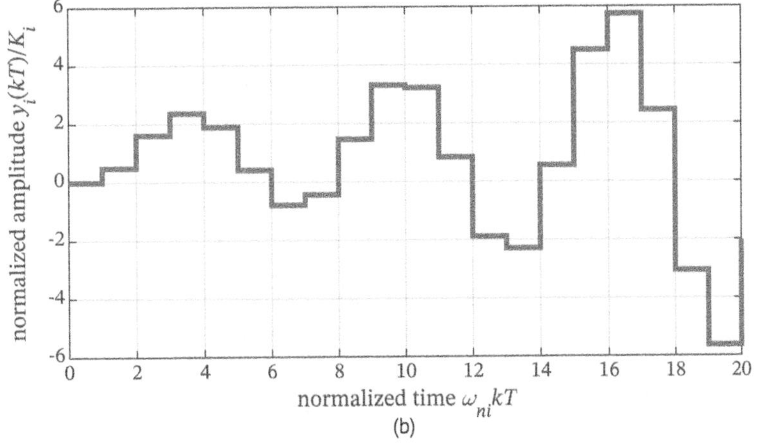

Figure 4.5 Examples of second-order discrete-time unit step response. (a) $\zeta_i = 0.1$. (b) $\zeta_i = -0.1$.

The portion of Equation 4.26 associated with a real root $r_i = \alpha_i + j0$ is

$$\frac{Y_i(z)}{X(z)} = \frac{q_i}{z - \alpha_i} \tag{4.36}$$

where q_i is real. The corresponding difference equation is

$$y_i(kT) = \alpha_i y_i((k-1)T) + q_i x((k-1)T) \tag{4.37}$$

which has the following well-known solution for $k \geq 0$, input $x(kT) = 0$, and initial condition $y_i(0)$:

$$y_i(kT) = y_i(0)(\alpha_i)^k \tag{4.38}$$

Hence, the portion of the response of output $y_i(kT)$ associated with real root $r_{i,} = \alpha_i + j0$ decays towards zero with increasing kT when $|\alpha_i| < 1$; conversely, the portion of response $y(t)$ associated with real root $r_{i,} = \alpha_i + j0$ approaches infinity with increasing time kT when $|\alpha_i| > 1$.

The portion of Equation 4.26 associated with a pair of complex conjugate roots $r_{i,} = \alpha_i + j\beta_i$ and $r_{i+1} = \alpha_i - j\beta_i$ is

$$\frac{Y_i(z)}{X(z)} = \frac{q_i}{z - r_i} + \frac{q_{i+1}}{z - r_{i+1}} \tag{4.39}$$

The corresponding difference equation is

$$\begin{aligned} y_i(kT) = {} & 2\alpha_i y_i((k-1)T) - (\alpha_i^2 + \beta_i^2) y_i((k-2)T) + 2\sigma_i x((k-1)T) \\ & - 2(\alpha_i\sigma_i + \beta_i\gamma_i) x((k-2)T) \end{aligned} \tag{4.40}$$

where q_i and q_{i+1} are complex conjugates with real and imaginary parts defined in Equations 4.15 and 4.16.

The well-known solution of this difference equation for $k \geq 0$, input $x(kT) = 0$, and initial conditions $y_i(0)$ and $y_i(-T)$ is

$$y_i(kT) = |r_i|^k \left(y_i(0)\cos(\theta_i k) - \left(\frac{y_i(-T)|r_i| - y_i(0)\cos(\theta_i)}{\sin(\theta_i)} \right) \sin(\theta_i k) \right)$$

where

$$\theta_i = \tan^{-1}\left(\frac{\beta_i}{\alpha_i}\right) \tag{4.41}$$

Hence, the portion of the response of output $y(kT)$ associated with conjugate roots $r_{i,} = \alpha_i + j\beta_i$ and $r_{i+1} = \alpha_i - j\beta_i$ decays towards zero with increasing kT when $|r_i| < 1$, and the amplitude of oscillations approaches infinity with increasing time kT when $|r_i| > 1$.

Regardless of whether roots are real or complex, the discrete-time criterion for system or component stability therefore is

$$|r_i| < 1 \text{ for all roots } r_i \tag{4.42}$$

Example 4.8 Relationship Between Proportional Decision-Rule Parameter and Stability of a Discrete-Time First-Order Production System

K_p days^{-1} is an adjustable decision-making parameter in the discrete-time closed-loop production system shown in Figure 4.6, and it is of interest to determine the relationship between values chosen for K_p and the stability of the closed-loop system. The block diagram in Figure 4.7 shows the topology of relationships in the production system, in which the following discrete-time capacity adjustment rule is used to adjust production capacity $r_o(kT)$ hours/day in proportion to the work in progress (WIP) $w_w(kT)$ hours at an intervals of T days:

$$r_o(kT) = K_p w_w(kT)$$

The following transfer function relates measured work output $w_o(kT)$ hours to work input $w_i(kT)$ hours:

$$\frac{W_o(z)}{W_i(z)} = \frac{K_p \mathcal{Z}\left\{\left(\frac{1-e^{-Ts}}{s}\right)\frac{1}{s}\right\}}{1+K_p \mathcal{Z}\left\{\left(\frac{1-e^{-Ts}}{s}\right)\frac{1}{s}\right\}} = \frac{K_p T z^{-1}}{1-(1-K_p T)z^{-1}}$$

The characteristic equation is

$$z-(1-K_p T)=0$$

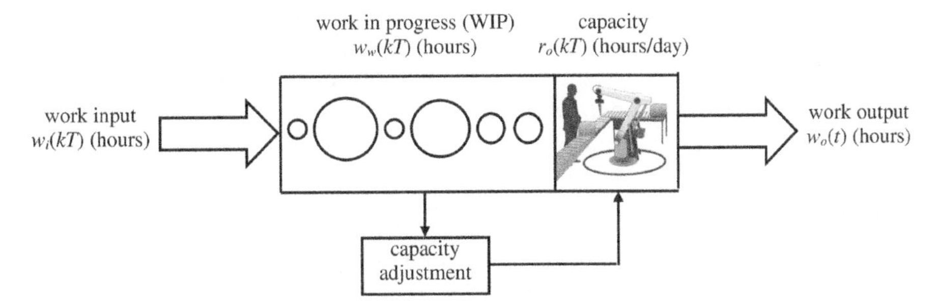

Figure 4.6 Production system with discrete-time capacity adjustment.

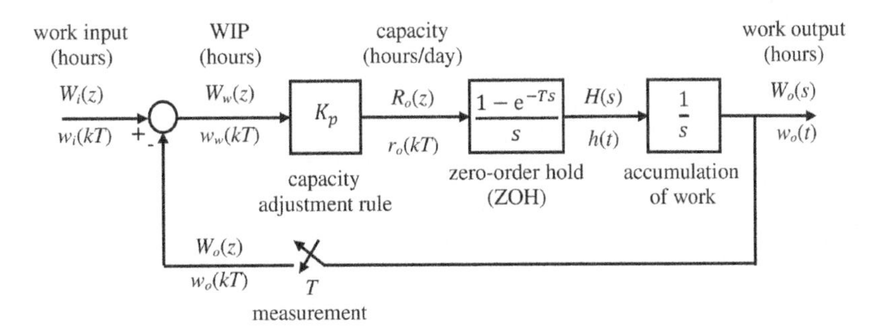

Figure 4.7 Block diagram for proportional discrete-time capacity adjustment.

and the root is

$$r = (1 - K_p T)$$

From Equation 4.42, capacity adjustment is stable when

$$|1 - K_p T| < 1$$

or

$$-1 < K_p T < 1$$

Example 4.9 Stability of Discrete-Time Actuator Position Control

A block diagram for computer control of actuator position is shown in Figure 4.8. The following discrete-time decision rule is used to adjust desired actuator velocity $v_c(kT)$ cm/second in proportion to the difference between actuator position command $p_c(kT)$ cm and actuator position $p(kT)$ cm:

$$v_c(kT) = K_p(p_c(kT) - p(kT))$$

K_p is an adjustable decision-making parameter, and it is desired to investigate stability and fundamental dynamic characteristics for $K_p = 2$, 10, and 400 seconds^{-1}.

The discrete-time actuator transfer function is

$$\frac{P(z)}{V_c(z)} = \mathcal{Z}\left\{\left(\frac{1 - e^{-Ts}}{s}\right)\left(\frac{1}{\tau_v s + 1}\right)\frac{1}{s}\right\} = \frac{b_1 z + b_0}{z^2 + a_1 z + a_0}$$

where coefficients a_1, a_0, b_1 and b_0 can be obtained for actuator time constant τ_v using function c2d. The closed-loop transfer function is

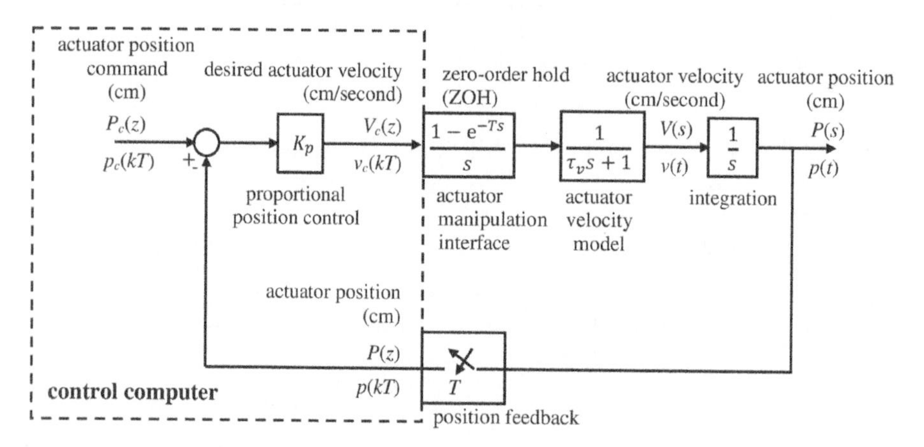

Figure 4.8 Block diagram for computer control of actuator position.

$$\frac{P(z)}{P_c(z)} = \frac{K_p \mathcal{Z}\left\{\left(\frac{1-e^{-Ts}}{s}\right)\left(\frac{1}{\tau_v s+1}\right)\frac{1}{s}\right\}}{1+K_p \mathcal{Z}\left\{\left(\frac{1-e^{-Ts}}{s}\right)\left(\frac{1}{\tau_v s+1}\right)\frac{1}{s}\right\}} = \frac{K_p b_1 z + K_p b_0}{z^2+(a_1+K_p b_1)z+(a_0+K_p b_0)}$$

and the characteristic equation is

$$z^2+(a_1+K_p b_1)z+(a_0+K_p b_0)=0$$

which has two roots that can be found and analyzed using function `damp`.

When position control parameter K_p = 2 seconds^{-1}, two real roots are found using Program 4.2 for period $T = 0.01$ seconds and time constant $\tau_v = 0.05$ seconds. With this choice of K_p, the characteristic equation has two real roots: $0.977 < 1$ and $0.839 < 1$; therefore, the closed-loop actuator position control system is not fundamentally oscillatory and, from the discrete-time stability criterion in Equation 4.42, the closed-loop system is stable in this case. The time constants associated with these roots are 0.438 seconds and 0.0571 seconds.

When K_p = 10 seconds^{-1}, the roots are complex conjugates and the closed-loop system tends to be oscillatory. The magnitude of the roots is $0.91 < 1$ and, from the discrete-time stability criterion in Equation 4.42, the closed-loop system is stable in this case. The damping ratio associated with the roots is 0.671 and the time constant associated with the roots is 0.106 seconds.

When K_p = 400 seconds^{-1}, the roots are complex conjugates with magnitude $1.08 > 1$. From the discrete-time stability criterion in Equation 4.42, the closed-loop system is not stable in this case. The damping ratio associated with the roots is -0.0927 and the time constant associated with the roots is -0.128 seconds.

Program 4.2 Calculation of discrete-time dynamic characteristics

```
T=0.01;  % period (seconds)
tauv=0.05;  % actuator velocity time constant (seconds)

Gpz=c2d(tf(1,[tauv, 1, 0]),T)  % actuator transfer function
```

```
  Gpz =

       0.0009365 z + 0.0008762
    ----------------------------
       z^2 - 1.819 z + 0.8187

  Sample time: 0.01 seconds
  Discrete-time transfer function.
```

```
Kp=2;  % parameter for two stable real roots (seconds^-1)
Gclz=feedback(Kp*Gpz,1)  % closed-loop transfer function
```

```
  Gclz =

      0.001873 z + 0.001752
    -----------------------
      z^2 - 1.817 z + 0.8205

  Sample time: 0.01 seconds
  Discrete-time transfer function.
```

```
damp(Gclz)
```

```
      Pole          Magnitude       Damping    Time Constant
                                                 (seconds)
   9.77e-01         9.77e-01        1.00e+00     4.38e-01
   8.39e-01         8.39e-01        1.00e+00     5.71e-02
```

```
Kp=10;  % parameter for two stable complex roots (seconds^-1)
Gclz=feedback(Kp*Gpz,1)  % closed-loop transfer function
```

```
Gclz =

      0.009365 z + 0.008762
    --------------------------
      z^2 - 1.809 z + 0.8275

Sample time: 0.01 seconds
Discrete-time transfer function.
```

```
damp(Gclz)
```

```
          Pole                 Magnitude     Damping      Time Constant
                                                            (seconds)
   9.05e-01 + 9.51e-02i        9.10e-01      6.71e-01        1.06e-01
   9.05e-01 - 9.51e-02i        9.10e-01      6.71e-01        1.06e-01
```

```
Kp=400;  parameter for two unstable complex roots (seconds^-1)
Gclz=feedback(Kp*Gpz,1)  % closed-loop transfer function
```

```
Gclz =

       0.3746 z + 0.3505
    ----------------------
     z^2 - 1.444 z + 1.169

Sample time: 0.01 seconds
Discrete-time transfer function.
```

```
damp(Gclz)
```

```
          Pole                 Magnitude     Damping      Time Constant
                                                            (seconds)
   7.22e-01 + 8.05e-01i        1.08e+00      -9.27e-02      -1.28e-01
   7.22e-01 - 8.05e-01i        1.08e+00      -9.27e-02      -1.28e-01
```

4.2 Characteristics of Time Response

As illustrated in Figure 4.9, time response to an input such as a step input often is considered to consist of an initial transient portion and a final steady-state portion. The transient response is associated with initial conditions and, when a production system or component is stable, the effects of the initial conditions decay as time becomes relatively large. The steady-state response remains after the transient response has decayed over this relatively long period. Characteristics of time response are defined for specific inputs and for step inputs, characteristics of the transient response include the settling time, percent overshoot, and variously defined rise times, while final value is a characteristic of the steady-state response. These measures can be

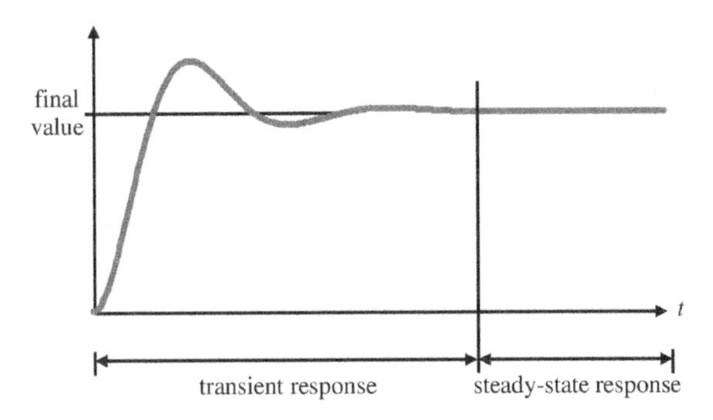

Figure 4.9 Transient and steady-state portions of time response.

calculated using control system engineering software and are of significant interest in assessing whether the dynamic behavior of a production system is favorable and guiding design of decision making.

4.2.1 Calculation of Time Response

When the relationships between the inputs and outputs of a production system are modeled using differential or difference equations, it is possible to solve those equations for specific input functions. The solutions then can be used to plot the output response as a function of time and the response can be analyzed to determine, for example, how quickly the response converges to the steady state. However, it can be difficult to solve differential equations or difference equations when inputs vary in a general way with time or when the differential or difference equations are not of low order. An alternative that is commonly used in control system engineering software is to calculate time response using the use difference equation

$$\begin{aligned} y(kT) = & -\frac{a_{n-1}}{a_n} y((k-1)T) - \cdots - \frac{a_2}{a_n} y((k-n+2)T) - \frac{a_1}{a_n} y((k-n+1)T) \\ & -\frac{a_0}{a_n} y((k-n)T) + \frac{b_m}{a_n} x((k-n+m)T) + \cdots + \frac{b_2}{a_n} x((k-n+2)T) \\ & + \frac{b_1}{a_n} x((k-n+1)T) + \frac{b_0}{a_n} x((k-n)T) \end{aligned} \quad (4.43)$$

which corresponds to discrete-time transfer function

$$\frac{Y(z)}{X(z)} = \frac{b_m z^m + \cdots + b_2 z^2 + b_1 z + b_0}{a_n z^n + \cdots + a_2 z^2 + a_1 z + a_0} \quad (4.44)$$

Equation 4.43 can be used when $m \leq n$ to find the current value of the output y(kT) from past values of output y(kT) and given values of input $x(kT)$. With known input sequence $x(kT)$, this equation can be recursively applied for k = 1,2,3,... to obtain output sequence $y(kT)$. Initial conditions can be specified or can be assumed to be zero.

For a continuous-time system modeled using transfer function $Y(s)/X(s)$, an equivalent discrete-time transfer function $Y(z)/X(z)$ can be obtained as described in Section 3.9 by choosing period T so that it is relatively short compared to the time constants in the continuous-time model and so that changes in input $x(t)$ during period T are relatively small; the latter permits the approximation $x(t) = x(kT)$ for $kT \le t < (k+1)T$. A sequence of samples $x(kT)$ of the input $x(t)$ is used as the input and output sequence $y(kT)$ then can be calculated using Equation 4.43. With relatively short period T, linear interpolation between successive values $y(kT)$ and $y((k+1)T)$ tends to result in a good approximation of the continuous response $y(t)$.

Example 4.10 Response of Continuous-Time Capacity Adjustment to a Unit Step in Work Input

It is desired to calculate and plot the response of work output, $w_o(t)$ hours in the continuous-time production system described in Example 4.3 to a step change in work input $w_i(t)$ hours. The following continuous-time transfer function was obtained for proportional capacity adjustment modeled as shown in the block diagram in Figure 3.22:

$$\frac{W_o(s)}{W_i(s)} = \frac{K_p}{s + K_p}$$

Response to a unit step input can be calculated and plotted using

`step(G)`	step input; continuous-time transfer function `G`
`step(G,tFinal)`	`tFinal` is the final time
`[y,t]=step(...)`	results saved in vectors: `y` is the response vector; `t` is the time vector; no plot is drawn

Function `step` assumes zero initial conditions by default, and uses straight-line interpolation when plotting continuous-time response and staircase interpolation when plotting discrete-time response.

Use of function `step` to plot the response to a unit step in work input, $w_i(t) = 1$ hour for$t \ge 0$, and $K_p = 0.4$ days^{-1} is demonstrated in Program 4.3. The result is shown in Figure 4.10. The final value of work output is $w_o(\infty) = 1$ hour. The time constant is $\tau = 1/K_p$ hours and, from a practical perspective, this final value is reached in approximately $4\tau = 10$ days.

Program 4.3 Calculation of continuous-time step response

```
Kp=0.4;   % decision-making parameter (days^-1)
s=tf('s','TimeUnit','days');

step(Kp/(s+Kp),15);   % plot step response - Figure 4.10
xlabel('time t'); ylabel('work output w_o(t) (hours)')
```

Example 4.11 Response of Discrete-Time Capacity Adjustment to a Unit Step in Work Input

In Example 4.8 the following discrete-time transfer function was obtained that relates work output $w_o(kT)$ hours to work input $w_i(kT)$ hours in a production system with pro-

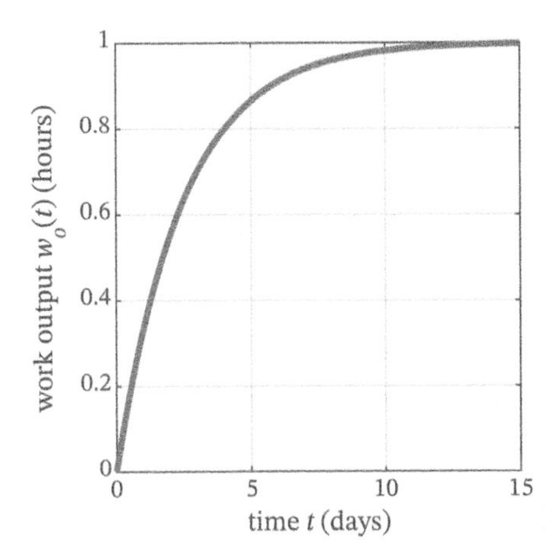

Figure 4.10 Response of continuous-time work output to a unit step in work input.

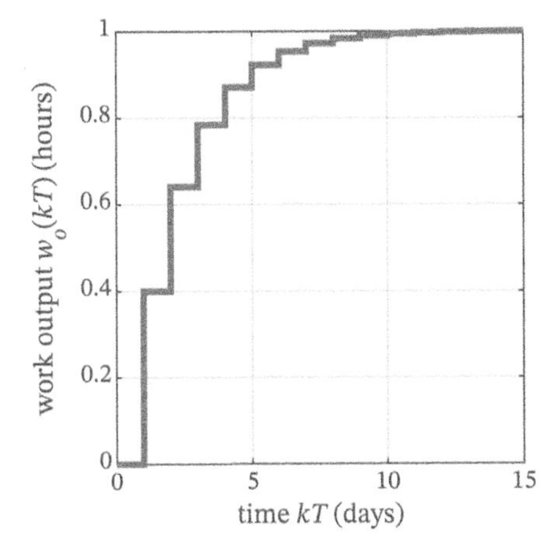

Figure 4.11 Response of discrete-time work output to a unit step in work input.

portional capacity adjustments modeled as shown in the block diagram in Figure 4.7:

$$\frac{W_o(z)}{W_i(z)} = \frac{K_p T z^{-1}}{1-(1-K_p T)z^{-1}}$$

The unit step response shown in Figure 4.11 can be plotted for capacity adjustment period $T = 1$ day and $K_p = 0.4$ days^{-1} using function `step` as is demonstrated in Program 4.4. The results are similar to those obtained in continuous-time Example 4.10, but here only daily values of work output $w_o(kT)$ hours are represented by the discrete-time model.

Example 4.12 Response of Discrete-Time Actuator Position Control to Constant Velocity Command

It is desired to calculate the response of the position control described in Example 4.9 when position command $p_c(kT)$ has a constant rate of change (constant velocity) of

Program 4.4 Calculation of discrete-time step response

```
T=1;  % period (days)
Kp=0.4;  % decision-making parameter (days^-1)
z=tf('z','TimeUnit','days');

step(Kp*T/(z-(1-Kp*T)),15);  % plot step response - Figure 4.11
xlabel('time kT'); ylabel('work output w_o(kT) (hours)')
```

Program 4.5 Calculation of response to given input function of time

```
T=0.01;  % period (seconds)
tauv=0.05; % actuator time constant (seconds)
Kp=10; % proportional control parameter (seconds^-1)

v=5;  % 5 cm/second velocity
pc(1)=0;  % initial position command
for k=1:1/T  % create input for 1-second simulation duration
    t(k+1)=k*T;
    if k<0.4/T
        pc(k+1)=pc(k)+v*T;  % position ramp for first 0.4 seconds
    else
        pc(k+1)=5*0.4;  % constant position for last 0.6 seconds
    end
end

Gpz=c2d(tf(1,[tauv 1 0]),T);  % actuator transfer function
Gclz=feedback(Kp*Gpz,1);  % closed-loop transfer function
damp(Gclz)  % two stable complex roots

        Pole                Magnitude     Damping      Time Constant
                                                         (seconds)
  9.05e-01 + 9.51e-02i      9.10e-01     6.71e-01       1.06e-01
  9.05e-01 - 9.51e-02i      9.10e-01     6.71e-01       1.06e-01

p=lsim(Gclz,pc,t);  % actuator position in response to input pc(kT)
stairs(t,pc); hold on; stairs(t,p); hold off  % response - Figure 4.12
xlabel('time kT (seconds)'); ylabel('position (cm)')
legend('actuator position command p_c(kT)','actuator position p(kT)')
```

5 cm/second for 0.4 seconds. The position command is constant 2 cm after 0.4 seconds.

The following function calculates and plots response to an arbitrary input sequence:

`lsim(G,u,t)`	transfer function `G` (continuous-time or discrete-time); `u` is the input sequence regularly spaced in time; `t` is the time sequence
`y=lsim(G,u,t)`	response sequence saved in vector `y;` no plot is drawn

As demonstrated in Program 4.5, function `lsim` can be used to obtain the response of actuator position $p(kT)$ cm to a constant rate of change in actuator position command $p_c(kT)$ cm. The input function and the results are shown in Figure 4.12 for period

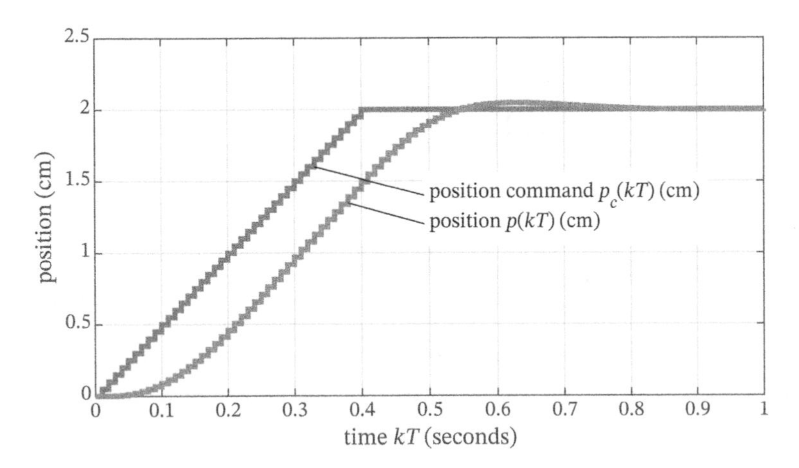

Figure 4.12 Response of actuator position to a constant velocity (constant rate of change in position) command of 5 cm/second for 0.4 seconds.

$T = 0.01$ seconds, actuator time constant $\tau_v = 0.05$ seconds, and proportional position control parameter $K_p = 10$ seconds^{-1}. With this value of K_p the closed-loop system is stable and the time constant associated with the two roots of the characteristic equation, calculated using function `damp`, is 0.106 seconds. The damping ratio is 0.671, and this results in some overshoot of actuator position beyond the final value of 2 cm as expected from Figure 4.2a. When a steady actuator velocity (constant rate of change in actuator position) has been reached at approximately 0.4 seconds, the actual position of the actuator differs from the commanded position by 0.5 cm. This steady-state following error is inversely proportional to K_p and may not be acceptable depending upon the application. If not, an alternative position control decision rule must be designed; this topic is discussed in Chapter 6.

4.2.2 Step Response Characteristics

Analysis of the response of a production system or component to a step input yields key characteristics such as the final value of the response, how long it takes for the response to complete, and whether the response overshoots the final value. These are referred to as the final value, settling time, and percent overshoot, respectively, and are illustrated in Figure 4.13. These are important characteristics of dynamic behavior and, among other measures, are useful as performance goals in design of decision making for production systems.

The overshoot in responses such as those shown in Figure 4.2a can be calculated theoretically for some models, as can final values, settling time, percent overshoot and other step response characteristics. However, these calculations are difficult for higher-order models, discrete-time systems and systems with internal delays. An alternative method is to calculate and analyze a sufficiently long step response sequence and, as illustrated in Figure 4.13, use the last value in the sequence as a

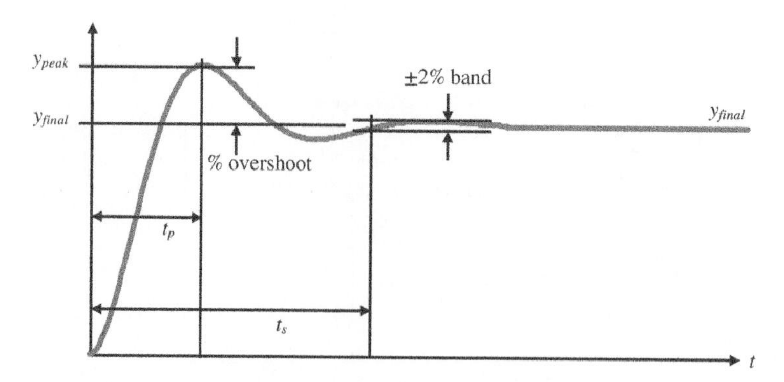

Figure 4.13 Characteristics of step response: final value y_{final}, settling time t_s and percent overshoot at peak time t_p.

practical measure of the final value y_{final},[5] define the settling time t_s as the time after which the response remains within 2% of the final value, and measure the percent overshoot using the amount by which by which the peak in the response y_{peak} exceeds the final value y_{final}.

Example 4.13 Characteristics of Step Response of Continuous-Time Capacity Adjustment

Figure 4.10 in Example 4.10 shows the response of the work output $w_o(t)$ hours of a production system, modeled by the following continuous-time transfer function, to a unit step in work input $w_i(t)$ hours when capacity adjustment parameter $K_p = 0.4$ days^{-1}:

$$\frac{W_o(s)}{W_i(s)} = \frac{K_p}{s + K_p}$$

The following function can be used to obtain the characteristics of this step response:

`stepinfo(G)` calculates a step response and computes performance characteristics; transfer function `G` (continuous-time or discrete-time); the steady-state value used is the last value of the calculated step response:

`RiseTime`	rise time of the response
`SettlingTime`	settling time of the response
`SettlingMin`	minimum value after the response has risen
`SettlingMax`	maximum value after response has risen

5 There is a final value theorem that can be used to calculate the theoretical final value of response to a given input if this final value exists, and there is a companion initial value theorem. Also, as noted in Section 5.3.1, the zero-frequency magnitude (DC gain) is the ratio between the steady-state output and amplitude of the input; hence, it is the final value of the response to a unit step input if this final value exists. The reader is referred to the Bibliography and many other control theory publications for additional information regarding calculating final values.

	`Overshoot`	percentage overshoot (relative to steady-state)
	`Undershoot`	percentage undershoot
	`Peak`	peak absolute value of response
	`PeakTime`	time at which peak response is reached
`stepinfo(y,t)`	computes performance characteristics from step response data `y` versus time `t`; the steady-state value used is the last value of `y`	
`S=stepinfo(...)`	`S` is a structure containing the performance characteristics	

As demonstrated in Program 4.6, the settling time calculated using function `stepinfo` is 9.8 days. No overshoot is expected, and the calculated % overshoot is zero. Because there is no overshoot, the peak value is the same as the last value of the calculated response, which is the approximate final value $w_o(\infty)$.

Program 4.6 Calculation of step response characteristics

```
Kp=0.4;  % decision-making parameter (days^-1)
s=tf('s','TimeUnit','days');

stepinfo(Kp/(s+Kp))  % step response characteristics

  ans =
          RiseTime:  5.4925
      SettlingTime:  9.7802
       SettlingMin:  0.9045
       SettlingMax:  1.0000
         Overshoot:  0
        Undershoot:  0
              Peak:  1.0000
          PeakTime:  26.3646
```

Example 4.14 Step Response Characteristics of Discrete-Time Temperature Regulation

It is desired to predict how long it will take for mixture temperature to reach the desired mixture temperature in a portion of a production process in which the temperature of a mixture flowing out of a pipe is regulated as shown in Figure 3.5 using a heater at the beginning of the pipe and a temperature sensor placed at the end of the pipe. Other time-response characteristics including the settling time and whether there is any temperature overshoot also are of interest.

The discrete model for mixture heating and temperature sensing is

$$H_o(z) = K_a \mathcal{Z}\left\{\left(\frac{1-e^{-Ts}}{s}\right)\frac{K_h}{\tau s+1}e^{-Ds}\right\}V_c(z) + \mathcal{Z}\left\{e^{-Ds}H_i(s)\right\}$$

where $h_o(kT)$ °C is the outlet mixture temperature, $h_i(t)$ °C is the inlet mixture temperature, and $v_c(kT)$ is the heater voltage manipulation calculated by a control computer. Mixture heating time constant τ = 49.8 seconds and mixture heating parameter K_h = 0.4°C/V were found using experimental data. K_a = 20 V/V is the heater voltage amplification parameter and delay D = 12 seconds is the time required for the mixture to flow through the pipe from the heater to the temperature sensor.

The discrete-time temperature regulation decision rule chosen, which is implemented in the control computer, is

$$v_c(kT) = v_c((k-1)T) + K_c T(h_c(kT) - h_o(kT))$$

or

$$V_c(z) = \frac{K_c T z^{-1}}{1 - z^{-1}}(H_c(z) - H_o(z))$$

where $h_c(kT)$ is the desired outlet temperature, K_c = 0.1 (V/s)/°C is the chosen value of the temperature regulation parameter, and T = 5 seconds is the period between heater voltage adjustments. The block diagram in Figure 4.14 shows the topology of this mixture regulation system.

As demonstrated in Program 4.7, function `stepinfo` can be used to characterize the response of the mixture heating to a 60°C step in desired mixture temperature $h_c(kT)$. The resulting mixture temperature $h_o(kT)$ °C is shown in Figure 4.15. For convenience, the mixture inlet temperature $h_i(t)$ °C and initial conditions are assumed to be zero, but the results obtained are valid for other initial temperatures $\Delta h(0)$ °C and other constant mixture inlet temperatures $h_i(t)$ °C. The mixture output temperature reaches and remains at the desired temperature; regardless of the initial conditions. The setting time found using function `stepinfo` is 485 seconds, which is the time after which the response remains within ±2% of the final value. The response has an overshoot in outlet temperature of approximately 4.1%.

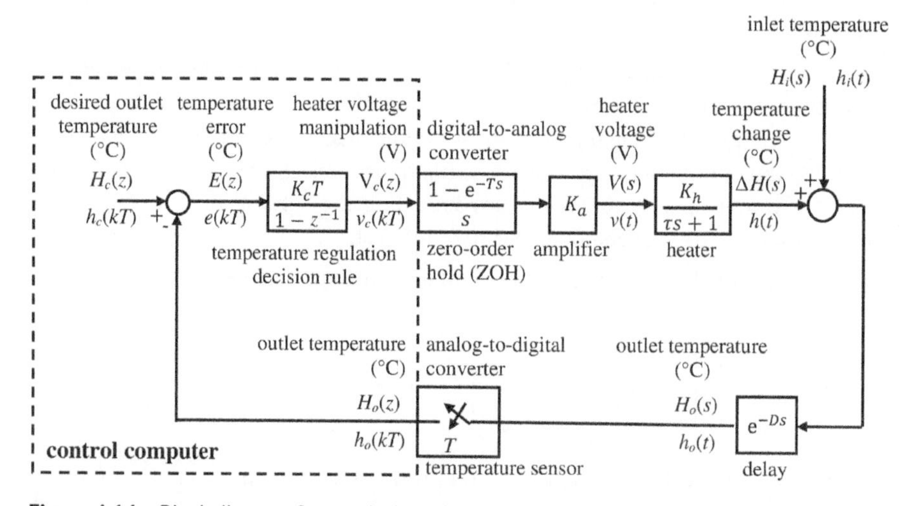

Figure 4.14 Block diagram for regulation of mixture temperature.

Program 4.7 Calculation of step response characteristics of discrete-time temperature regulation

```
tau=49.8;  % mixture heating time constant (seconds)
Kh=0.4;  % mixture heating parameter (°C/V)
D=12;  % pipe delay (seconds)
Ka=20;  % voltage amplification parameter (V/V)
Kc=0.001;  % temperature regulation parameter ((V/s)/°C)
A=60;  % 60 °C step in desired mixture temperature (°C)

T=5;  % period between heater voltage adjustments (seconds)
Ghz=c2d(tf(Kh,[tau, 1],'iodelay',D),T,'zoh')  % transfer function
```

```
Ghz =

                   0.02338 z + 0.01483
    z^(-3) * ---------------------------
                      z - 0.9045

Sample time: 5 seconds
Discrete-time transfer function.
```

```
Gcz=tf([Kc*T 0],[1, -1],T)  % temperature regulation transfer function
```

```
Gcz =

     0.005 z
    ----------
      z - 1

Sample time: 5 seconds
Discrete-time transfer function.
```

```
Gclz=minreal(feedback(Gcz*Ka*Ghz,1))  % closed-loop transfer function
```

```
Gclz =

                     0.002338 z + 0.001483
    ---------------------------------------------------
     z^4 - 1.904 z^3 + 0.9045 z^2 + 0.002338 z + 0.001483

Sample time: 5 seconds
Discrete-time transfer function.
```

```
step(A*Gclz,800)  % plot step response of temperature - Figure 4.15
xlabel('time kT'); ylabel('mixture temperature h_o(kT) (°C)')

stepinfo(A*Gclz)  % step response characteristics
```

```
ans =
         RiseTime: 175
     SettlingTime: 485
      SettlingMin: 54.5470
      SettlingMax: 62.4796
        Overshoot: 4.1326
       Undershoot: 0
             Peak: 62.4796
         PeakTime: 365
```

```
[y,t]=step(A*Gclz,800);   % step response data
y(end)  % approximate final value

    ans = 59.9291
```

Example 4.15 Step Response Characteristics of a System with Two Time Constants

The following second-order model of the dynamic behavior of a continuous-time production system has two factors with time constants $\tau_1 > 0$ and $\tau_2 > 0$:

$$G(s) = \frac{Y(s)}{X(s)} = K\left(\frac{1}{\tau_1 s + 1}\right)\left(\frac{1}{\tau_2 s + 1}\right)$$

Figure 4.16 shows response $y(t)$ to a unit step input $x(t)$ for several time constant ratios τ_2/τ_1. The associated settling times are listed in Table 4.1 and are the time after which the response remains within 2% of the final value. Settling times become longer as time constant τ_2 becomes more significant with respect to τ_1.

When $\tau_2 = \tau_1$ the transfer function is

$$G(s) = \frac{Y(s)}{X(s)} = K\frac{1}{(\tau_1 s + 1)^2}$$

The settling time in this case is $5.8\tau_1$. A goal in designing decision making for production systems often is to avoid oscillatory response, and a decision rule that yields a second-order system transfer function with two equal real roots often results in a fast settling time that is achieved with little or no overshoot. This will be discussed in more detail in Chapter 6.

The case where $\tau_2 = 0$ also is shown in Figure 4.16 and listed in Table 4.1. In this case, the settling time is approximately $4\tau_1$. Comparing this result with the results when τ_2 is relatively small with respect to τ_1 yields the simplification

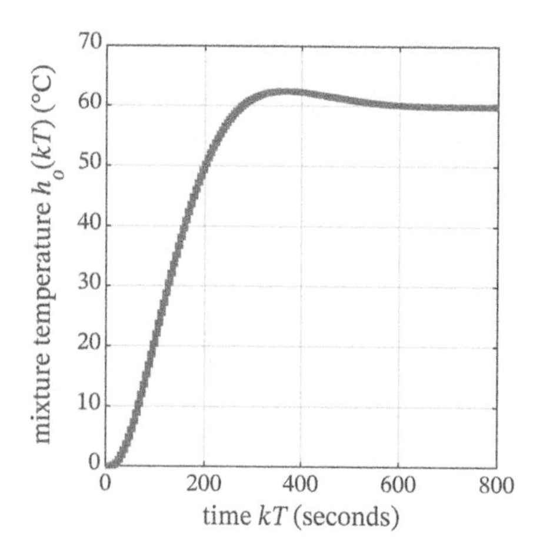

Figure 4.15 Response of mixture outlet temperature to a 60°C step in desired mixture temperature.

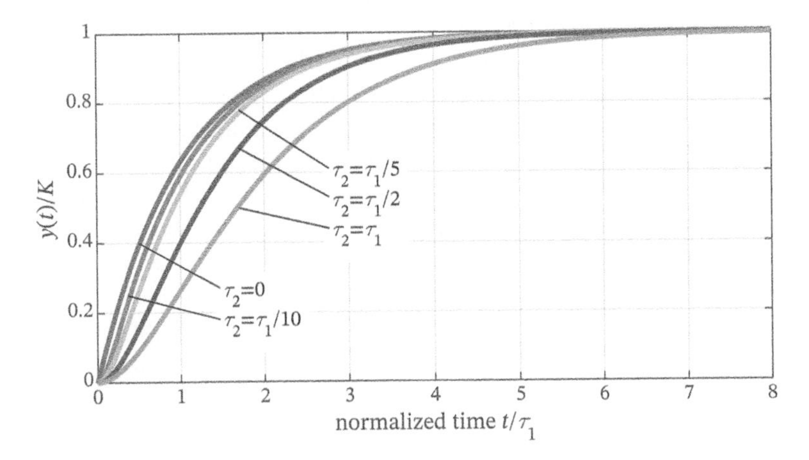

Figure 4.16 Unit step response with various ratios of time constants τ_1 and τ_2.

Table 4.1 Settling times for various ratios of time constants τ_1 and τ_2.

	settling time
$\tau_2 = 0$	$3.91\tau_1$
$\tau_2 = \frac{\tau_1}{10}$	$4.02\tau_1$
$\tau_2 = \frac{\tau_1}{5}$	$4.16\tau_1$
$\tau_2 = \frac{\tau_1}{2}$	$4.60\tau_1$
$\tau_2 = \tau_1$	$5.83\tau_1$

$$G(s) \approx K\left(\frac{1}{\tau_1 s + 1}\right)$$

when $\tau_2 \leq \tau_1/5$. When transfer functions are of higher order, factors that are associated with relatively small time constants (relatively quick response) often can be ignored when the goal is to understand fundamental dynamic behavior and design decision rules for production systems and components.

4.3 Summary

It has been shown in this chapter that the transfer functions that describe the dynamic behavior of production systems and components can be analyzed to directly determine the fundamental dynamic characteristics of these systems and components. These characteristics include time constants, damping ratios, natural frequencies and stability; they represent general tendencies that are true regardless of specific inputs over time or operating conditions. They describe how quickly the outputs of a system or component tend to respond to changing inputs, whether these responses tend to be oscillatory, and whether the system or component is stable or unstable. Production engineers often are familiar with models and simulations that predict how production systems and their components respond to specific inputs, and the use of transfer functions to calculate and analyze step response and other input functions of time has been described in this chapter. Characteristics of step response and response to other inputs of interest can significantly facilitate understanding of the dynamic behavior of production systems and components. In subsequent chapters it will be shown how fundamental dynamic characteristics influence frequency response, which is response to sinusoidal inputs, and how fundamental dynamic characteristics, time response and frequency response can be used to design decision making for production systems.

5

Frequency Response

Fundamental dynamic characteristics of production systems and components such as time constants, damping ratios, natural frequencies, and stability were defined in Chapter 4, along with important characteristics of time response such as final value, settling time, and percent overshoot. The response of production systems and components to sinusoidal inputs also is of significant interest because decision rules often must be designed so that the production system reacts differently in various ranges of frequency. For example, production systems may be required to meet goals in the presence of relatively slowly changing disruptions such as fluctuations in demand, but it may not be physically or economically practical to meet the same goals in the presence of relatively quickly changing disruptions. Adjustments made using decision rules therefore may be relatively large when responding to lower frequency fluctuations in production system inputs but relatively small when responding at higher frequency fluctuations in inputs. Response to sinusoidal inputs over a range of frequencies is referred to as frequency response.

In this chapter, important fundamental dynamic characteristics associated with frequency response will be defined. These include the magnitude, phase, zero-frequency magnitude, bandwidth, unity magnitude crossing frequency, and -180° phase crossing frequency. The latter facilitate calculation of gain and phase margins, which are measures of relative stability. Calculation of the frequency response of continuous-time and discrete-time systems will be described, and examples will be presented that illustrate the frequency response characteristics of production systems and components. Analysis of frequency response leads to important additional ways of characterizing and understanding dynamic behavior and guiding design of decision-making for production systems and components.

5.1 Frequency Response of Continuous-Time Systems

An example of the response of a production system or component to a sinusoidal input that commences at time $t = 0$ is shown in Figure 5.1. There is a transient period during which the effects of initial conditions decay with time. The response then converges to a steady-state sinusoidal behavior. It is the steady-state relationship between the

Control Theory Applications for Dynamic Production Systems: Time and Frequency Methods for Analysis and Design, First Edition. Neil A. Duffie.

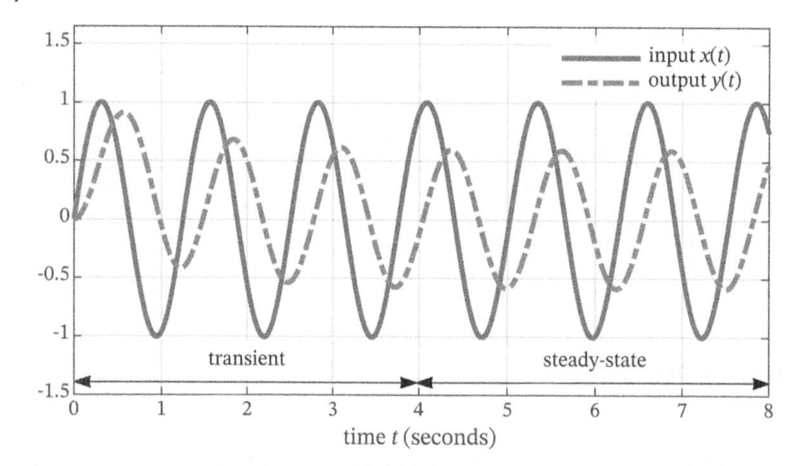

Figure 5.1 Example of transient and steady-state portions of response of a production system to a sinusoidal input.

amplitudes of the input and output sinusoids, and the shift of the output sinusoid in time with respect to the input sinusoid that are referred to as frequency response.

When an input to a stable, linear continuous-time production system or component has an oscillating portion with amplitude A,

$$x(t) = A\sin(\omega t) \tag{5.1}$$

the corresponding steady-state portion of the output is

$$y(t) = M(\omega)A\ \sin(\omega t + \phi(\omega)) \tag{5.2}$$

where $M(\omega)$ is the magnitude of the frequency response, and $\phi(\omega)$ is the phase in radians of the frequency response at radian frequency $\omega \geq 0$ (radians/second, radians/day, etc.).

The differential equation that models the relationship between $x(t)$ and $y(t)$ can be solved for the sinusoidal input in Equation 5.1. At $t \gg 0$, the portion of the solution in the form of Equation 5.2 then represents the frequency response. Magnitude and phase as a function of frequency are fundamental dynamic characteristics of a production system or component, and initial conditions and response to non-oscillating portions of inputs such as average values are not considered when analyzing frequency response.

It is not necessary to solve differential equations for sinusoidal inputs to determine the magnitude and phase of frequency response. Instead, as described in Section 3.8.6, the continuous-time transfer function that represents the production system or component

$$G(s) = \frac{Y(s)}{X(s)} = \frac{b_m s^m + \ldots + b_2 s^2 + b_1 s + b_0}{a_n s^n + \ldots + a_2 s^2 + a_1 s + a_0} e^{-Ds} \tag{5.3}$$

can be factored into the form

$$G(s)=\frac{Y(s)}{X(s)}=K_s\frac{(s-c_1)(s-c_2)\dots(s-c_m)}{(s-r_1)(s-r_2)\dots(s-r_n)}e^{-Ds} \tag{5.4}$$

It then can be shown that after substituting $j\omega$ for s in transfer function $G(s)$,[1]

$$G(j\omega)=K_s\frac{(j\omega-c_1)(j\omega-c_2)\dots(j\omega-c_m)}{(j\omega-r_1)(j\omega-r_2)\dots(j\omega-r_n)}e^{-Dj\omega} \tag{5.5}$$

the magnitude of the frequency response of a stable continuous-time system ($\alpha_i < 0$ for all $r_i = \alpha_i + j\beta_i$) is

$$M(\omega)=|K_s|\frac{|j\omega-c_1||j\omega-c_2|\dots|j\omega-c_m|}{|j\omega-r_1||j\omega-r_2|\dots|j\omega-r_n|} \tag{5.6}$$

and the phase angle[2] of the frequency response is

$$\begin{aligned}\phi(\omega)=&\angle(K_s)+\angle(j\omega-c_1)+\angle(j\omega-c_2)+\ldots+\angle(j\omega-c_m)\\&-\angle(j\omega-r_1)-\angle(j\omega-r_2)-\ldots-\angle(j\omega-r_n)-D\omega\end{aligned} \tag{5.7}$$

Frequency also can be expressed as cycles of oscillation per time unit (cycles/second, cycles/day, etc.):

$$f=\frac{\omega}{2\pi} \tag{5.8}$$

---

1 One approach is to use Laplace transform methods to solve for the complete response and then to recognize that the sinusoidal steady-state portion of the solution is obtained by substituting $j\omega$ for s. The reader is referred to the Bibliography and many control theory publications that treat this development.

2 Phase angle often varies beyond $\pm\pi$ radians ($\pm 180°$) for higher-order systems and use of Equation 5.7 can be preferable to substituting $j\omega$ for s in the unfactored numerator and denominator of the transfer function of Equation 5.3, which yields the real and imaginary parts of the frequency response for each frequency ω:

$$G(j\omega)=\frac{b_m(j\omega)^m+\ldots+b_2(j\omega)^2+b_1 j\omega+b_0}{a_n(j\omega)^n+\ldots+a_2(j\omega)^2+a_1 j\omega+a_0}e^{-Dj\omega}$$

$$G(j\omega)=\mathrm{Re}(G(j\omega))+j\mathrm{Im}(G(j\omega))$$

Magnitude can be calculated using

$$\sqrt{\mathrm{Re}(G(j\omega))^2+(\mathrm{Im}(G(j\omega)))^2}$$

and phase angle can be calculated using

$$\tan^{-1}\left(\frac{\mathrm{Im}(G(j\omega))}{\mathrm{Re}(G(j\omega))}\right)$$

considering all four quadrants. Angles may need to be "unwrapped" to obtain results that are outside the range $\pm\pi$ radians. Control system engineering software typically supports unwrapping of calculated phase angles when models are of higher order.

The corresponding period of oscillation (seconds/cycle, days/cycle, etc.) is

$$\delta = \frac{2\pi}{\omega} = \frac{1}{f} \tag{5.9}$$

Often, a logarithmic frequency scale is used when plotting frequency response, with the magnitude plotted in decibels (dB) and the phase angle plotted in degrees:

$$dB(\omega) = 20\log_{10} M(\omega) \tag{5.10}$$

and

$$\deg(\omega) = \phi(\omega)\frac{180°}{\pi} \tag{5.11}$$

The magnitude and phase characteristics of common continuous-time models are discussed in the following sections, and examples are presented that illustrate important aspects of frequency response as well as calculation and plotting of frequency response.

5.1.1 Frequency Response of Integrating Continuous-Time Production Systems or Components

A production system or component with a transfer function in the form of Equation 5.4 that has $(s - 0)$ as one of the factors in its denominator often is referred to as an integrating system or component, and $(s - 0)^2$ in the denominator often is referred to as double integrating. Integration often is an important physical aspect of models, and several examples with integration have been presented in previous chapters. While the dynamic behavior of a production system or component represented by a transfer function containing integration is not stable according to the stability criterion of continuous-time systems presented in Section 4.1.5, the magnitude and phase of response to sinusoidal inputs is well-defined: the amplitude of oscillation does not grow with time beyond a steady-state value.

Consider the steady-state response of a continuous-time system or component with a single integration:

$$y(t) = K\int_0^t x(t)dt \tag{5.12}$$

For the sinusoidal input in Equation 5.1, the response is

$$y(t) = K\left(-\frac{A}{\omega}\cos(\omega t) + \frac{A}{\omega}\cos(0t)\right) = \frac{KA}{\omega}\sin\left(\omega t - \frac{\pi}{2}\right) + \frac{KA}{\omega} \tag{5.13}$$

This result is plotted in Figure 5.2.

As previously noted, the magnitude and phase as a function of frequency are fundamental dynamic characteristics that are not related to the non-sinusoidal portions of the response. In this case, the constant portion of the response is ignored. The magnitude of the frequency response then is

$$M(\omega)=\frac{\frac{KA}{\omega}}{A}=\frac{K}{\omega} \tag{5.14}$$

and the phase is

$$\phi(\omega)=-\frac{\pi}{2} \tag{5.15}$$

The magnitude and phase are plotted versus frequency ω on a linear scale in Figure 5.3. The time unit is seconds. The magnitude is infinite at frequency $\omega = 0$ and zero at $\omega = \infty$ radians/second. The phase is $-\pi/2$ radians at all frequencies ω.

Alternatively, the transfer function for a single integration is

$$G(s)=\frac{Y(s)}{X(s)}=\frac{K}{s} \tag{5.16}$$

Substituting $j\omega$ for s,

$$G(j\omega)=\frac{K}{j\omega} \tag{5.17}$$

For $K > 0$, the magnitude of the frequency response is calculated using Equation 5.6:

$$M(\omega)=|G(j\omega)|=\left|\frac{K}{j\omega}\right|=\frac{K}{\omega} \tag{5.18}$$

and the phase of the frequency response is calculated using the angles in Equation 5.7

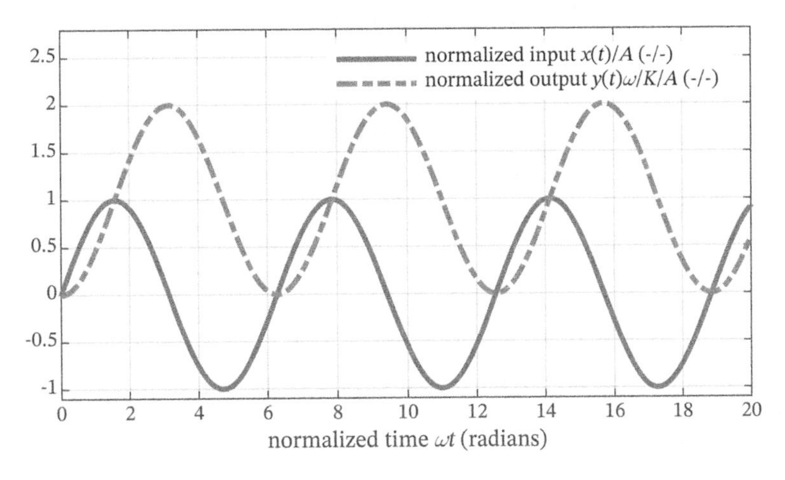

Figure 5.2 Example of response of integrating continuous-time system or component when the input is a sinusoid.

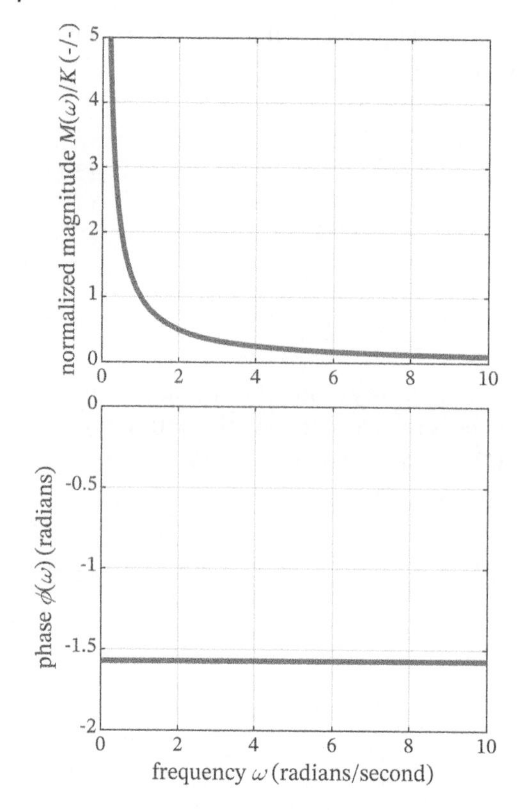

Figure 5.3 Frequency response of an integrating continuous-time production system or component plotted on a linear frequency scale.

$$\phi(\omega) = \angle G(j\omega) = \angle(K) - \angle(j\omega) = -\frac{\pi}{2} \tag{5.19}$$

These results for magnitude and phase are identical to those obtained above using the integral.

Example 5.1 Frequency Response of Backlog in a Production System

The following continuous-time model represents the accumulation of backlog $w_b(t)$ orders in the production system shown in Figure 2.3:

$$w_b(t) = w_d(t) + \int_0^t (r_i(t) - r_o(t))dt$$

where $w_d(t)$ orders represents rush orders and cancelled orders, $r_i(t)$ orders/day is the order input rate and $r_o(t)$ orders/day is the order release rate. The order input rate varies sinusoidally:

$$r_i(t) = \bar{r}_i + A\sin(\omega t)$$

Where $\bar{r}_i$ orders/day is the average order input rate and A orders/day is the amplitude of variation. Both a relatively long period of oscillation and a relatively short period of oscillation, 50 days and 5 days, respectively, are of particular interest in terms of their effect on accumulation of backlog. The general trend of magnitude of oscillation in backlog as a function of frequency of oscillation in order input rate also is of interest, and the magnitude and phase of its frequency response characterize this behavior.

The frequency response can be calculated using the transform

$$W_b(s) = W_d(s) + \frac{1}{s}\left(R_i(s) - R_o(s)\right)$$

From Equations 5.6 and 5.7, the magnitude and phase are

$$M(\omega) = \left|\frac{W_b(j\omega)}{R_i(j\omega)}\right| = \left|\frac{1}{j\omega}\right| = \frac{1}{\omega}$$

$$\phi(\omega) = \angle\left(\frac{W_b(j\omega)}{R_i(j\omega)}\right) = -\frac{\pi}{2}$$

Program 5.1 demonstrates calculation and plotting of the magnitude and phase of the frequency response of backlog for fluctuating order input rate. Functions that are useful in calculating and plotting frequency response include

`bode(G,options)`	draw Bode plot; transfer function `G` (continuous-time or discrete-time); options allow selection of frequency units, logarithmic scales, etc.
`bode(G,W)`	draw Bode plot for frequencies in vector `W`
`[M,P,W]=bode(...)`	obtain the magnitude and phase of the frequency response (in this case, no plot is drawn); `M` is the magnitude vector (absolute); `P` is the phase vector (degrees); `W` is the frequency vector
`[Re,Im,W]=nyquist(...)`	obtain the real and imaginary parts of the frequency response; `Re` is the real part vector; `Im` is the imaginary part vector; `W` is the frequency vector
`db(M)`	convert magnitude to `dB`
`rad2deg(R)`	convert angle from radians to degrees
`deg2rad(P)`	convert angle from degrees to radians

In Program 5.1 the magnitude and phase of the frequency response for periods of oscillation of $\delta = 50$ and $\delta = 5$ days/cycle (frequencies $f = 0.02$ and 2 cycles/day, $\omega = 0.04\pi$ and 0.4π radians/day) are calculated using function `bode`. The magnitudes found are 7.958 and 0.796, respectively, illustrating that when fluctuations in order input rate have longer periods, they tend to result in larger fluctuations in backlog.

Figure 5.4 is a plot of the magnitude and the phase on a logarithmic frequency scale; this often is called a Bode plot. In Program 5.1, `options` are used to set the units of frequency to cycles/day, plot magnitude on an absolute scale in Figure 5.4a and on a dB scale in Figure 5.4b, as well as phase in degrees in Figure 5.4c. The amplitude of variation in backlog is high when the frequency of variation in order input rate is low, and the

Program 5.1 Calculation of frequency response of backlog

```
s=tf(1,[1, 0],'TimeUnit','days');   % backlog transfer function

delta=50;  % 50 day period (frequency 0.02 cycles/day)
w=2*pi/delta;   % 2π/50 radians/day
[M,phi,w]=bode(Gs,w)   % calculate magnitude and phase

    M = 7.9577
    phi = -90
    w = 0.1257

delta=5;   % 5 day period (frequency 0.02 cycles/day)
w=2*pi/delta;   % 2π/5 radians/day
[M,phi,w]=bode(Gs,w)   % calculate magnitude and phase

    M = 0.7958
    phi = -90
    w = 1.2566

options=bodeoptions; options.FreqUnits='cycles/day';
W=logspace(-2,0,100)*2*pi';   % frequency range (radians)

% log magnitude, absolute units
options.MagScale='log'; options.MagUnits='abs';
bode(Gs,W,options)   % frequency response - Figure 5.4a and c

% linear magnitude, dB units
options.MagScale='linear'; options.MagUnits='db';
bode(Gs,W,options)   % frequency response - Figure 5.4b and c
```

amplitude of variation in backlog is low when the frequency of variation in order input rate is high. The phase is -90° for all frequencies of variation in order input rate.

5.1.2 Frequency Response of 1st-order Continuous-Time Production Systems or Components

Consider a production system or component with the transfer function

$$G(s)=\frac{Y(s)}{X(s)}=\frac{K}{\tau s+1} \tag{5.20}$$

where τ is the time constant and K is the constant that relates the units of $y(t)$ to the units of $x(t)$. Substituting $\mathrm{j}\omega$ for s,

$$G(\mathrm{j}\omega)=\frac{K}{\tau \mathrm{j}\omega+1} \tag{5.21}$$

The magnitude of the frequency response is calculated using Equation 5.6:

$$M(\omega)=|G(\mathrm{j}\omega)|=\frac{|K|}{|\mathrm{j}\tau\omega+1|}=\frac{|K|}{\sqrt{1+(\tau\omega)^2}} \tag{5.22}$$

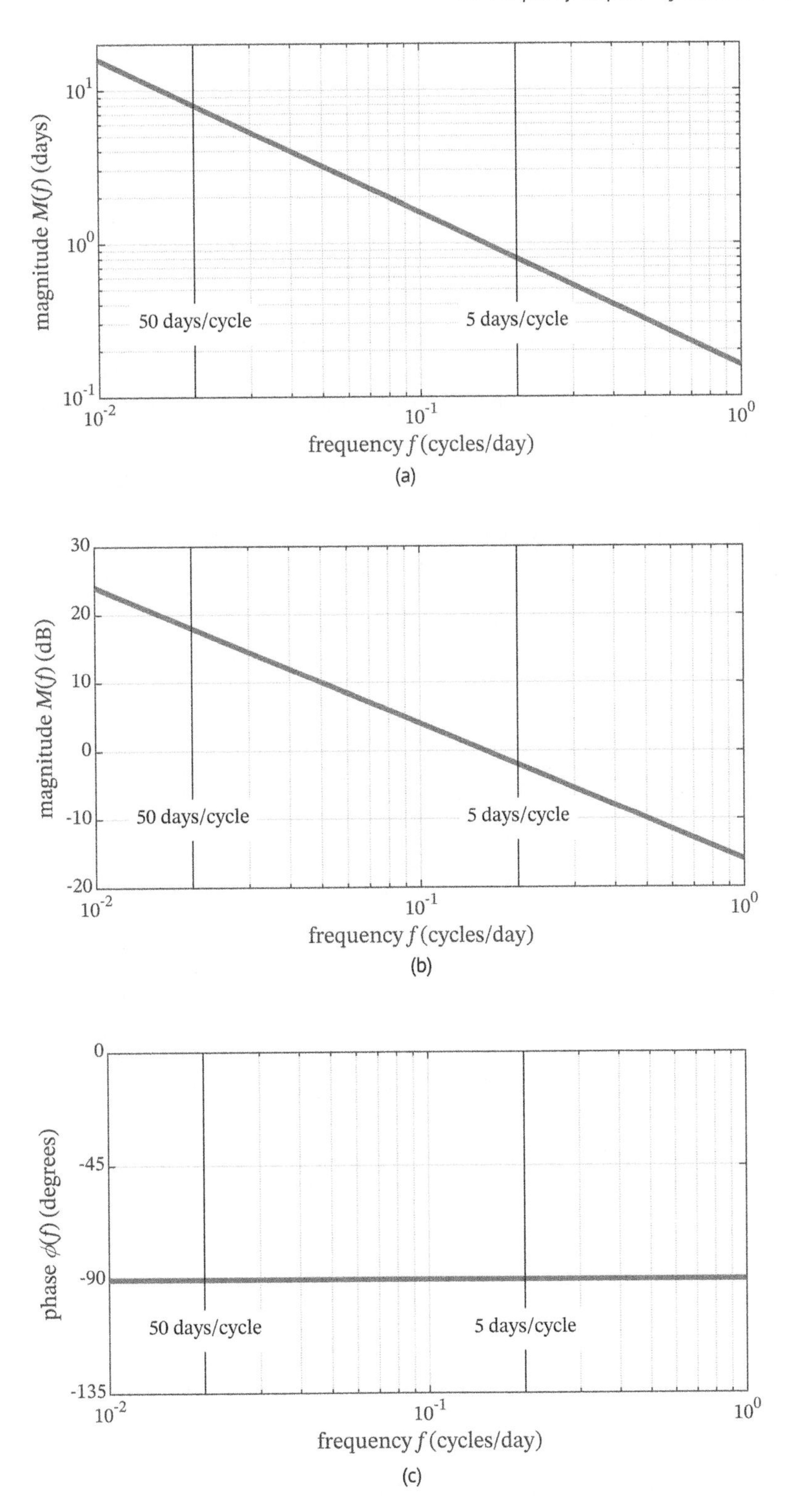

Figure 5.4 Bode plot of frequency response of backlog with logarithmic frequency scale. (a) Logarithmic magnitude. (b) dB magnitude. (c) Phase.

and the phase of the frequency response is calculated using the angles in Equation 5.7

$$\phi(\omega) = \angle G(j\omega) = \angle(K) - \angle(\tau j\omega + 1) \tag{5.23}$$

For $K > 0$,

$$\phi(\omega) = -\tan^{-1}(\tau\omega) \tag{5.24}$$

The magnitude and phase are plotted versus normalized frequency $\omega\tau$ on a logarithmic scale in Figure 5.5. The magnitude is approximately constant at relatively low frequencies, approaching $M(\omega) = K$ at $\tau\omega = 0$, which corresponds to $20\log_{10}K$ dB. At relatively high frequencies, the magnitude decreases by a factor of 10 per decade of frequency, which corresponds to -20 dB per decade. The phase is approximately 0° at relatively low frequencies and approximately -90° at relatively high frequencies.

As shown in Figure 5.5, the normalized frequency $\tau\omega = 1$ is a convenient quantitative border between low-frequency dynamic behavior and high-frequency dynamic behavior. At $\tau\omega = 1$, the phase is $\phi(1/\tau) = -45°$ and the magnitude had decreased to $M(1/\tau) = 0.707\ K$, a reduction in magnitude of -3 dB, and the magnitude of the frequency response is 70.7% of the magnitude at $\omega\tau = 0$.

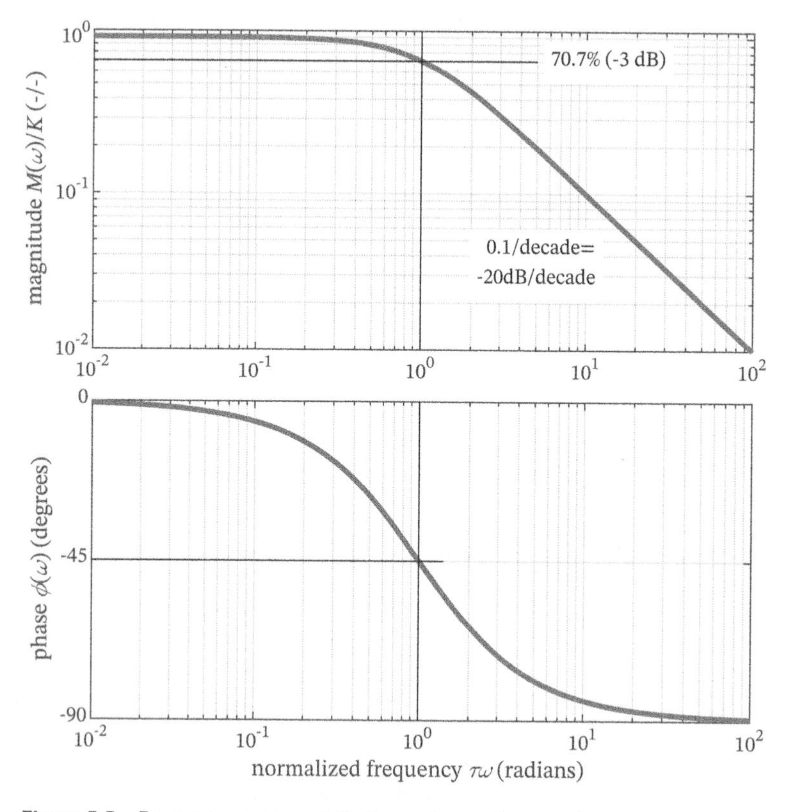

Figure 5.5 Frequency response of a 1st-order continuous-time system.

Example 5.2 Frequency Response of Continuous-Time Capacity Adjustment

The block diagram in Figure 5.6 shows the topology of a production system with capacity adjustment such as that illustrated in Figure 3.21. It is desired to characterize oscillation in the production system's work output rate $r_o(t)$ hours/day and WIP $w_w(t)$ hours as a function of frequency of oscillation in the rate of work input $r_i(t)$ hours/day. Trends in the dynamic behavior of the production system as a function of frequency then can be identified by analyzing its frequency response.

WIP is modelled using

$$\frac{dw_w(t)}{dt} = r_i(t) - r_o(t)$$

and work output rate is adjusted in proportion to WIP using the following decision rule that reacts to increasing WIP by increasing work output rate:

$$r_o(t) = K_p w_w(t)$$

where proportional decision parameter $K_p > 0$ days^{-1}. The relationship between work output rate and work input rate is

$$\frac{R_o(s)}{R_i(s)} = \frac{K_p}{s + K_p} = \frac{1}{\frac{1}{K_p}s + 1}$$

and the relationship between WIP and work input rate is

$$\frac{W_w(s)}{R_i(s)} = \frac{1}{s + K_p} = \frac{\frac{1}{K_p}}{\frac{1}{K_p}s + 1}$$

The frequency response of work output rate $r_o(t)$ to oscillation in work input rate $r_i(t)$ of frequency ω radians/day and amplitude A hours/day around average $\bar{r}_i$ hours/day

$$r_i(t) = \bar{r}_i + A\sin(\omega t)$$

can be calculated using Equations 5.22 and 5.24 where time constant $\tau = 1/K_p$:

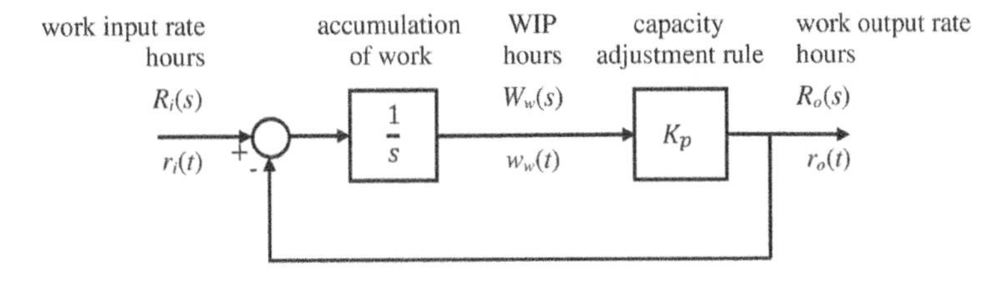

Figure 5.6 Production system with capacity adjustment.

$$M(\omega)=\left|\frac{R_i(j\omega)}{R_i(j\omega)}\right|=\left|\frac{K_p}{j\omega+K_p}\right|=\frac{1}{\sqrt{1+\left(\frac{\omega}{K_p}\right)^2}}$$

$$\phi(\omega)=\angle\left(\frac{R_i(j\omega)}{R_i(j\omega)}\right)=-\tan^{-1}\left(\frac{\omega}{K_p}\right)$$

The magnitude of the frequency response of WIP $w_w(t)$ with respect to oscillation in work input rate $r_i(t)$ is $M(\omega)/K_p$ because of the decision rule used in capacity adjustment, and the phase of the frequency response of WIP is $\phi(\omega)$.

Program 5.2 demonstrates calculation and plotting of the magnitude and phase of the frequency response of capacity adjustment for $K_p = 0.4$ days^{-1}, which corresponds to time constant $\tau = 2.5$ days. The results are shown in Figure 5.7. Work output rate is approximately the same as work input rate, $M(\omega) \approx 1$, when the period of oscillation is significantly longer than $\delta = 2\pi/K_p = 15.7$ days/cycle (frequency $\omega = K_p$ radians/day, $f = K_p/2\pi = 0.064$ cycles/day). On the other hand, oscillations in work output rate are relatively small when the period of oscillation in work input rate is significantly shorter than $\delta = 2\pi/K_p = 15.7$ days/cycle. Hence, work output rate is effectively adjusted to match work input rate at relatively low-frequency fluctuations in input rate but significant adjustments are not made in work output rate in response to relatively high-frequency fluctuations in work input rate.

Program 5.2 Calculation of capacity adjustment frequency response

```
Kp=0.4;  % capacity adjustment decision parameter (days^-1)
Gs=tf(Kp, [1, Kp],'TimeUnit','days')  % work output rate
```

```
Gs =

    0.4
  -------
  s + 0.4

Continuous-time transfer function.
```

```
options=bodeoptions; options.FreqUnits='cycles/day';
options.MagScale='linear';
bode(Gs,options)  % input rate frequency response - Figure 5.7a and c
bode(Gs/Kp,options)  % WIP frequency response - Figure 5.7b and c
```

5.1.3 Frequency Response of 2nd-order Continuous-Time Production Systems or Components

Consider a production system or component with the transfer function

$$G(s)=\frac{Y(s)}{X(s)}=\frac{K}{\frac{1}{\omega_n^2}s^2+\frac{2\zeta}{\omega_n}s+1} \tag{5.25}$$

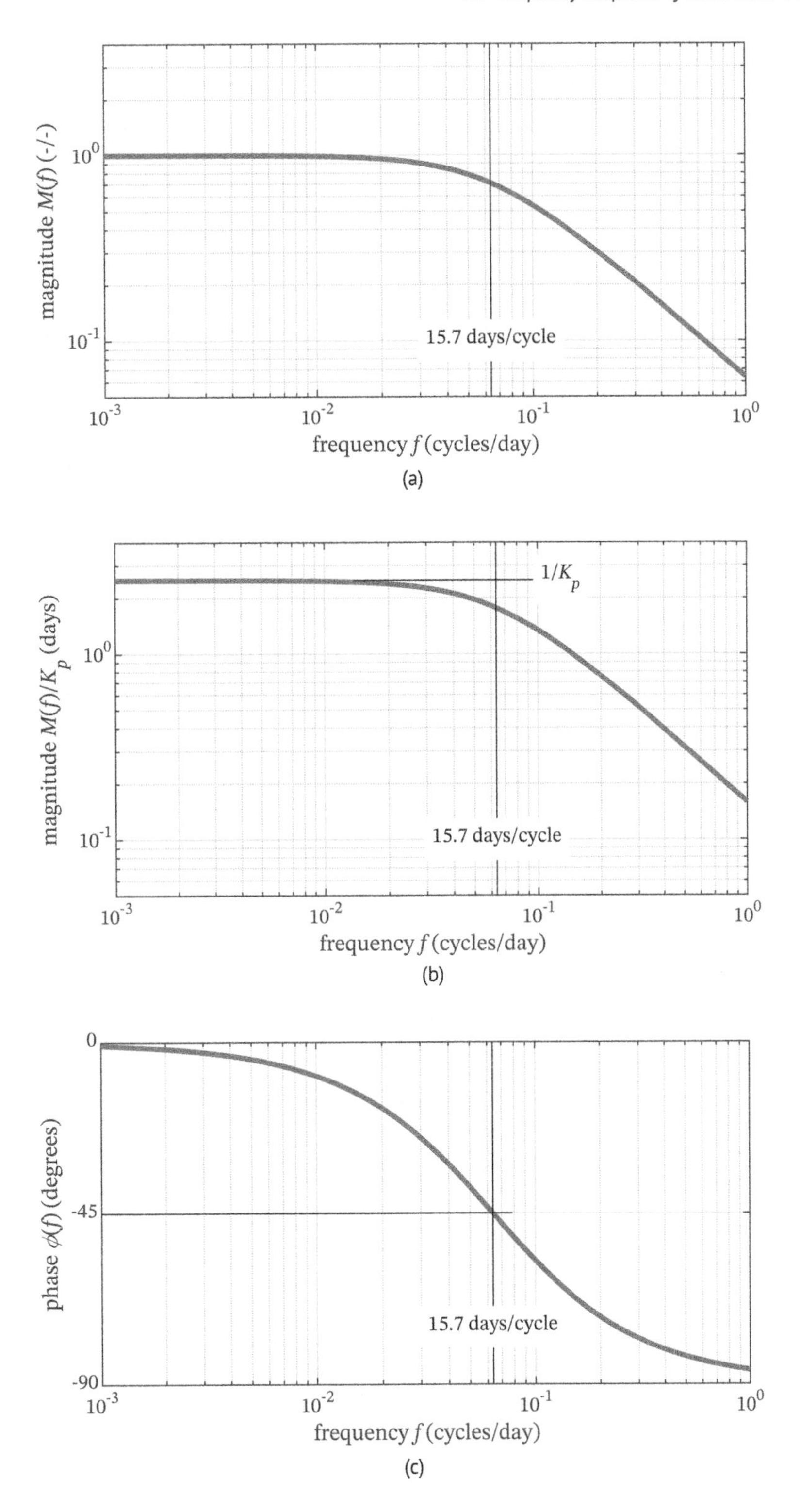

Figure 5.7 Frequency response of capacity adjustment. (a) Magnitude for work output rate. (b) Magnitude for WIP. (c) Phase for work output rate and WIP.

where ζ *is* the damping ratio, ω_n is the natural frequency and K is the constant that relates the units of $y(t)$ to the units of $x(t)$. From Equations 5.4 through 5.6, the magnitude of the frequency response is

$$M(\omega)=\frac{K}{\sqrt{\left(1-\left(\frac{\omega}{\omega_n}\right)^2\right)^2+4\zeta^2\left(\frac{\omega}{\omega_n}\right)^2}} \tag{5.26}$$

and for $K > 0$ the phase is

$$\phi(\omega)=-\tan^{-1}\left(\frac{2\zeta\left(\frac{\omega}{\omega_n}\right)}{1-\left(\frac{\omega}{\omega_n}\right)^2}\right) \tag{5.27}$$

The magnitude and phase of this 2nd-order continuous-time frequency response are plotted versus normalized frequency ω/ω_n in Figure 5.8 for several values of damping ratio ζ. The magnitude is approximately constant at relatively low frequencies, approaching $M(\omega) = K$ which corresponds to $20\log_{10}K$ dB at $\omega/\omega_n = 0$. At relatively

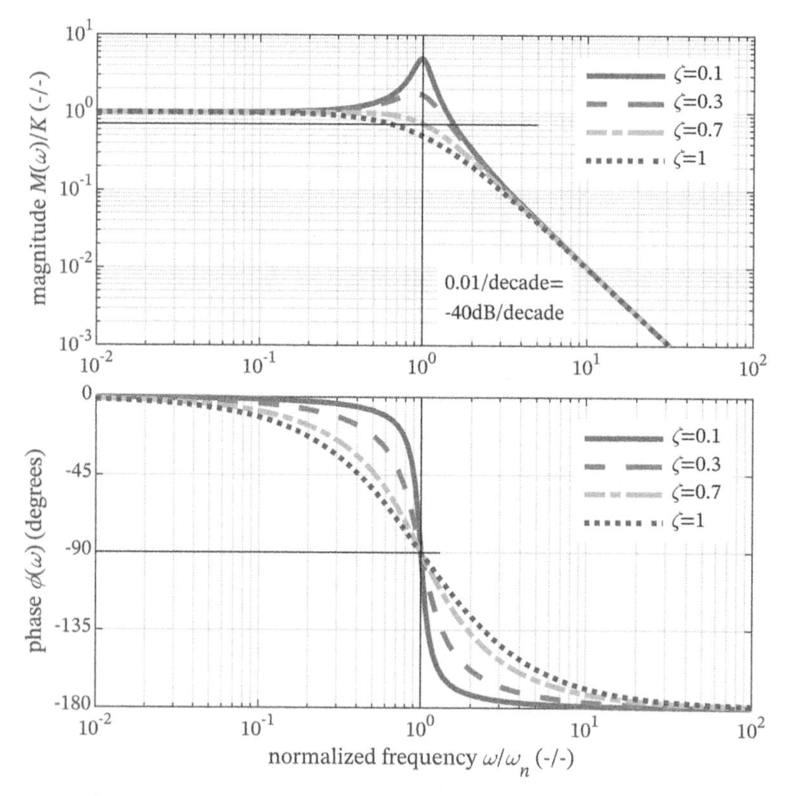

Figure 5.8 2nd-order continuous-time frequency response magnitude and phase plotted versus normalized frequency on a logarithmic frequency scale.

high frequencies, the magnitude decreases by a factor of 100 per decade of frequency, which corresponds to -40 dB per decade. The phase is approximately 0° at relatively low frequencies and approximately -180° at relatively high frequencies. As shown in Figure 5.8, the normalized frequency $\omega/\omega_n = 1$ is a convenient quantitative border between low-frequency dynamic behavior and high-frequency dynamic behavior. At $\omega/\omega_n = 1$, the phase is $\phi(\omega_n) = -90°$. The magnitude has a peak in the vicinity of this frequency for $\zeta < 0.707$.

Example 5.3 Frequency Response of Mixture Temperature Regulation

It is desired to calculate the frequency response of mixture temperature regulation using a heater as shown in Figure 2.4. It is expected that temperature regulation will be effective and outlet temperature will approximately equal the desired temperature when fluctuations in mixture inlet temperature are of relatively low frequency. Conversely, it is expected that mixture temperature regulation will be less effective when fluctuations in mixture inlet temperature are of relatively high frequency.

As described in Example 2.3, the change in temperature of the mixture $\Delta h(t)$ °C due to heating is

$$\Delta h(t) = h_o(t) - h_i(t)$$

where $h_o(t)$ °C is the outlet temperature and $h_i(t)$ °C is the inlet temperature. The change in temperature of the mixture is characterized by the relationship

$$\tau \frac{d\Delta h(t)}{dt} + \Delta h(t) = K_h v(t)$$

where time constant τ minutes characterizes how quickly temperature difference $\Delta h(t)$ changes in response to changes in heater voltage $v(t)$ and constant of proportionality K_h °C/V relates the temperature difference to the applied heater voltage.
The decision rule used for continuous-time temperature regulation is

$$\frac{dv(t)}{dt} = K_c\left(h_c(t) - h_o(t)\right)$$

where $h_c(t)$ °C is the desired outlet temperature and K_c V/second/°C is the voltage adjustment decision parameter.

The transfer function that describes the dynamic relationship between mixture outlet temperature and desired mixture outlet temperature is

$$\frac{H_o(s)}{H_c(s)} = \frac{K_h K_c}{\tau s^2 + s + K_h K_c}$$

and the transfer function that describes the dynamic relationship between mixture outlet temperature and mixture inlet temperature is

$$\frac{H_o(s)}{H_i(s)} = \frac{(\tau s + 1)s}{\tau s^2 + s + K_h K_c}$$

The magnitude and phase of the frequency response of mixture temperature regulation is calculated in Program 5.3 for $\tau = 0.83$ minutes, $K_h = 0.4$°C/V and $K_c = 0.03$ (V/second)/°C. The frequency response for desired mixture outlet temperature is shown in Figure 5.9a and the frequency response for mixture inlet temperature is shown in Figure 5.9b. For frequencies of variation in mixture inlet temperature that are significantly lower than the natural frequency $f_n = 0.00247$ cycles/second (0.0155 radians/second, 405 seconds/cycle), the magnitude of response to fluctuation in mixture inlet temperature becomes small; hence, outlet temperature tends to remain nearly equal to desired outlet temperature when the latter is constant.

On the other hand, for frequencies that are significantly higher than 0.00247 cycles/second, the magnitude of response to fluctuation in mixture inlet temperature in Figure 5.9b is approximately 1; hence, at these frequencies, mixture outlet temperature tends to fluctuate with mixture inlet temperature rather than remain at the desired mixture outlet temperature. Additionally, the magnitude of the frequency response to fluctuation in mixture inlet temperature has a peak of 1.31 at frequency 0.00320 cycles/second (313 seconds/cycle, 0.0201 radians/second). This peak of magnitude $1.31 > 1$ often is referred to as magnification. With the decision rule used in this example, the amplitude of fluctuation mixture outlet temperature therefore is 31% greater than the amplitude of fluctuation in mixture inlet temperature at frequencies in the vicinity of this peak.

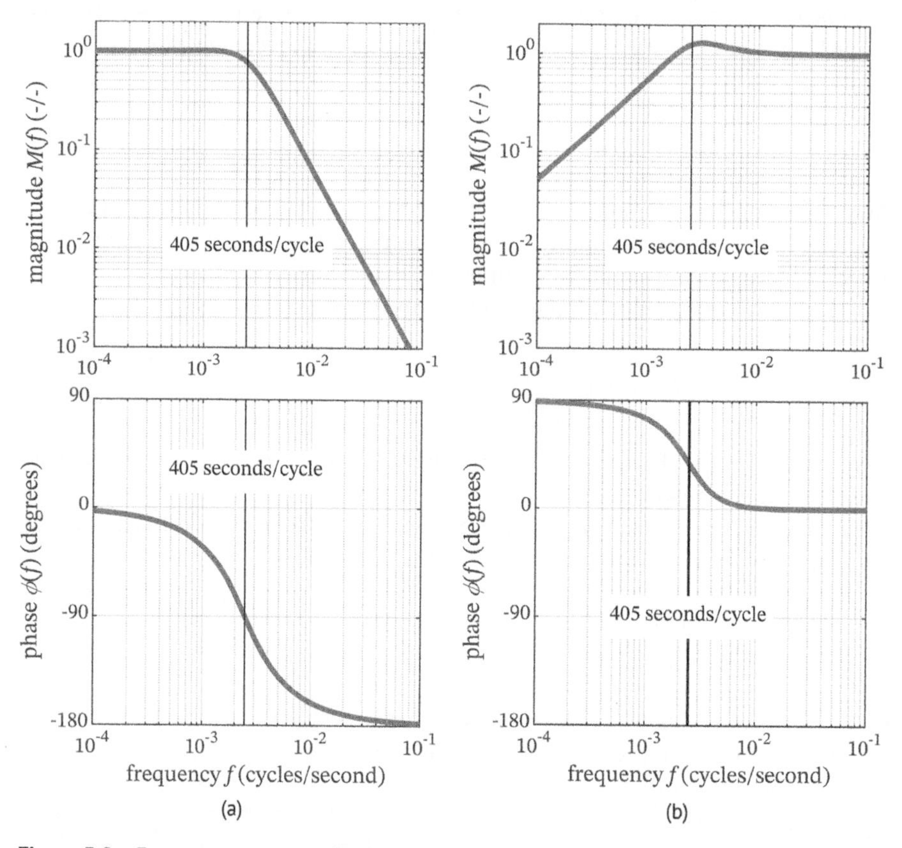

Figure 5.9 Frequency response of mixture temperature regulation. (a) Relationship between outlet temperature and desired outlet temperature. (b) Relationship between outlet temperature and inlet temperature.

Program 5.3 Frequency response of mixture temperature regulation

```
tau=49.8;  % mixture heating time constant (seconds)
Kh=0.4;  % mixture heating parameter (°C/V)

Ghs=tf(Kh,[tau 1])  % mixture heating transfer function
```

```
    Ghs =

          0.4
      ----------
      49.8 s + 1

    Continuous-time transfer function.
```

```
Kc=0.03;  % temperature regulation parameter ((V/second)/°C)
Gcs=tf(Kc,[1, 0])  % integral control transfer function
```

```
    Gcs =

      0.03
      ----
       s

    Continuous-time transfer function.
```

```
GHoHc=feedback(Gcs*Ghs,1)  % desired outlet temperature
```

```
    GHoHc =

               0.012
      --------------------
      49.8 s^2 + s + 0.012

    Continuous-time transfer function.
```

```
GHoHi=feedback(1,Gcs*Ghs)  % transfer function for inlet temperature
```

```
    GHoHi =

          49.8 s^2 + s
      --------------------
      49.8 s^2 + s + 0.012

    Continuous-time transfer function.
```

```
options=bodeoptions;
options.FreqUnits='Hz'; options.MagScale='log'; options.MagUnits='abs';

% frequency response for desired outlet temperature - Figure 5.9a
bode(GHoHc,options)

% frequency response for inlet temperature - Figure 5.9b
bode(GHoHi,options)
```

5.1.4 Frequency Response of Delay in Continuous-Time Production Systems or Components

Time delays are common in production systems and components. They often are related to physical or logistical limitations such as transportation times and delays in making and implementing decisions. Delay usually adversely affects production system performance because fluctuations in inputs and disturbances cannot be responded to immediately. Furthermore, delays within closed-loop production system topologies can significantly affect stability and limit the magnitude of manipulations that can be implemented. The frequency response of delay is readily calculated without approximation and provides important information that can help explain dynamic behavior and guide design of decision-making.

The continuous-time delay D that was included in Equation 5.3 has the transfer function

$$G(s) = e^{-Ds} \tag{5.28}$$

Substituting $j\omega$ for s

$$G(j\omega) = e^{-Dj\omega} \tag{5.29}$$

The magnitude is constant:

$$M(\omega) = 1 \tag{5.30}$$

and the phase in radians is

$$\phi(\omega) = -D\omega \tag{5.31}$$

This result was included in Equation 5.7. The frequency response of delay D is plotted on a linear normalized frequency scale in Figure 5.10. The phase becomes more negative as frequency increases, and phase decreases at a greater linear rate with increasing delay D.

Example 5.4 Frequency Response of Delay in a 2-Company Production System

As illustrated in Figure 2.12, Company A obtains unfinished orders from a supplier, Company B, and then performs the work required to finish them. Unfinished orders are shipped from distant Company B to Company A, and the time between an order leaving Company B and arriving at Company A is constant D days. In Example 2.7 the following continuous-time model was developed for this 2-company production system with lead time and transportation delays:

$$r_o(t) = r_i(t - L_B - D - L_A)$$

where the input rate to Company B is demand $r_i(t)$ orders/day, the output rate from Company A is $r_o(t)$ orders/day, Companies A and B have lead times L_A and L_B days, respectively, and the time between an order leaving Company B and arriving at Company A is D days. The corresponding transfer function is

$$\frac{R_o(s)}{R_i(s)} = e^{-(L_B + D + L_A)s}$$

From Equation 5.31, phase in degrees as a function of frequency f cycles/day is

$$\phi(f) = -360(L_B + D + L_A)f$$

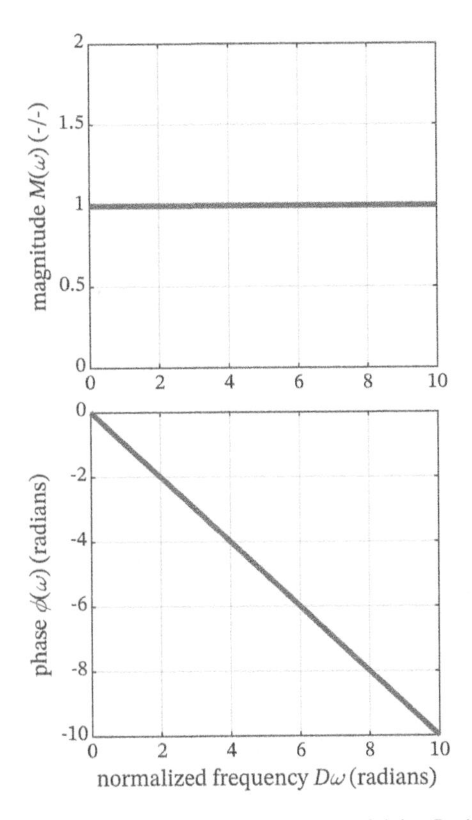

Figure 5.10 Frequency response of delay D plotted with a linear frequency scale.

Program 5.4 demonstrates how the frequency response of this model with delay can be calculated and plotted. The result is shown in Figure 5.11 for lead time delays $L_A = 7$ and $L_B = 11$ days and transportation delay $D = 4$ days.

Program 5.4 Calculation of frequency response of delay

```
LA=7;  % Company A lead time delay (days)
LB=11;  % Company B lead time delay (days)
D=4;  % transportation delay (days)

% 2-company delay transfer function
Gs=tf(1,1,'TimeUnit','days','OutputDelay',LA+D+LB)
```

```
Gs =
  exp(-22*s) * (1)
Continuous-time transfer function.
```

```
options=bodeoptions;
options.FreqUnits='cycles/day'; options.MagScale='log';
options.MagUnits='abs';
W=logspace(-3,-1,100)*2*pi;  % frequency range in radians/day

bode(Gs,W,options)  % frequency response - Figure 5.11
```

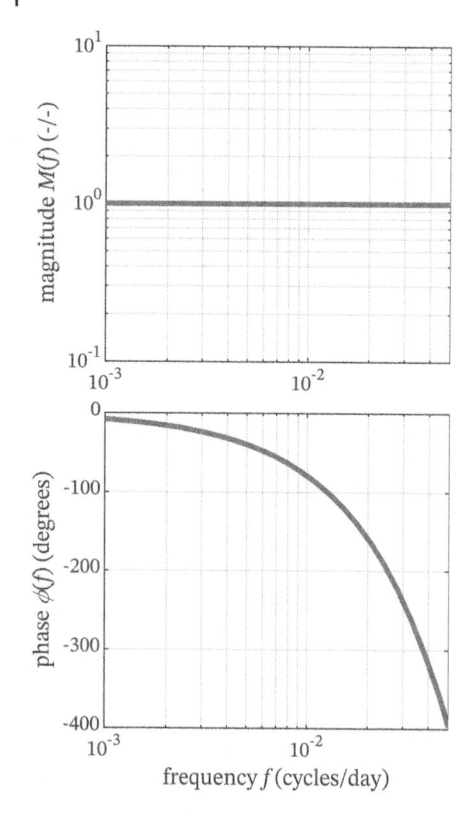

Figure 5.11 Frequency response of a 2-company production system when lead time and transportation delays total 22 days, plotted with a logarithmic frequency scale.

5.2 Frequency Response of Discrete-Time Systems

When the input to a stable, linear discrete-time system has an oscillating portion with amplitude A:

$$x(kT) = A\sin(\omega kT) \tag{5.32}$$

the corresponding steady-state portion of the output is

$$y(kT) = M(\omega)A\ \sin(\omega kT + \phi(\omega)) \tag{5.33}$$

where $M(\omega)$ is the magnitude of the frequency response and $\phi(\omega)$ is the phase in radians of the frequency response at radian frequency $\omega > 0$ (radians/second, radians/day, etc.).

Similar to the case for continuous-time production systems or components, it is not necessary to solve difference equations for sampled sinusoidal inputs to determine the magnitude and phase of frequency response; instead, the frequency response can be calculated directly from the transfer function that represents the discrete-time production system or component.

In a manner similar to that described for continuous-time systems and components in Section 5.1, the discrete-time transfer function that represents a production system or component

$$G(z)=\frac{Y(z)}{X(z)}=\frac{b_m z^m+\ldots+b_2 z^2+b_1 z+b_0}{a_n z^n+\ldots+a_2 z^2+a_1 z+a_0}z^{-d} \tag{5.34}$$

can be factored into the form

$$G(z)=\frac{K(z-c_1)(z-c_2)\ldots(z-c_m)}{(z-r_1)(z-r_2)\ldots(z-r_n)}z^{-d} \tag{5.35}$$

It then can be shown that the discrete-time frequency response can be calculated by substituting jω for s in the definition $z = e^{Ts}$:

$$G\left(e^{Tj\omega}\right)=\frac{K\left(e^{Tj\omega}-c_1\right)\left(e^{Tj\omega}-c_2\right)\ldots\left(e^{Tj\omega}-c_m\right)}{\left(e^{Tj\omega}-r_1\right)\left(e^{Tj\omega}-r_2\right)\ldots\left(e^{Tj\omega}-r_n\right)}e^{-Tj\omega d} \tag{5.36}$$

The magnitude of the frequency response of a stable discrete-time system (all $|r_i| < 1$ for all r_i) is

$$M(\omega)=\frac{|K|\left|e^{Tj\omega}-c_1\right|\left|e^{Tj\omega}-c_2\right|\ldots\left|e^{Tj\omega}-c_m\right|}{\left|e^{Tj\omega}-r_1\right|\left|e^{Tj\omega}-r_2\right|\ldots\left|e^{Tj\omega}-r_n\right|} \tag{5.37}$$

and the phase of the frequency response is

$$\begin{aligned}\phi(\omega)=&\angle(K)+\angle\left(e^{Tj\omega}-c_1\right)+\angle\left(e^{Tj\omega}-c_2\right)+\ldots+\angle\left(e^{Tj\omega}-c_m\right)-\angle\left(e^{Tj\omega}-r_1\right)\\&-\angle\left(e^{Tj\omega}-r_2\right)-\ldots-\angle\left(e^{Tj\omega}-r_n\right)-Td\omega\end{aligned} \tag{5.38}$$

Applying Euler's formulas,

$$M(\omega)=\frac{|K|\left|\cos(T\omega)+j\sin(T\omega)-c_1\right|\ldots\left|\cos(T\omega)+j\sin(T\omega)-c_m\right|}{\left|\cos(T\omega)+j\sin(T\omega)-r_1\right|\ldots\left|\cos(T\omega)+j\sin(T\omega)-r_n\right|} \tag{5.39}$$

and

$$\begin{aligned}\phi(\omega)=&\angle(K)+\angle\left(\cos(T\omega)+j\sin(T\omega)-c_1\right)+\ldots+\angle\left(\cos(T\omega)+j\sin(T\omega)-c_m\right)\\&-\angle\left(\cos(T\omega)+j\sin(T\omega)-r_1\right)-\ldots-\angle\left(\cos(\omega T\omega)+j\sin(T\omega)-r_n\right)\\&-Td\omega\end{aligned} \tag{5.40}$$

Because of the periodic nature of sin($T\omega$) and cos($T\omega$), the magnitude and phase functions in Equations 5.39 and 5.40 also are periodic.

5.2.1 Frequency Response of Discrete-Time Integrating Production Systems or Components

Similar to the case for continuous-time production systems and components described in Section 5.1.1, it is customary to use Equations 5.39 and 5.40 to calculate the magnitude and phase of the frequency response of discrete-time systems that have one or more factors (z - 1) in the denominator of transfer function $G(z)$.

The transfer function for a production system or component modeled as discrete-time integration is

$$G(z)=\frac{Y(z)}{X(z)}=\frac{KT}{z-1} \tag{5.41}$$

Substituting $e^{Tj\omega}$ for z,

$$G\left(e^{Tj\omega}\right)=\frac{KT}{e^{Tj\omega}-1}=\frac{KT}{\cos(T\omega)+j\sin(T\omega)-1} \tag{5.42}$$

The magnitude and phase of the frequency response are calculated using Equations 5.39 and 5.40:

$$M(\omega)=\frac{|KT|}{|\cos(T\omega)+j\sin(T\omega)-1|}=\frac{|KT|}{\sqrt{(\cos(T\omega)-1)^2+\sin^2(T\omega)}} \tag{5.43}$$

$$\phi(\omega)=\angle KT-\angle\big((\cos(T\omega)-1)+j\sin(T\omega)\big)=\angle KT-\tan^{-1}\left(\frac{\sin(T\omega)}{\cos(T\omega)-1}\right) \tag{5.44}$$

The results for $K>0$ are plotted on a normalized linear frequency scale in Figure 5.12. Note that the magnitude at frequencies in the range $0.5/T \le f < 1/T$ mirrors the

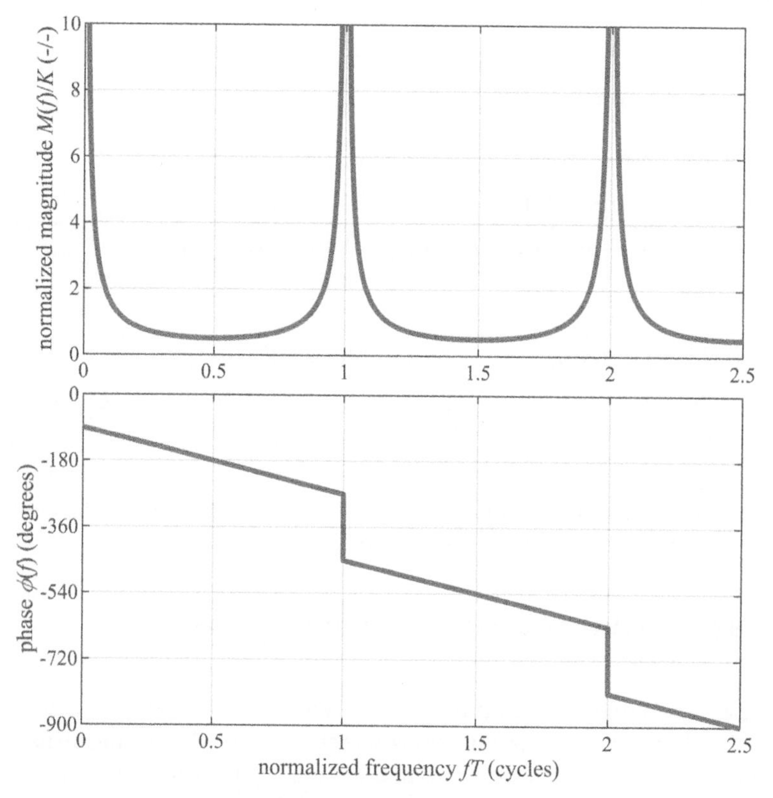

Figure 5.12 Frequence response of integrating discrete-time production system or component plotted on a linear frequency scale.

magnitude of frequencies in the range $0 \le f \le 0.5/T$. Also, the magnitude is replicated at integer multiples of frequency $f = 1/T$. This periodicity in the frequency response is a manifestation of aliasing. As will be discussed in Section 5.2.3, a sinusoidal sequence of higher frequency $f > 0.5/T$ can be identical to a sinusoidal sequence of lower frequency $f < 0.5/T$. If this sequence is input to a production system or component, then the corresponding sinusoidal sequence at the output of the production system or component also is identical for the two frequencies and the frequency response is the same for the two frequencies.

Example 5.5 Frequency Response of Warehouse Inventory

It is desired to calculate and plot the discrete-time frequency response of the warehouse modeled as shown in the block diagram in Figure 5.13. The relationship between inventory $x(kT)$ items, rate of items received $r_i(kT)$ and rate of items shipped $r_o(kT)$ is

$$X(z) = \frac{Tz^{-1}}{1 - z^{-1}}\left(R_i(z) - R_o(z)\right)$$

Equations 5.43 and 5.44 yield the magnitude and phase of the frequency response of warehouse inventory $x(kT)$ to sinusoidal oscillations in input $r_i(kT)$:

$$M(\omega) = \frac{T}{\sqrt{\left(\cos(T\omega) - 1\right)^2 + \sin^2(T\omega)}}$$

and

$$\phi(\omega) = -\tan^{-1}\left(\frac{\sin(T\omega)}{\cos(T\omega) - 1}\right)$$

Calculation and plotting of the frequency response of warehouse inventory is demonstrated in Program 5.5 for $T = 1$ day. The results are shown in the Bode plot in Figure 5.14. As noted above, the results for frequencies $f > 0.5/T$ cannot be distinguished in practice from results for lower frequencies $f < 0.5/T$. Hence, discrete-time frequency response commonly is plotted only for frequencies in the range $0 \le f \le 0.5/T$.

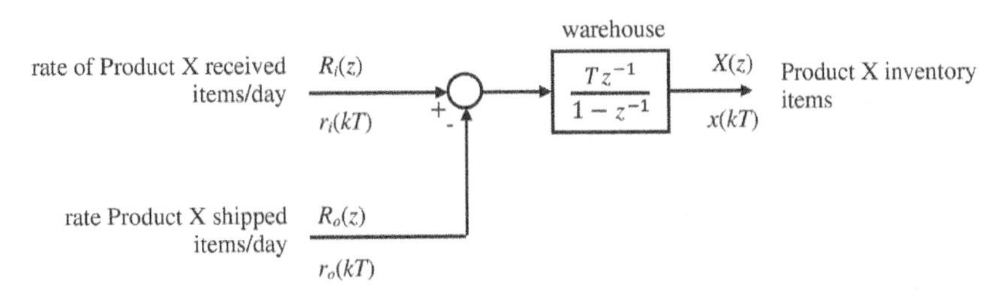

Figure 5.13 Block diagram for discrete-time model of warehouse.

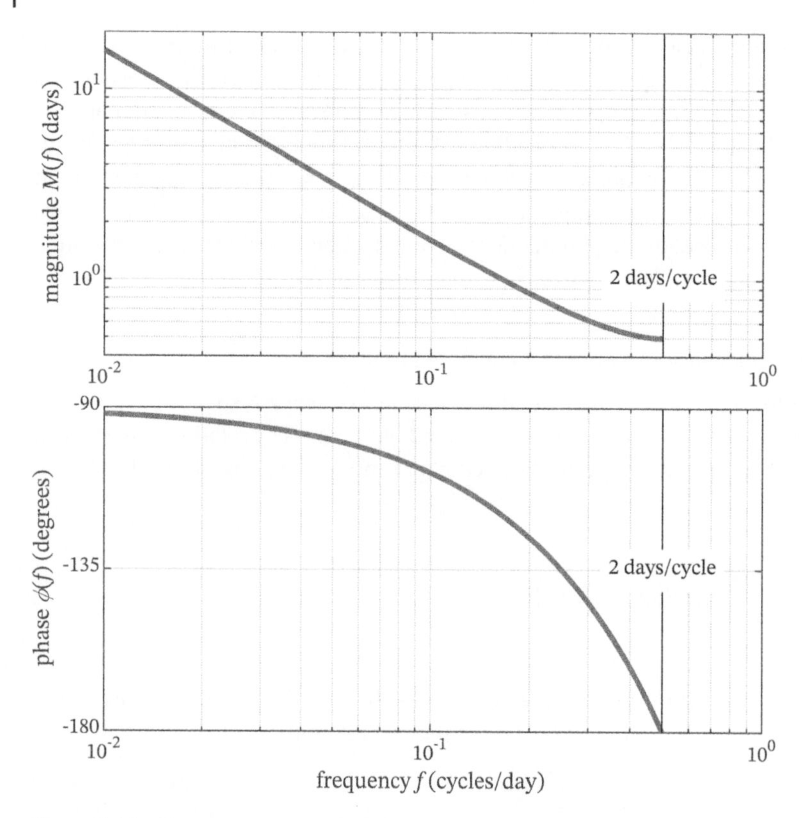

Figure 5.14 Frequency response of discrete-time warehouse inventory for T = 1 day plotted with a logarithmic frequency and magnitude scales.

Program 5.5 Frequency response of discrete-time warehouse inventory

```
T=1;  % period (days)
Gz=tf(1,[1, -1],T,'TimeUnit','days')  % warehouse inventory
```

```
Gz =

    1
  -----
  z - 1

Sample time: 1 days
Discrete-time transfer function.
```

```
options=bodeoptions;
options.FreqUnits='cycles/day'; options.MagScale='log';
options.MagUnits='abs';

bodeplot(Gz,options);  % frequency response of inventory - Figure 5.14
```

5.2.2 Frequency Response of Discrete-Time 1st-Order Production Systems or Components

Consider a discrete-time system represented by the transfer function

$$G(z)=\frac{Y(z)}{X(z)}=\frac{K\left(1-e^{-T/\tau}\right)}{z-e^{-T/\tau}} \tag{5.45}$$

where τ is the time constant and K is the constant relating the units of input $x(kT)$ to the units of output $y(kT)$. Substituting $e^{Tj\omega}$ for z,

$$G\left(e^{j\omega T}\right)=\frac{K\left(1-e^{-T/\tau}\right)}{e^{Tj\omega}-e^{-T/\tau}}=\frac{K\left(1-e^{-T/\tau}\right)}{\cos(T\omega)+j\sin(T\omega)-e^{-T/\tau}} \tag{5.46}$$

The magnitude and phase of the frequency response are obtained using Equations 5.39 and 5.40:

$$M(\omega)=\frac{\left|K\left(1-e^{-T/\tau}\right)\right|}{\left|\cos(T\omega)+j\sin(T\omega)-e^{-T/\tau}\right|}=\frac{\left|K\left(1-e^{-T/\tau}\right)\right|}{\sqrt{\left(\cos(T\omega)-e^{-T/\tau}\right)^{2}+\sin^{2}(T\omega)}} \tag{5.47}$$

$$\begin{aligned}\phi(\omega)&=\angle K\left(1-e^{-T/\tau}\right)-\angle\left(\cos(\omega T)+j\sin(T\omega)-e^{-T/\tau}\right)\\&=\angle K\left(1-e^{-T/\tau}\right)-\tan^{-1}\left(\frac{\sin(T\omega)}{\cos(T\omega)-e^{-T/\tau}}\right)\end{aligned} \tag{5.48}$$

The magnitude and phase of the frequency response are plotted on a normalized logarithmic frequency scale in Figure 5.15 for $K > 0$ and several ratios of T/τ. As period T becomes shorter with respect to time constant τ, both the magnitude and the phase of the frequency response become more similar to the magnitude and phase of a continuous-time system or component with the same time constant. On the other hand, as period T becomes longer with respect to time constant τ, the phase decreases rapidly with frequency. As will be discussed in Chapter 6, stability and achievable performance of production systems and components with closed-loop topologies tends to be degraded when phase is more negative; hence, choice of period T is an important aspect of design of decision-making.

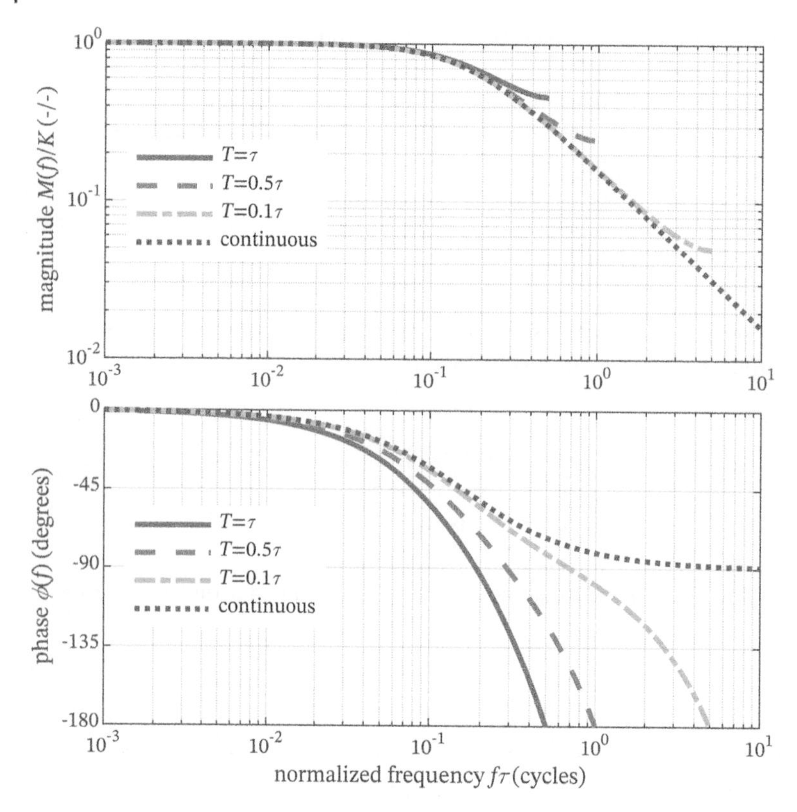

Figure 5.15 Frequency response of 1st-order discrete-time production system or component for various ratios T/τ.

Example 5.6 Discrete-Time Frequency Response of Capacity Provided by Permanent Workers

The block diagram in Figure 3.15 shows the topology of a model of a production system in which the capacity provided by permanent workers should not be adjusted at high frequencies; furthermore, there is a delay of dT days in implementing permanent worker adjustment decisions where T is the period between decisions and d is a positive integer. The remaining capacity needed to meet demand is provided by cross-trained workers; this portion can be adjusted immediately. Frequency response can be calculated for both adjustments in permanent workers and adjustments in cross-trained workers, and the results can be used to assess choice of parameter α for the filter that reduces higher frequency adjustments in capacity provided by permanent workers.

The transfer function that relates permanent worker capacity adjustment decisions $r_p(kT)$ orders/day to order input rate $r_i(kT)$ orders/day is

$$\frac{R_p(z)}{R_i(z)} = \frac{\alpha}{1-(1-\alpha)z^{-1}} z^{-d}$$

where $0 < \alpha \leq 1$ is the adjustable parameter of the exponential filter. Furthermore, the transfer function that relates cross-trained worker capacity adjustment decisions $r_c(kT)$ orders/day to order input rate $r_i(kT)$ is

$$\frac{R_c(z)}{R_i(z)} = 1 - \frac{\alpha}{1-(1-\alpha)z^{-1}} z^{-d} = \frac{1-(1-\alpha)z^{-1} - \alpha z^{-d}}{1-(1-\alpha)z^{-1}}$$

Program 5.6 can be used to calculate the frequency response of adjustments in permanent and cross-trained workers. The results are shown in Figure 5.16 for $\alpha = 0.2$. Order input rate is measured weekly, $T = 5$ days, $dT = 10$ days, and permanent and cross-trained worker capacity adjustment decisions are made weekly. The magnitude of adjustments in permanent worker capacity is $M(f) \approx 1$ for relatively low frequencies of oscillation f (relatively long periods of oscillation $\delta = 1/f$), while the magnitude of adjustments in cross-trained worker capacity is relatively small at those frequencies. On the other hand, the magnitude of adjustments in permanent worker capacity is

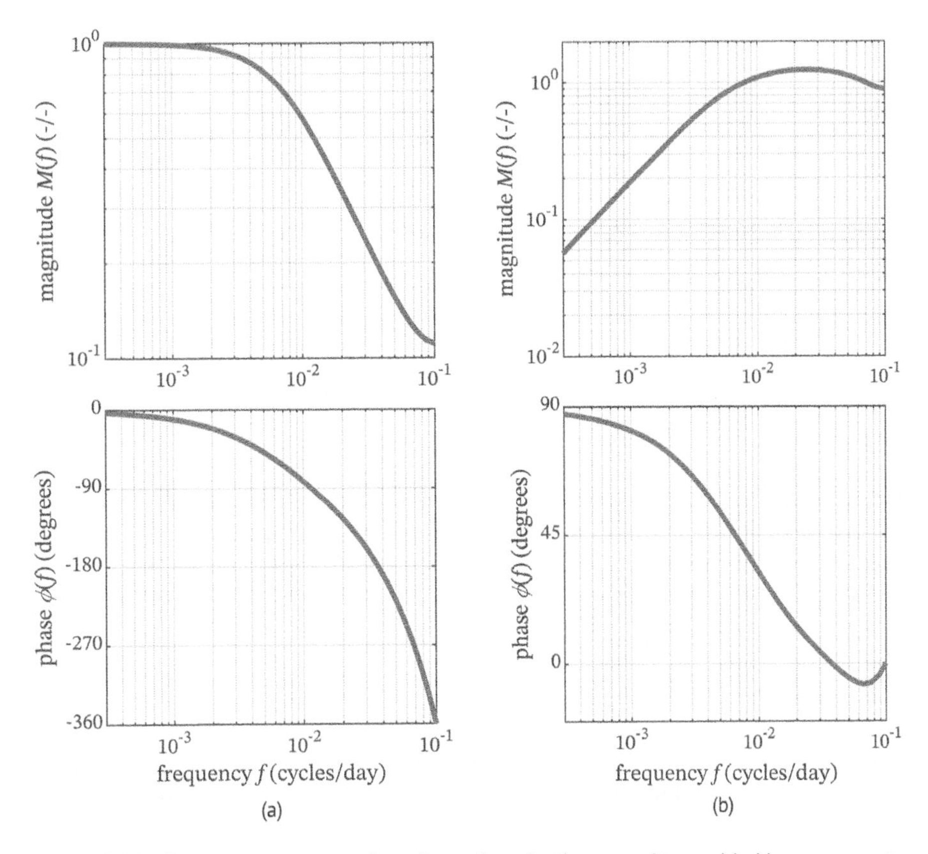

Figure 5.16 Frequency response of portions of production capacity provided by permanent and cross-trained workers. (a) Permanent workers. (b) Cross-trained workers.

Program 5.6 Calculation of frequency response of permanent and cross-trained worker capacity adjustments

```
T=5;   % period (days)
d=2;   % delay dT=10 (days)
alpha=0.2;   % filter parameter

% transfer funtion for permanent worker capacity
Gpz=tf([alpha, 0],[1, -(1-alpha)],T,'OutputDelay',d,'TimeUnit','days')
```

```
    Gpz =

                 0.2 z
      z^(-2) * -------
                z - 0.8

    Sample time: 5 days
    Discrete-time transfer function.
```

```
Gcz=minreal(1-Gpz)   % transfer funtion for cross-trained worker capacity
```

```
    Gcz =

      z^2 - 0.8 z - 0.2
      -----------------
         z^2 - 0.8 z

    Sample time: 5 days
    Discrete-time transfer function.
```

```
options=bodeoptions;
options.FreqUnits='cycles/day'; options.MagScale='log';
options.MagUnits='abs';

% frequency response of permanent workers - Figure 5.16a
bode(Gpz,options);

% frequency response of cross-trained workers - Figure 5.16b
bode(Gcz,options);
```

reduced by nearly 90% at frequency $f = 0.1$ cycles/day ($\delta = 10$ days/cycle), while the magnitude of adjustments in cross-trained workers at that frequency is $M(f) \approx 1$. The magnitude of adjustments in cross-trained workers is $M(f) > 1$ for some frequencies f because of the phase characteristics of the frequency response of adjustments in permanent worker capacity.

5.2.3 Aliasing Errors

Aliasing errors occur when sinusoidal functions of frequency $f \geq 0.5/T$ (period of oscillation $\delta \leq 2T$, frequency $\omega \geq \pi/T$) are represented using a sequence with period T. Sinusoidal sequences of frequencies higher than $f = 0.5/T$ cannot be distinguished in practice from sinusoidal sequences of frequencies lower than $f = 0.5/T$. This is illustrated in Figure 5.17, which shows a sinusoidal sequence that is the same for frequencies of oscillation $f_1 = 0.08$ cycles/day (12.5 days/cycle), $2f_1 = 0.16$ cycles/day (6.25 days/cycle) and $f_2 = 0.12$ cycles/day (8.33 days/cycle), the latter with a delay of $0.5/f_2$ days ($-\pi$ radians phase)

Another straightforward example of aliasing error is illustrated in Figure 5.18. Work in progress (WIP) is measured daily in a production work system but WIP varies during the day for reasons such as work shifts, deliveries of work, and inflow of work from upstream work systems. The WIP sequence obtained from measurements at 6:00 AM is shown, which is different from the WIP sequence obtained from measurements at 12:00 noon, which also is shown; they are different because of aliasing errors. WIP may vary more from day to day than is shown in Figure 5.18, and it may be necessary to measure WIP significantly more often and then to use a filter[3] to obtain a reliable daily measurement.

Aliasing errors are sensitive to small changes in frequency and timing of measurements. They can be difficult to predict and detect, and are difficult to correct using

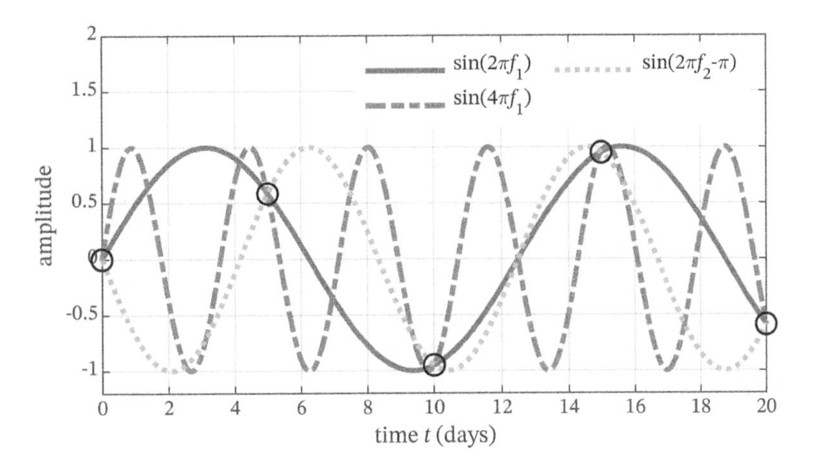

Figure 5.17 Aliasing errors result in identical sequences for three different frequencies.

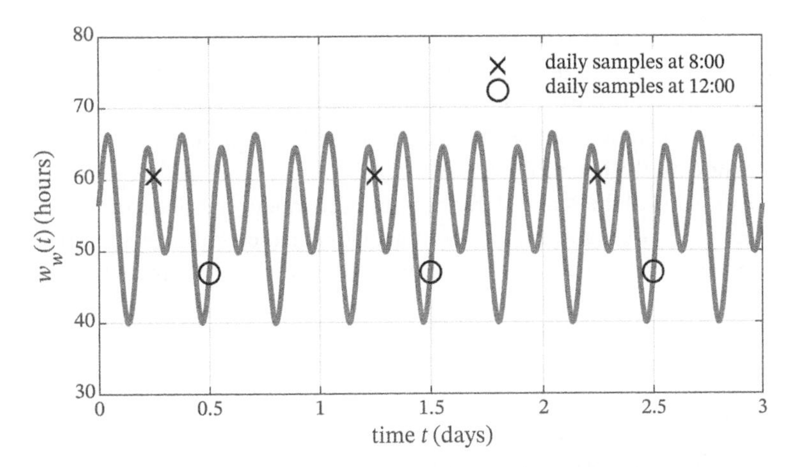

Figure 5.18 Aliasing errors in daily measurement of work in progress (WIP).

3 There are many types of filters, including finite impulse response filters such as averaging and infinite impulse response filters such as exponential moving average filters. The reader is referred to the Bibliography and many other signal processing publications for more information on this topic.

software; hence, they need to be avoided or reduced to an insignificant level. The following recommendations can be of assistance:

1) Period T should be chosen to be short enough to ensure that content at frequencies above $f = 0.5/T$ is of sufficiently small amplitude so as to not cause significant aliasing errors.[4]
2) Measurements acquired with sufficiently short period T, without significant aliasing errors, can be filtered with a discrete-time transfer function to reduce higher frequency content. The filtered measurements then can be decimated (downsampled) if desired, without introducing significant aliasing errors, to obtain a new sequence with longer period T.
3) When continuous-time signals are sampled, a continuous-time anti-aliasing filter should be used before sampling to reduce frequency content at and above $f = 0.5/T$ to a sufficiently low amplitude to avoid significant aliasing errors.

5.3 Frequency Response Characteristics

The frequency response of a production system or component can be analyzed, and key characteristics can be calculated that are useful in understanding and guiding design of dynamic behavior. Important frequency response characteristics include the zero-frequency magnitude, bandwidth, unity magnitude crossing frequency, and -180° phase crossing frequency. The latter are used to determine magnitude and phase margins, which are measures of relative stability that are obtained from the open-loop frequency response but guide the design of closed-loop frequency response.[5] These characteristics will be defined in the following sections and examples will be presented to illustrate them.

5.3.1 Zero-Frequency Magnitude (DC Gain) and Bandwidth

The relationship between input and output amplitude at relatively low frequencies often is of particular interest. The zero-frequency magnitude, often called the DC gain, is the magnitude of the frequency response at frequency $\omega = 0$.[6] It is a straightforward way to characterize behavior at relatively low frequencies. For integrating production systems and components, the magnitude $M(0) = \infty$.

Bandwidth is a measure of the range of frequencies over which a production system or component responds with significant magnitude. Often, bandwidth is determined as the lowest frequency where the magnitude of the frequency response is decreasing

4 Fourier analysis can be helpful in identifying frequency content.

5 In Chapter 6 it will be shown how open-loop frequency response can be used to design decision making for production systems with closed-loop topologies.

6 The zero-frequency magnitude is the ratio between the steady-state output and amplitude of the input; hence, it is the final value of the response to a unit step input if this final value exists.

with frequency and is 70.7% of the zero-frequency magnitude (3 dB below the zero-frequency magnitude).[7] Bandwidth is not defined when zero-frequency magnitude $M(0) = \infty$.

Example 5.7 Zero-Frequency Magnitude and Bandwidth

The following mixture heating transfer function was found in Example 3.7 for the portion of a production process shown in Figure 3.5 in which the temperature of a mixture flowing out of a pipe is regulated using a heater and a temperature sensor:

$$G(s) = \frac{\Delta H(s)}{V(s)} = \frac{K_h}{\tau s + 1} e^{-Ds}$$

where mixture heating time constant $\tau = 49.8$ seconds, mixture heating parameter $K_h = 0.4$°C/V, and delay $D = 12$ seconds. The following functions can be used to determine the zero-frequency magnitude and bandwidth of the frequency response represented by this transfer function:

`dcgain(G)`	calculate the zero-frequency magnitude (magnitude of the frequency response at frequency $\omega = 0$) of transfer function G
`bandwidth(G)`	calculate the lowest radian frequency at which the magnitude of the frequency response of transfer function G is 70.7% of (3 dB below) the magnitude at $\omega = 0$

The zero-frequency magnitude and bandwidth can be found for mixture heating transfer function $G(s)$ as demonstrated in Program 5.7, and the results are shown in the plot of frequency response magnitude in Figure 5.19. The zero-frequency magnitude (DC

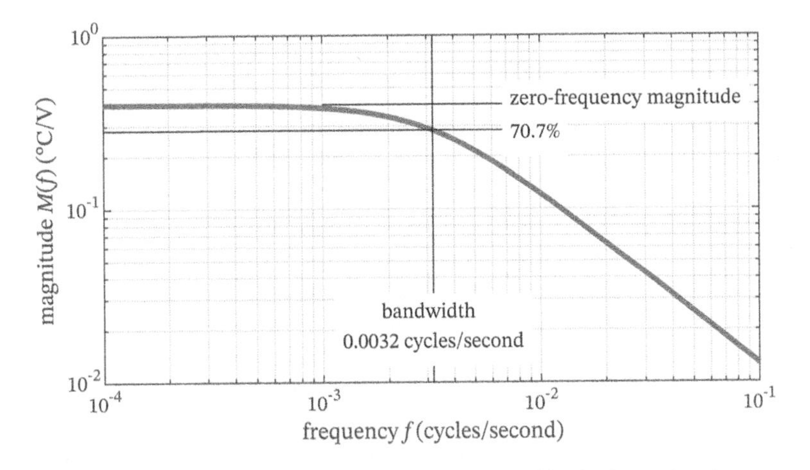

Figure 5.19 Zero-frequency magnitude and bandwidth of mixture heating.

7 This measure often is referred to as baseband bandwidth.

Program 5.7 Calculation of zero-frequency magnitude and bandwidth

```
tau=49.8;  % mixture heating time constant (seconds)
Kh=0.4;   % mixture heating parameter (°C/V)
D=12;   % transportation delay (seconds)

Gs=tf(Kh,[tau, 1],'OutputDelay',D)  % mixture heating transfer function
```

```
   Gs =

                      0.4
     exp(-12*s) * ----------
                   49.8 s + 1

   Continuous-time transfer function.
```

```
dcgain(Gs)  % zero-frequency magnitude (°C/V)
```

```
   ans = 0.4000
```

```
bandwidth(Gs)/(2*pi)  % bandwidth (cycles/second)
```

```
   ans = 0.0032
```

```
options=bodeoptions; options.PhaseVisible='off';
options.FreqUnits='Hz'; options.MagScale='log'; options.MagUnits='abs';

bode(Gs,options);  % frequency response magnitude - Figure 5.19
```

gain) is 0.4°C/V. A constant input of 100 V therefore can be expected to change mixture temperature by 40°C. The bandwidth is 0.0032 cycles/second (period 313 seconds/cycle). Significant response of mixture heating therefore can be expected when heater voltage fluctuates with frequencies significantly below 0.0032 cycles/second. The delay of 12 seconds does not affect the magnitude of the frequency response.

5.3.2 Magnitude (Gain) Margin and Phase Margin

Many production systems and components have a closed-loop topology and design of decision-making components in these closed-loop systems and components can be guided by the characteristics of open-loop frequency response, which is the combined frequency response of production and decision-making components in the loop excluding the effects of feedback. For the general closed-loop topology shown in Figure 5.20a, the corresponding open-loop topology is shown in Figure 5.20b. The latter has frequency response $G(j\omega)H(j\omega)$ for continuous-time production systems and components and frequency response $G(e^{Tj\omega})H(e^{Tj\omega})$ for discrete-time production systems and components. It can be shown that the trajectory, as a function of frequency ω, that $G(j\omega)H(j\omega)$ or $G(e^{Tj\omega})H(e^{Tj\omega})$ follows with respect to reference point $-1 + j0$ provides a test for whether the closed-loop system or component will be stable[8]

8 The reader is referred to the Bibliography and many other control-theory publications for information regarding the Nyquist stability criterion and polar-coordinate approaches to analyzing open-loop frequency response. These publications also address limitations in assessment of stability and relative stability using magnitude (gain) and phase margins; fortunately, these limitations usually do not diminish the utility of these margins in guiding design of decision making as described in Chapter 6.

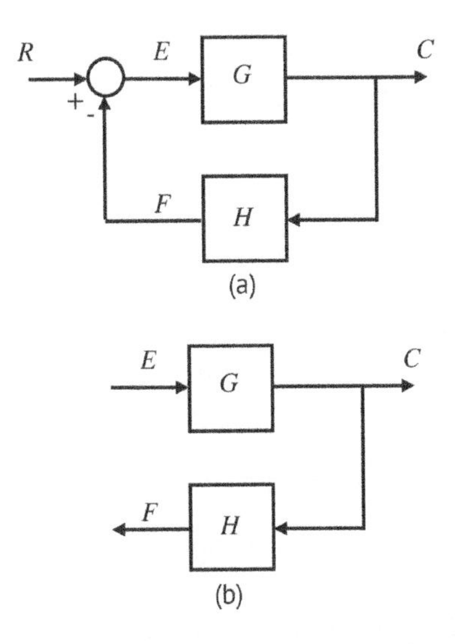

Figure 5.20 Continuous-time or discrete-time production system or component with feedback. (a) Closed-loop transfer function $G(s)/(1+G(s)H(s))$ or $G(z)/(1+G(z)H(z))$. (b) Open-loop transfer function $G(s)H(s)$ or $G(z)H(z)$.

and also measures of relative stability. The magnitude margin, often referred to as the gain margin, and phase margin are two such measures of relative stability. The use of these measures in guiding design of decision-making for production systems and components will be described in Chapter 6. Here, the magnitude margin and phase margin are defined, and examples are given that illustrate how they can be calculated.

The magnitude $M_{ol}(\omega)$ of the open-loop frequency response often is greater than 1 (dB > 0) at lower frequencies and less than 1 (dB < 0) at higher frequencies. Between these extremes there is a frequency ω_{cp} where the magnitude $M_{ol}(\omega_{cp}) = 1$ (dB = 0).[9] This frequency is the unity-magnitude (0 dB) crossing frequency and is where the phase margin is measured. Also, the phase $\phi_{ol}(\omega)$ of the open-loop frequency response often is greater at lower frequencies than it is at higher frequencies. If there is a frequency ω_{cg} at which the phase of the open-loop frequency response $\phi_{ol}(\omega_{cg}) = -180$, then this frequency is the -180° crossing frequency and is where the magnitude margin is measured.

The magnitude margin is defined as the reciprocal of $M_{ol}(\omega_{cg})$, if ω_{cg} exists:

$$G_m = \frac{1}{M_{ol}\left(\omega_{cg}\right)} \tag{5.49}$$

This is the factor by which the magnitude of the open-loop frequency response can be increased at the -180° crossing frequency to make the magnitude $M_{ol}(\omega_{cg}) = 1$ (0 dB). Increasing the magnitude by this factor or more results in a closed-loop system or component that is not stable. (If $G_m < 1$, the magnitude of the open-loop frequency response must be decreased to obtain a stable closed-loop system.) The magnitude margin therefore is a measure of relative stability.

9 If there is more than one unity-magnitude (0 dB) gain crossing frequency or more than one -180° crossing frequency, margins usually are determined using the lowest crossing frequency.

The phase margin is calculated by adding 180° to the phase angle at unity magnitude (0 dB) crossing frequency ω_{cp}, if ω_{cp} exists:

$$P_m = \deg(\phi_{ol}(\omega_{cp})) + 180^\circ \tag{5.50}$$

This is the amount by which the phase can be decreased at the unity-magnitude (0 dB) crossing frequency to make the phase $\phi_{ol}(\omega_{cp}) = -180^\circ$. Decreasing the phase by this amount or more than this amount results in a closed-loop system that is not stable. (If $P_m < 0$, the phase of the open-loop frequency response must be increased to obtain in a stable closed-loop system.) The phase margin therefore also is a measure of the relative stability.

Example 5.8 Open-Loop Magnitude Margin and Phase Margin of Capacity Adjustment

The block diagram in Figure 5.21a shows the closed-loop topology of the production system with capacity adjustment that is described in Example 4.8, except there is a delay dT in implementing capacity adjustments where d is a positive integer and T is the period between capacity adjustments. Figure 5.21b shows the corresponding open-loop topology, which excludes feedback. The combination of the transfer functions in the open loop is

$$\frac{W_o(z)}{W_w(z)} = K_p z^{-d} Z\left\{\left(\frac{1-e^{-Ts}}{s}\right)\frac{1}{s}\right\} = K_p \frac{Tz^{-1}}{1-z^{-1}} z^{-d}$$

The magnitude and phase margins can be obtained from this open-loop transfer function using

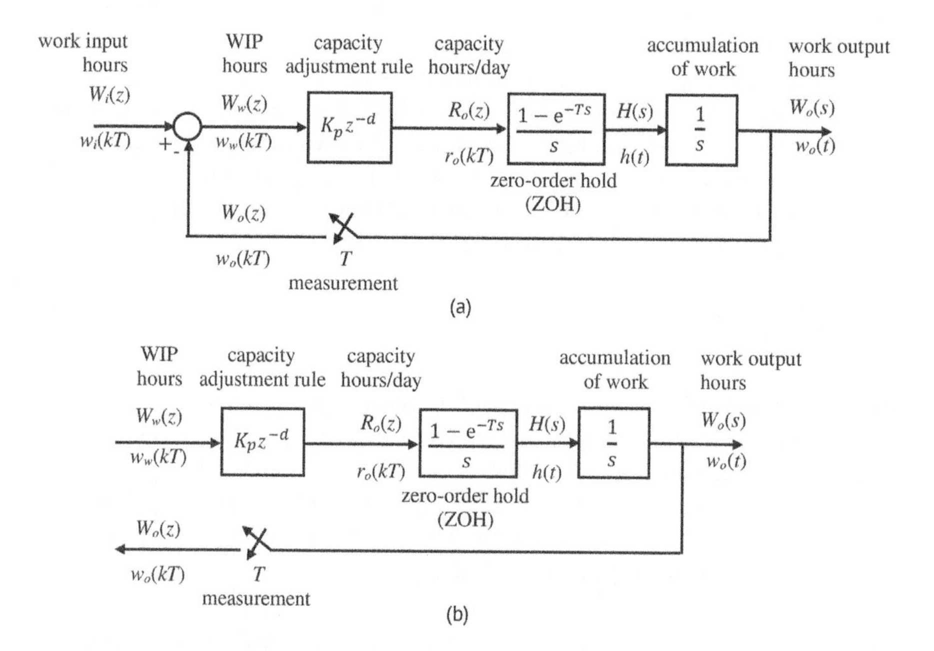

Figure 5.21 Block diagrams for discrete-time capacity adjustment with delay dT in implementation. (a) Closed-loop. (b) Open-loop.

`margin(G)`	plot the frequency response of transfer function `G` and calculate and display the magnitude (gain) margin and phase margin, as well as the -180° crossing frequency and the unity-magnitude (0 dB) crossing frequency
`[Gm,Pm,Wcg,Wcp]=margin(G)`	calculate the magnitude (gain) margin `Gm`, phase margin `Pm`, -180° crossing frequency `Wcg`, and unity magnitude (0 dB) crossing frequency `Wcp`

This is demonstrated for two cases of delay dT in Program 5.8. Results are shown in Figure 5.22a for capacity adjustment period $T = 1$ day, delay $dT = 2$ days and proportional decision-making parameter $K_p = 0.2$ days^{-1}. The magnitude margin is $G_m = 3.09$ and the phase margin is $P_m = 61.2°$. In this case it is expected that closed-loop capacity adjustment will be stable because $G_m > 1$ and $P_m > 0°$.

Results are shown in Figure 5.22b for capacity adjustment period $T = 5$ days, delay $dT = 10$ days and proportional decision-making parameter $K_p = 0.2$ days^{-1}. The

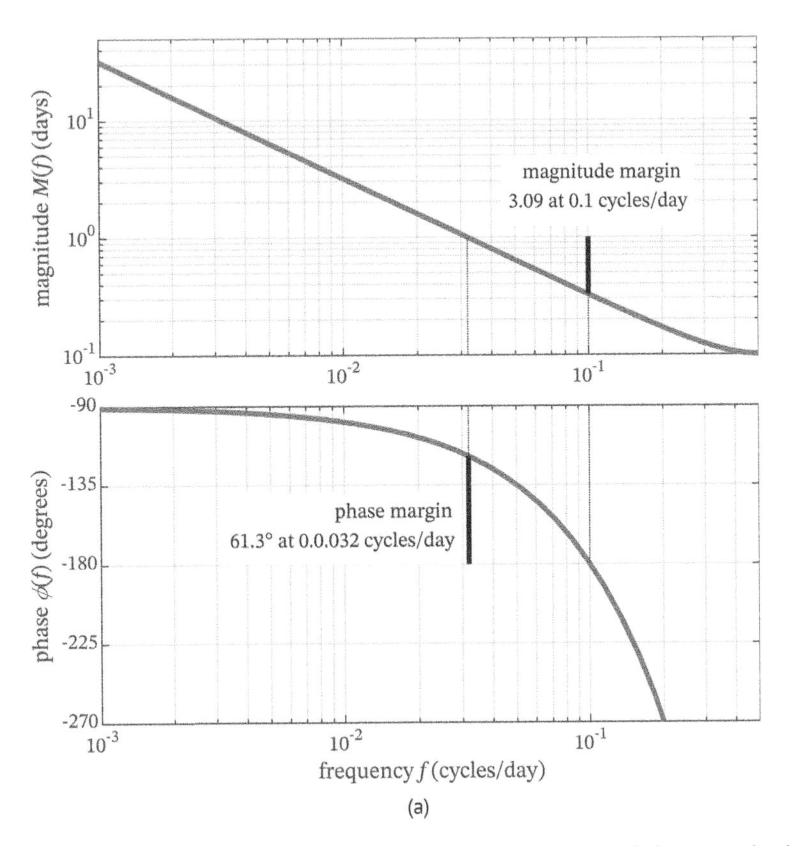

Figure 5.22 Open-loop frequency response and magnitude and phase margins for discrete-time capacity adjustment. (a) $T = 1$ day, $dT = 2$ days, $Kp = 0.2$ days-1 (stable closed-loop capacity adjustment).

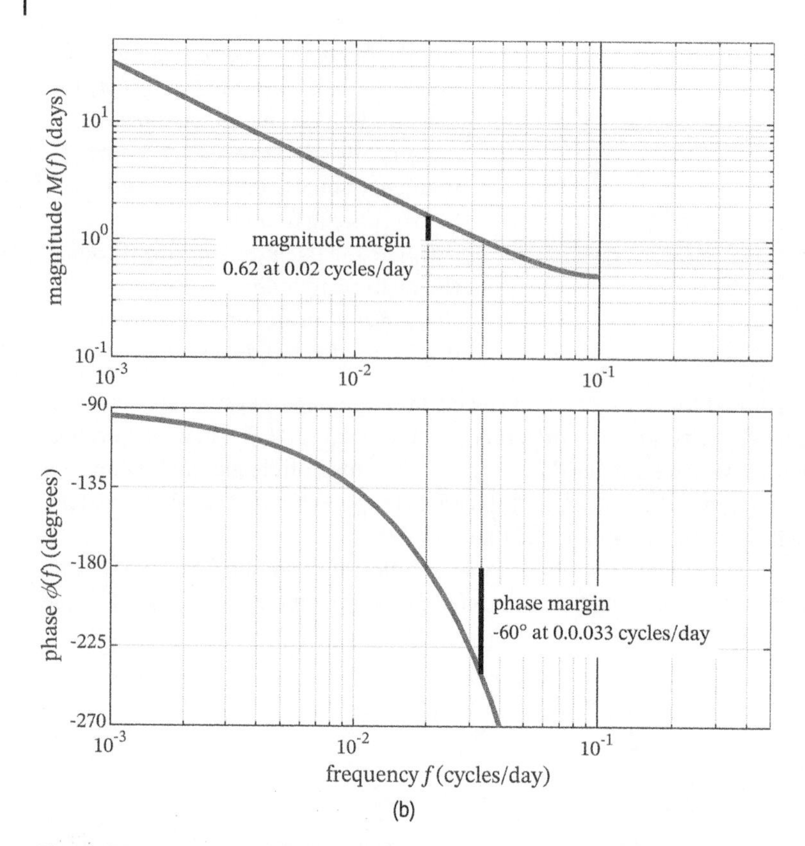

Figure 5.22 (b) $T = 5$ *days*, $dT = 10$ *days*, $Kp = 0.2$ *days*-1 (unstable closed-loop capacity adjustment).

Program 5.8 Calculation of magnitude margin and phase margin

```
Kp=0.2;  % proportional decision-making parameter (days^-1)
options=bodeoptions;
options.FreqUnits='cycles/day'; options.MagScale='log';
options.MagUnits='abs';

T2=1;  % period (days)
d=2;  % delay dT=2 (days)

% open-loop transfer function for capacity adjustment with dT=2 days
Gz2=tf(Kp*T2,[1 -1],T2,'OutputDelay',d,'TimeUnit','days')
```

```
Gz2 =

              0.2
  z^(-2) * -----
            z - 1

Sample time: 1 days
Discrete-time transfer function.
```

```
bode(Gz2,options)  % frequency response - Figure 5.22a

[Gm2,Pm2,Wcg2,Wcp2]=margin(Gz2)  % margins and frequencies
```

```
Gm2 = 3.0902
Pm2 = 61.3040
Wcg2 = 0.6283
Wcp2 = 0.2003
```

```
T10=5;  % period (days)
d=2;  % delay dT=10 (days)

% open-loop transfer function for capacity adjustment with dT=10 days
Gz10=tf(Kp*T10,[1 -1],T10,'OutputDelay',d,'TimeUnit','days')
```

```
Gz10 =

               1
  z^(-2) * -----
           z - 1

Sample time: 5 days
Discrete-time transfer function.
```

```
bode(Gz10,options)  % frequency response - Figure 5.22b

[Gm10,Pm10,Wcg10,Wcp10]=margin(Gz10)  % magnitude and phase margins
```

```
Gm10 = 0.6180
Pm10 = -60.0308
Wcg10 = 0.1257
Wcp10 = 0.2095
```

magnitude margin is $G_m = 0.62$ and the phase margin is $P_m = -60°$. In this case it is expected that closed-loop capacity adjustment will be unstable because $G_m < 1$ and also because $P_m < 0°$.

5.4 Summary

Fundamental dynamic characteristics associated with frequency response have been defined in this chapter. These include the magnitude, phase, zero-frequency magnitude, bandwidth, magnitude margin and phase margin. The latter are measures of relative stability that are obtained from the open-loop frequency response but anticipate closed-loop frequency response and closed-loop dynamic behavior. In Chapter 6 it will be shown how these fundamental characteristics can be used to guide design of decision-making. The magnitude and phase of the frequency response are important information that augments time-response measures, providing insight into and understanding of the dynamic behavior of production systems and components.

6

Design of Decision-Making for Closed-Loop Production Systems

As illustrated in Figure 6.1, production systems often have closed-loop topologies in which the performance, status, and condition of key components are observed and data collected are analyzed to detect trends, errors, and deviations from desired goals, commands, and targets. Decisions then are made using appropriate rules and algorithms; these can be made manually, automatically or in a hybrid combination depending upon how quickly decisions must be made, how much human experience and judgment is required, and the level of automation in monitoring and analysis. Action then is taken to implement decisions with the objective of manipulating the production system to effectively respond to trends and eliminate deviations from desired behavior. The closed-loop cycle illustrated in Figure 6.1 can take place continuously, or nearly so, or observation, decision-making, and implementation of decisions can take place at discrete intervals, with a period between decisions that can range from milliseconds to months.

Decisions in production systems often are made using algorithms and rules that can consist of algebraic or statistical calculations, detection of trends and response to them, if-then evaluations, and limits. These often are developed with the assistance of experience and intuition. Using the methods discussed in Chapter 2, the decision-making components of production systems can be modeled using linear differential, difference, and algebraic equations that represent their fundamental dynamic nature. Relationships between goals, feedback, and manipulations then can be represented using transfer functions. This facilitates analysis of dynamic behavior of production systems with embedded closed-loop decision-making and selection of decision-making parameters that can be expected to result in required dynamic behavior when decision-making components are implemented in practice with logic, AI, limits, exceptions, and other nonlinear relationships.

This chapter is focused on design of decision-making to meet dynamic behavior requirements for closed-loop production systems, which usually involves the following steps:

1. Set dynamic behavior requirements: These can include accurate regulation, settling time after changes in demand or disturbances, tendency to oscillate, and fundamental measures of dynamic behavior such as time constants, damping ratios, and bandwidth.

Control Theory Applications for Dynamic Production Systems: Time and Frequency Methods for Analysis and Design, First Edition. Neil A. Duffie.

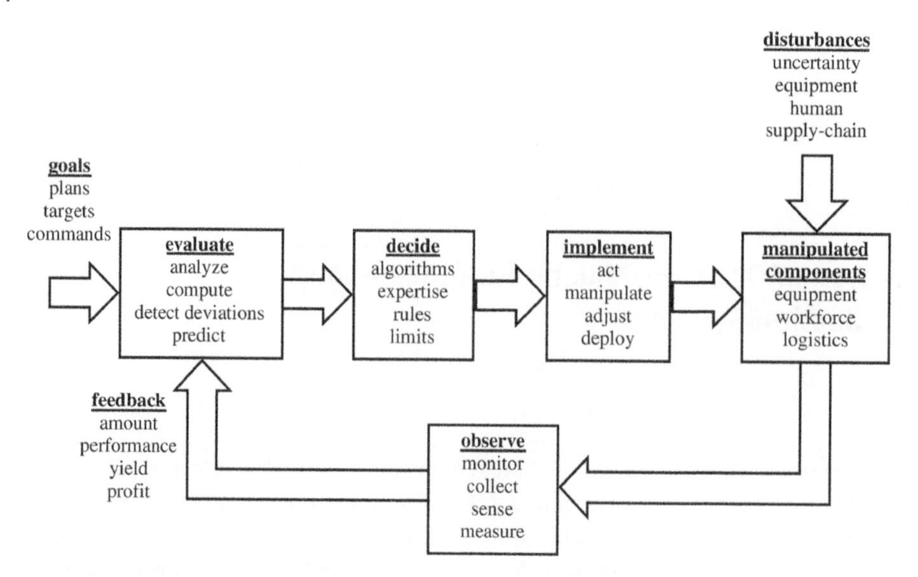

Figure 6.1 Closed-loop decision-making in production systems.

2. Choose closed-loop decision-making topology: There are many possible topologies for closed-loop decision-making. Proportional plus integral plus derivative (PID) control is perhaps the most well-known topology, but there are many alternatives for locating decision-making components in the closed-loop and adding decision-making components that can significantly improve performance.
3. Choose decision-making parameter values: Closed-loop decision-making components have parameters for which values need to be found that result in dynamic behavior that satisfies requirements. Design of these components and their parameters can be aided by analysis of time and frequency response, as well as by direct computation using transfer functions.
4. Assess design and modify if necessary: Often, an initially chosen closed-loop decision-making topology is found to be unsuitable. It may not be possible to meet all requirements, and alternative topologies may need to be investigated. On the other hand, requirements may be infeasible, and they may need to be modified, or components of a production system may need to be modified to make requirements feasible. For example, delays in making observations and implementing decisions may need to be reduced to make requirements for time response feasible. The dynamic behavior that results from a chosen topology and parameter values may be unacceptably sensitive to uncertainties in production system behavior including variation in behavior during operation and modeling inaccuracies.

Decision-making component designs obtained that meet dynamic requirements can be used to guide further design and eventual implementation of decision-making for production systems that incorporates more detailed consideration of physical, operational and logistical requirements.

In this chapter, decision-making using proportional, integral, derivative control, and common combinations is described first because these easily can be related to the need to act based on the magnitude errors or deviations, the persistence of errors or deviations, and the rate at which errors or deviations are increasing or decreasing. Then, three basic approaches are presented for choosing parameter values for decision-making components: design using time response, design directly from transfer functions, and design using frequency response. While design using time response and frequency response can be readily performed using models of any order with the assistance of control system engineering software, direct design using transfer functions usually is focused on low-order production system models. Simplification of models by eliminating insignificant dynamic terms therefore is discussed. Furthermore, models with a limited number of easily recognizable dynamic terms can be helpful in taking advantage of the designer's intuition and experience.

Choices for decision-making components and their placement in closed-loop production systems are highly dependent on the application and dynamic requirements being considered. Some of the most common and important alternatives will be described next including the proportional plus integral plus derivative (PID) topology, which is perhaps the most well-known, placement of decision-making components in the forward and feedback paths, cascade topologies, addition of feedforward control, and the Smith predictor topology, which is an example of incorporation of an internal model in closed-loop decision-making.

Calculation of sensitivity is discussed at the end of this chapter. The closed-loop behavior that results from use of designed decision-making can vary in actual production system operation because of modeling inaccuracies. Parameters in dynamic models of production system components often are assumed to be constant in the design process but are variable in reality, and models often are simplified and omit many physical, operational, and logistical details. Even if variations or uncertainty in parameter values are small, it is useful to be able to assess the sensitivity of key measures of dynamic behavior and performance to variations in system parameters. The results can be important in verifying the suitability of chosen decision-making topologies and parameters.

6.1 Basic Types of Continuous-Time Control

Many closed-loop production systems have the topology shown in Figure 6.2a. For continuous-time production systems, $G_m(s)$ and $H_m(s)$ are models of one or more manipulated components and $G_c(s)$ and $H_c(s)$ are models of one or more decision-making components. $R(s)$ represents the goal, target, or command and $C(s)$ represents the production system output that results from manipulations $M(s)$ and disturbances $P(s)$. $F(s)$ is the feedback, which can differ from output $C(s)$, often because of scaling or dynamic behavior of sensing and measurement components.

The open-loop transfer function contains all of the components that are within the closed-loop in Figure 6.2a:

$$G_{ol}(s) = \frac{B(s)}{E(s)} = G_c(s)G_m(s)H_m(s)H_c(s) \tag{6.1}$$

This transfer function and the corresponding topology in Figure 6.2b are important in guiding the choice of a closed-loop decision-making topology for a production system and analysis required to choose values of decision-making parameters, particularly in design using frequency response.

The decision-making rules and algorithms used in production systems with the closed-loop topology shown in Figure 6.1 often can be approximately represented using straightforward dynamic relationships between feedback and manipulations. These can be proportional where manipulations are greater in magnitude when conditions such as errors or deviations are greater in magnitude, accumulative where manipulations are greater in magnitude when conditions such as errors or deviations persist over time, and rate-based where manipulations are greater in magnitude when conditions such as errors or deviations are increasing rapidly thereby anticipating larger errors or deviations. These relationships are referred to as proportional, integral,

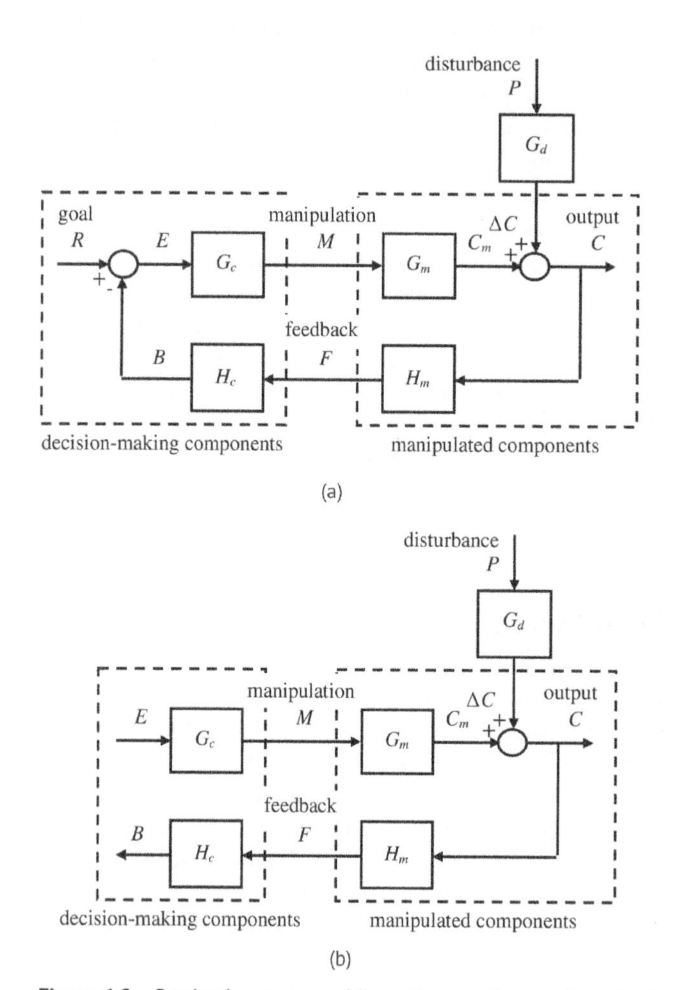

Figure 6.2 Production system with continuous-time or discrete-time decision-making components G_c and H_c and manipulated components G_m and H_m. (a) Closed-loop. (b) Open-loop.

and derivative, respectively. Integral or derivative decision-making often is combined with proportional decision-making.

Decision-making often is designed so that $H_m(s)H_c(s) = 1$ in the feedback path in Figure 6.2, and proportional, integral, or derivative control is the forward path in $G_c(s)$. In this case, $E(s)$ is the error or deviation between the production system goal $R(s)$ and the production system output $C(s)$:

$$E(s) = R(s) - C(s) \tag{6.2}$$

This is assumed to be the case in the rest of this section where continuous-time proportional, integral and derivative control and some common combinations will be defined and the effects they tend to have on dynamic behavior will be discussed. However, distributing decision-making components between $G_c(s)$ and $H_c(s)$ in other ways can be advantageous, and this will be discussed in Section 6.6.2.

6.1.1 Continuous-Time Proportional Control

It often is desirable to include proportional decision-making in design of closed-loop dynamic behavior to meet requirements. This is referred to as proportional control and with $H_c(s)H_m(s) = 1$ in Figure 6.2, the relationship between manipulation $m(t)$ and error $e(t)$ is

$$m(t) = K_p e(t) \tag{6.3}$$

where a value needs to be chosen for parameter K_p. The transfer function for proportional control then is

$$G_c(s) = K_p \tag{6.4}$$

Increasing proportional control parameter K_p often tends to

- decrease settling time
- increase closed-loop bandwidth
- decrease closed-loop damping ratio, making a closed-loop production system more oscillatory
- reduce error

However, results are highly dependent on the dynamic characteristics of the specific closed-loop production system being considered and may differ, even having opposite trends, over the range of possible values of K_p.

6.1.2 Continuous-Time Proportional Plus Derivative Control

It often is desirable to generate manipulations $m(t)$ that tend to counteract rate of change $dc(t)/dt$ in production system output, anticipating growth of errors and deviations, and increasing damping in a closed-loop production system, which tends to make its dynamic behavior less oscillatory. Derivative decision-making usually cannot be used alone because constant or slowly changing errors and disturbances do not generate required manipulations. Derivative control therefore usually is combined with proportional control. With $H_c(s)H_m(s) = 1$ in Figure 6.2, the relationship between manipulation $m(t)$ and error $e(t)$ is

$$m(t) = K_p e(t) + K_d \frac{de(t)}{dt} \tag{6.5}$$

where values need to be chosen for proportional control parameter K_p and derivative control parameter K_d. The transfer function for proportional plus derivative control is then

$$G_c(s) = (K_p + K_d s) \tag{6.6}$$

Increasing derivative control parameter K_d often tends to

- increase closed-loop damping ratio and make a closed-loop production system less oscillatory
- decrease error, but not affect steady-state error

Again, results are highly dependent on the dynamic characteristics of the specific closed-loop production system being considered and may differ, even having opposite trends, over the range of possible values of K_p and K_d.

An alternative form of proportional plus derivative control is

$$G_c(s) = K_c(\tau_c s + 1) \tag{6.7}$$

where values need to be chosen for parameters K_c and time constant τ_c, which are related to the parameters in Equation 6.6 by $K_c = K_p$ and $\tau_c = K_d/K_p$. The term $(\tau_c s + 1)$ often is referred to as lead control.

6.1.3 Continuous-Time Integral Control

A time-integral decision-making relationship often is desirable because manipulations can be generated that react to persistent deviations from production system goal $r(t)$. For example, when output $c(t)$ is perturbed by a constant but unknown disturbance that is modeled by input $p(t)$ in Figure 6.2, manipulation $m(t)$ can be increased with time until error is eliminated. With $H_c(s)H_m(s) = 1$ in Figure 6.2, the relationship between manipulation $m(t)$ and error $e(t)$ with integral control is

$$m(t) = K_i \int_0^t e(t)dt \tag{6.8}$$

where a value needs to be chosen for parameter K_i. The transfer function for integral control is then

$$G_c(s) = \frac{K_i}{s} \tag{6.9}$$

Increasing integral control parameter K_i often tends to

- decrease bandwidth and increase settling time
- decrease closed-loop damping ratio and make a closed-loop production system more oscillatory
- eliminate steady-state error.

However, results are highly dependent on the dynamic characteristics of the specific closed-loop production system being considered and may differ over various ranges of K_i, even having opposite trends.

6.1.4 Continuous-Time Proportional Plus Integral Control

Often, integral control cannot be used alone because it can result in a production system that reacts too slowly to disturbances, is too oscillatory, or is unstable. Integral control, therefore, often is combined with proportional control. With $H_c(s)H_m(s) = 1$ in Figure 6.2, the relationship between manipulation $m(t)$ and error $e(t)$ is

$$m(t) = K_p e(t) + K_i \int_0^t e(t)\,dt \tag{6.10}$$

where values need to be chosen for proportional control parameter K_p and integral control parameter K_i. The transfer function for proportional plus integral control then is

$$G_c(s) = K_p + \frac{K_i}{s} \tag{6.11}$$

In this case, increasing proportional control parameter K_p often tends to

- increase bandwidth and decrease settling time
- increase closed-loop damping ratio and make a closed-loop production system less oscillatory
- reduce error

Again, results are highly dependent on the dynamic characteristics of the specific closed-loop production system being considered and may differ over various ranges of K_p and K_p, even having opposite trends.

An alternative form of proportional plus integral control is

$$G_c(s) = \frac{K_c}{s}(\tau_c s + 1) \tag{6.12}$$

where values need to be chosen for parameters K_c and time constant τ_c. These are related to the parameters in Equation 6.11 by $K_c = K_i$ and $\tau_c = K_p/K_c$. Again, the term $(\tau_c s + 1)$ can be referred to as lead control, and the combination can be referred to as integral-lead control.

6.2 Basic Types of Discrete-Time Control

Many closed-loop production systems have discrete-time decision-making, and often they have the topology shown in Figure 6.2a. For discrete-time production systems, $G_m(z)$ and $H_m(z)$ are models of one or more manipulated components and $G_c(z)$ and $H_c(z)$ are models of one or more decision-making components. $R(z)$ represents the goal, target, or command, and $C(z)$ represents the production system output that results from manipulations $M(z)$ and disturbances $P(z)$. $F(z)$ is the feedback, which can differ from output $C(z)$, often because of scaling or dynamic behavior of sensing and measurement components.

The open-loop transfer function contains all of the components that are within the closed-loop in Figure 6.2a:

$$G_{ol}(z)=\frac{B(z)}{E(z)}=G_c(z)G_m(z)H_m(z)H_c(z) \tag{6.13}$$

This transfer function and the corresponding topology in Figure 6.2b are important in guiding the choice of a closed-loop decision-making topology for a production system and analysis required to choose values of decision-making parameters, particularly in design using frequency response.

As described in Section 6.1 for continuous-time decision-making, discrete-time dynamic relationships between feedback and manipulations are often designed to be proportional where manipulations are greater in magnitude when conditions such as errors or deviations are greater in magnitude, accumulative where manipulations are greater in magnitude when errors or deviation persist over time, and rate-based where manipulations are greater in magnitude when conditions such as errors or deviations are increasing rapidly thereby anticipating larger errors or deviations. Discrete-time decision-making often is designed so that $H_m(z)H_c(z) = 1$ in the feedback path in Figure 6.2, and proportional, integral, or derivative decision-making is in the forward path in $G_c(z)$. In this case, $E(z)$ is the error or deviation between the production system goal $R(z)$ and the production system output $C(z)$:

$$E(z)=R(z)-C(z) \tag{6.14}$$

This is assumed to be the case in the rest of this section where discrete-time proportional, integral, and derivative control and some common combinations will be defined. The effects they tend have on dynamic behavior are similar to those of continuous-time proportional, integral, and derivative control and will not be restated in this section; however, dynamic behavior is influenced by choice of period T. As is the case with continuous-time decision-making, distributing components between $G_c(z)$ and $H_c(z)$ in other ways can be advantageous, and this will be discussed in Section 6.6.2.

6.2.1 Discrete-Time Proportional Control

When only proportional control is present and $H_c(z)H_m(z) = 1$ in Figure 6.2, the discrete-time relationship between manipulation $m(kT)$ and error $e(kT)$ with proportional control is

$$m(kT)=K_p e(kT) \tag{6.15}$$

where a value needs to be chosen for parameter K_p. The transfer function for proportional control then is

$$G_c(z)=K_p \tag{6.16}$$

6.2.2 Discrete-Time Proportional Plus Derivative Control

It is desirable to generate manipulations $m(kT)$ that counteract observed changes in production system output $c(kT)$ from one instant to the next that can be relatively small but anticipate significant change in the output. Including derivative decision-making can increase damping in a closed-loop production system, making its dynamic behavior less oscillatory. Discrete-time approximation of continuous-time derivative control usually is combined with discrete-time proportional control because deviations from the goal must be responded to if they are non-zero but not changing. With $H_c(z)H_m(z) = 1$ in Figure 6.2, the relationship between manipulation $m(kT)$ and error $e(kT)$ is

$$m(kT) = K_p e(kT) + K_d\left(\frac{e(kT) - e((k-1)T)}{T}\right) \tag{6.17}$$

where values need to be chosen for proportional control parameter K_p and derivative control parameter K_d. The transfer function for proportional plus derivative control is then

$$G_c(z) = K_p + K_d\left(\frac{1 - z^{-1}}{T}\right) \tag{6.18}$$

An alternative form of proportional plus derivative control is

$$G_c(z) = K_c\left(\frac{1 - e^{-T/\tau_c} z^{-1}}{1 - e^{-T/\tau_c}}\right) \tag{6.19}$$

where values need to be chosen for parameters K_c and lead-control time constant τ_c. which is related to the parameters in Equation 6.18 by

$$\tau_c = \frac{-T}{\ln\left(\dfrac{K_d}{K_p T + K_d}\right)} \tag{6.20}$$

6.2.3 Discrete-Time Integral Control

An accumulating decision-making relationship often is desirable because manipulations can be generated that react to persistent differences between production system goal $r(kT)$ and production system output $c(kT)$. With $H_c(z)H_m(z) = 1$ in Figure 6.2, the discrete-time relationship between manipulation $m(kT)$ and error $e(kT)$ with integral control is

$$m(kT) = K_i \sum_{n=0}^{k} e(kT)T \tag{6.21}$$

where a value needs to be chosen for parameter K_i. For integer $k > 0$,

$$m(kT) = m((k-1)T) + K_i T e(kT) \tag{6.22}$$

and the transfer function for discrete-time integral control is

$$G_c(z) = K_i\left(\frac{T}{1-z^{-1}}\right) \tag{6.23}$$

6.2.4 Discrete-Time Proportional Plus Integral Control

As is the case with continuous-time decision-making, discrete-time integral control often cannot be used alone because it can result in a production system that reacts too slowly, is too oscillatory, or is unstable. Discrete-time integral control, therefore, often is combined with discrete-time proportional control. With $H_c(s)H_m(s) = 1$ in Figure 6.2, the relationship between manipulation $m(kT)$ and error $e(kT)$ is

$$m(kT) = K_p e(kT) + K_i \sum_{n=0}^{k} e(kT)T \tag{6.24}$$

where values need to be chosen for proportional control parameter K_p and integral control parameter K_i. For integer $k > 0$,

$$m(kT) = m((k-1)T) + (K_p + K_i T)e(kT) - K_p e((k-1)T) \tag{6.25}$$

and the transfer function for discrete-time proportional plus integral control is

$$G_c(z) = K_p + K_i\left(\frac{T}{1-z^{-1}}\right) \tag{6.26}$$

An alternative form of discrete-time proportional plus integral control is integral-lead control:

$$G_c(z) = K_c\left(\frac{T}{1-z^{-1}}\right)\left(\frac{1-e^{-T/\tau_c}z^{-1}}{1-e^{-T/\tau_c}}\right) \tag{6.27}$$

where values need to be chosen for parameters K_c and lead-control time constant τ_c, which is related to the parameters in Equation 6.26 by

$$\tau_c = \frac{-T}{\ln\left(\frac{K_p}{K_p + K_i T}\right)} \tag{6.28}$$

6.3 Control Design Using Time Response

Values can be chosen for the parameters in decision-making components in closed-loop production systems using the results of repeated simulations to search the parameter space for dynamic behavior with characteristics that satisfy requirements. Optimization and machine learning also can be applied. While this can be a trial and error approach, trends often can be observed as parameters are changed, and these trends can help guide design of decision-making to a satisfactory result. This approach, as well as selection of a closed-loop decision-making topology and type of control, often benefits from the designer's prior experience with similar systems and is a straightforward way to design decision-making that is low in complexity, for example, decision-making with one or two parameters such as those defined in Sections 6.1 and 6.2.

Example 6.1 Time Response Design of Continuous-Time Integral Control of Production Using Metal Forming

The influence of deviations in the quality of parts produced using metal forming and the use of integral control to adjust production can be modeled as illustrated in the block diagram in Figure 6.3. Forming equipment is modeled very simply as a proportional relationship between the number of presses that are used to produce the parts and the production rate. Physical phenomena that affect quality are modeled simply as a disturbance: the rate of rejection of parts produced because quality requirements are not met. As this rate increases, the production rate needs to be increased to maintain the rate of production of parts that meet quality requirements, and until otherwise corrected, this is done by increasing the number of presses that produce the parts. Integral control is used in planning and scheduling to decide how many presses are required. Adjustments in the number of presses are assumed to be continuous, and this number is not required to be an integer. The rate of production of parts that pass quality requirements is assumed to be continuously observed, and no other information is assumed to be available for use in decision-making. It is assumed that there are no delays in making observations and no delays such as setup times in implementing adjustments in the number of presses.

The simple continuous-time model of production using metal forming is

$$r_m(t) = K_m m(t)$$

$$r_o(t) = r_m(t) - r_d(t)$$

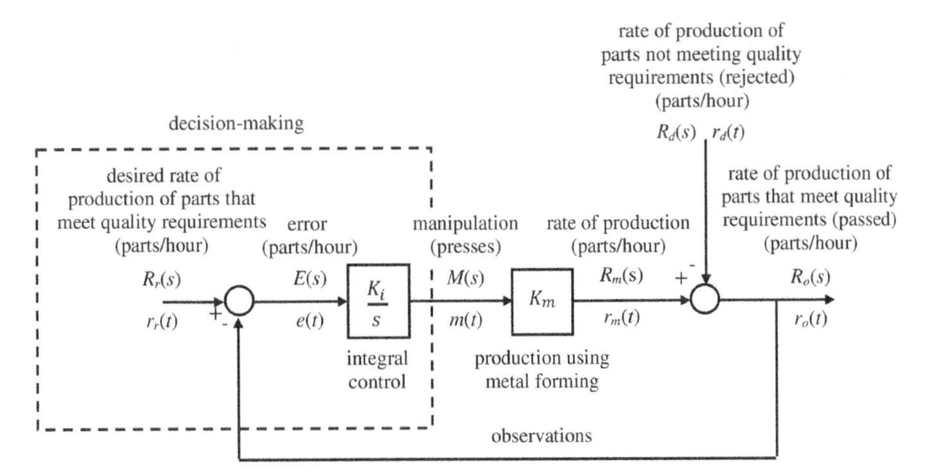

Figure 6.3 Block diagram for continuous-time integral control of production using metal forming.

where $m(t)$ is the number of presses, constant $K_m > 0$ relates the number of presses to the rate of production $r_m(t)$, $r_d(t)$ is the portion of the production rate that is rejected because parts do not meet quality requirements, and $r_o(t)$ is the remaining portion that does meet quality requirements.

The number of presses is adjusted using integral control:

$$e(t) = r_r(t) - r_o(t)$$

$$m(t) = K_i \int_0^t e(t)dt$$

where $r_r(t)$ parts/hour is the desired rate of production of parts that meet quality requirements, $e(t)$ parts/hour is the error between this desired rate of production and the actual rate of production that meets quality requirements, and $K_i > 0$ is the integral controller parameter that must be designed. Also,

$$\frac{m(t)}{dt} = K_i e(t)$$

When inputs $r_r(t)$ and $r_d(t)$ are constant in the closed-loop topology with integral control in Figure 6.3, steady-state error $e(\infty) = 0$ for $K_i > 0$ because the number of presses $m(t)$ increases or decreases until $r_o(t) = r_r(t)$.

It is expected that demand is relatively constant and hence desired rate of production $r_r(t)$ is changed infrequently. However, disturbances $r_d(t)$ often perturb production and the response of error $e(t)$ to this disturbance therefore is of significant interest; ideally, this error quickly will return to zero after a change in disturbance $r_d(t)$. The following closed-loop performance is desired:

- non-oscillatory response to step disturbance $r_d(t)$
- zero steady-state error for step disturbance $r_d(t)$
- closed-loop time settling time t_s not greater than 2 hours
- relatively low rate of increase or decrease in the number of presses $m(t)$.

A longer settling time represents a longer disruption in production of parts that meet quality requirements, while a shorter settling time represents more rapid recovery from such disruptions.

As noted in Section 3.8.4, the response of a production system to each of its inputs can be modeled separately using transfer functions. Desired production rate rarely changes and only disturbances $r_d(t)$ contribute significantly to dynamic response. The closed-loop transfer functions for error and manipulation are of interest:

$$\frac{E(s)}{R_d(s)} = \frac{\frac{1}{K_i K_m} s}{\frac{1}{K_i K_m} s + 1}$$

$$\frac{M(s)}{R_d(s)} = \frac{\frac{1}{K_m}}{\frac{1}{K_i K_m} s + 1}$$

The characteristic equation is

$$\frac{1}{K_i K_m} s + 1 = 0$$

Response to step disturbance inputs is not expected to be oscillatory because the characteristic equation is 1st-order and does not have complex conjugate roots.

It is desired to find a value for integral control parameter K_i that results in a closed-loop production system with dynamic behavior for production parameter $K_m = 200$ (parts/hour)/press that satisfies the requirements listed above. Program 6.1 can be used to calculate settling time for this production system and time response for step disturbance $r_d(t) = 300$ rejected parts/hour. Results are settling time $t_s = 3.91$, 1.96, and 0.98 hours for $K_i = 0.005$, 0.01, and 0.02 (presses/hour)/(parts/hour), respectively. The time response for the step disturbance and these values of K_i is shown in Figure 6.4.

Program 6.1 Calculation of settling times and step response for continuous-time integral control of production using metal forming

```
Km=200;  % production constant ((parts/hour)/press)
rd=300;  % production rate rejected due to quality (parts/hour)

s=tf('s','TimeUnit','hours');

Ki=0.005;  % integral control parameter ((presses/hour)/(parts/hour))
Ges1=(s/Ki/Km)/(1+s/Ki/Km);  % closed-loop transfer function for error
Gms1=(1/Km)/(1+s/Ki/Km);  % closed-loop transfer function for presses
stepinfo(Gms1)  % settling time
```

```
ans =
        RiseTime: 2.1970
    SettlingTime: 3.9121
     SettlingMin: 0.0045
     SettlingMax: 0.0050
       Overshoot: 0
      Undershoot: 0
            Peak: 0.0050
        PeakTime: 10.5458
```

```
Ki=0.01;  % integral control parameter ((presses/hour)/(parts/hour))
Ges2=(s/Ki/Km)/(1+s/Ki/Km);  % closed-loop transfer function for error
Gms2=(1/Km)/(1+s/Ki/Km);  % closed-loop transfer function for presses
stepinfo(Gms2)  % settling time
```

```
ans =
        RiseTime: 1.0985
    SettlingTime: 1.9560
     SettlingMin: 0.0045
     SettlingMax: 0.0050
       Overshoot: 0
      Undershoot: 0
            Peak: 0.0050
        PeakTime: 5.2729
```

```
Ki=0.02;  % integral control parameter ((presses/hour)/(parts/hour))
Ges3=(s/Ki/Km)/(1+s/Ki/Km);  % closed-loop transfer function for error
Gms3=(1/Km)/(1+s/Ki/Km);  % closed-loop transfer function for presses
stepinfo(Gms3)  % settling time
```

```
ans =
        RiseTime: 0.5493
    SettlingTime: 0.9780
     SettlingMin: 0.0045
     SettlingMax: 0.0050
       Overshoot: 0
      Undershoot: 0
            Peak: 0.0050
        PeakTime: 2.6365
```

```
step(rd*Ges1,rd*Ges2,rd*Ges3,4)  % response of error - Figure 6.4a
xlabel('time kT'); ylabel('error e(t) (parts/hour)');
legend('Ki=0.005','Ki=0.01','Ki=0.02')

step(rd*Gms1,rd*Gms2,rd*Gms3,4)  % response of presses - Figure 6.4b
xlabel('time kT'); ylabel('manipulation m(t) (presses)');
legend('Ki=0.005','Ki=0.01','Ki=0.02','location','east')
```

Closed-loop settling times decrease with increasing K_i, and $K_i = 0.01$ (presses/hour)/(parts/hour) satisfies the settling time requirement. Figure 6.4a confirms that steady-state error is zero for $K_i > 0$, and Figure 6.4b shows that the rate of increase in number of presses increases with K_i; hence, a lower value that meets requirements should be chosen. The responses are non-oscillatory, and all of the above requirements are satisfied with $K_i = 0.01$.

Example 6.2 Time Response Design of Discrete-Time Proportional Control of Actuator Position

Computer control of an actuator is illustrated in Figure 3.24, and a corresponding block diagram is shown in Figure 4.8. The discrete time transfer function for the actuator is

$$\frac{P(z)}{V_c(z)} = \mathcal{Z}\left\{\left(\frac{1-e^{-Ts}}{s}\right)\left(\frac{1}{\tau_v s+1}\right)\frac{1}{s}\right\}$$

where $v_c(kT)$ cm/second is the actuator velocity command calculated and output by the control computer and $p(kT)$ cm is the actuator position, which is measured as feedback by the control computer. τ_v is the actuator time constant, and T is the period between adjustments of desired actuator velocity.

Discrete-time proportional control is used to adjust desired actuator velocity as a function of the error between actuator position command $p_c(kT)$ and actuator position $p(kT)$:

$$V_c(z) = K_p\left(P_c(z) - P(z)\right)$$

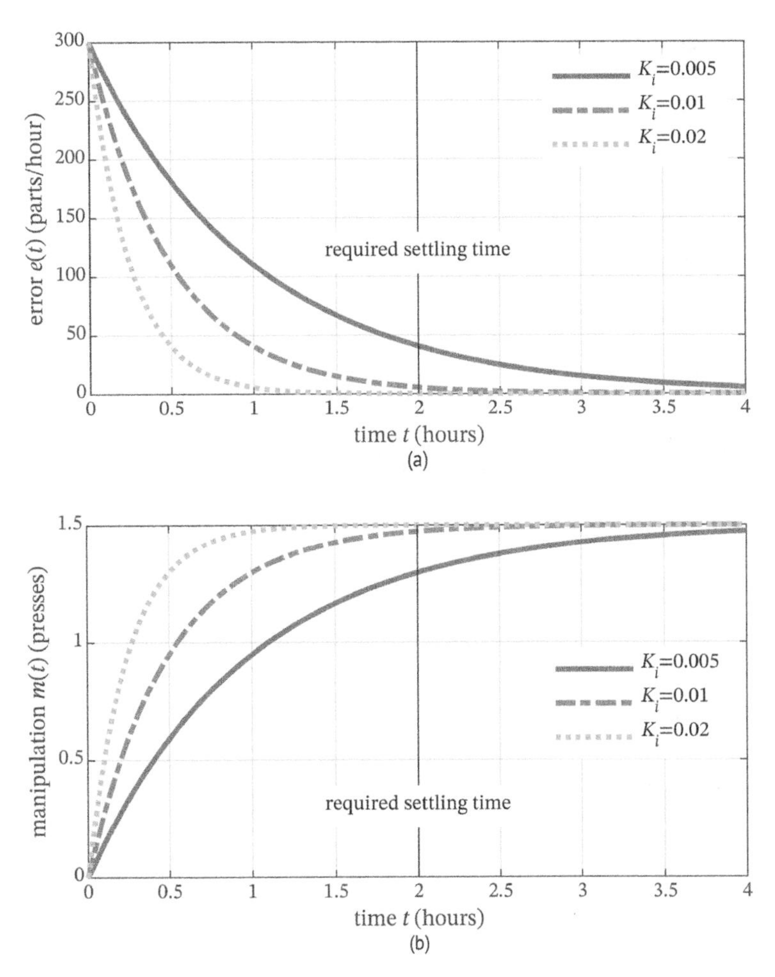

Figure 6.4 Response to a step disturbance of 300 rejected parts/hour for various values of integral control parameter K_i (presses/hour)/(parts/hour). (a) Error between desired and actual rate of production of parts that meet quality requirements. (b) Number of presses.

where K_p seconds^{-1} is the proportional control parameter. The actuator time constant is $\tau_v = 0.06$ seconds, and the period between adjustments in the velocity command is $T = 0.004$ seconds.

The following closed-loop performance is desired:

- zero steady-state error for step changes in position command $p_c(kT)$
- non-oscillatory response to step changes in position command $p_c(kT)$
- settling time $t_s \approx 0.2$ seconds.

It is desired to find a value for proportional control parameter K_p that results in closed-loop dynamic behavior that satisfies these requirements.

Program 6.2 can be used to calculate the time response for this system, and the results for K_p ranging from 2 to 16 seconds^{-1} are shown in Figure 6.5. By varying K_p it

Program 6.2 Proportional discrete-time control of actuator position

```
tauv=0.06;  % actuator time constant (seconds)
T=0.004;  % period between adjustments (seconds)

Gpz=c2d(tf(1,[tauv, 1, 0]),T);  % actuator transfer function

Kp=2;  % proportional control parameter
Gclz1=feedback(Kp*Gpz,1);  % closed-loop transfer function for position
stepinfo(Gclz1)
```

```
ans =
        RiseTime: 0.9640
    SettlingTime: 1.7520
     SettlingMin: 0.9005
     SettlingMax: 0.9999
       Overshoot: 0
      Undershoot: 0
            Peak: 0.9999
        PeakTime: 3.9560
```

```
Kp=4;  % proportional control parameter
Gclz2=feedback(Kp*Gpz,1);  % closed-loop transfer function for position
stepinfo(Gclz2)
```

```
ans =
        RiseTime: 0.4200
    SettlingTime: 0.7320
     SettlingMin: 0.9024
     SettlingMax: 0.9985
       Overshoot: 0
      Undershoot: 0
            Peak: 0.9985
        PeakTime: 1.1160
```

```
Kp=8;  % proportional control parameter
Gclz3=feedback(Kp*Gpz,1);  % closed-loop transfer function for position
stepinfo(Gclz3)
```

```
ans =
        RiseTime: 0.1880
    SettlingTime: 0.5160
     SettlingMin: 0.9032
     SettlingMax: 1.0421
       Overshoot: 4.2098
      Undershoot: 0
            Peak: 1.0421
        PeakTime: 0.3880
```

```
Kp=16;  % proportional control parameter
Gclz4=feedback(Kp*Gpz,1);  % closed-loop transfer function for position
stepinfo(Gclz4)
```

```
ans =
         RiseTime: 0.1000
     SettlingTime: 0.5000
      SettlingMin: 0.9159
      SettlingMax: 1.1679
        Overshoot: 16.7920
       Undershoot: 0
             Peak: 1.1679
         PeakTime: 0.2200
```

```
step(Gclz1,Gclz2,Gclz3,Gclz4,1)  % position response - Figure 6.5
xlabel('time kT'); ylabel('position p(kT) (cm)')
legend('Kp=2 seconds^-^1','Kp=4 seconds^-^1','Kp=8 seconds^-^1',...
    'Kp=16 seconds^-^1')
```

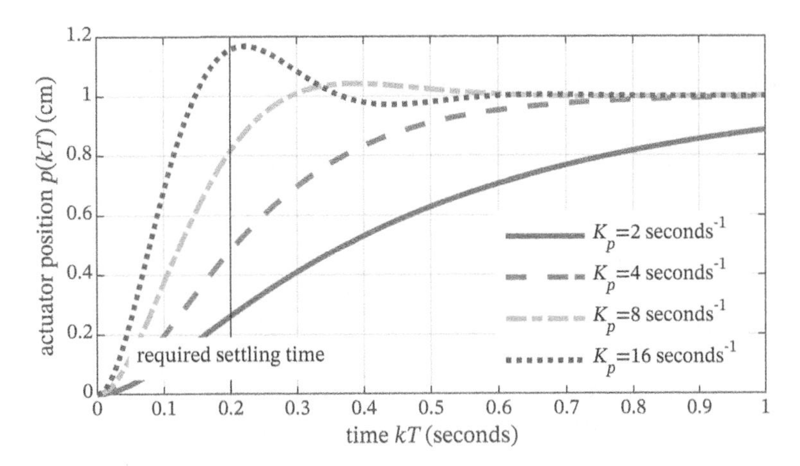

Figure 6.5 Step response with various values of controller parameter K_p.

is found that no value of K_p results in time response that meets the above requirements. Another control decision rule therefore needs to be chosen if the above requirements are to be met. Proportional plus derivative control can be designed to satisfy the above requirements as described in Example 6.3.

Example 6.3 Time Response Design of Discrete-Time Proportional Plus Derivative Control of Actuator Position

Consider computer control of an actuator as described in Example 6.2 except, as shown in the block diagram in Figure 6.6, decision-making uses a proportional plus derivative control with the transfer function in Equation 6.18. In this case, it is desired to find values of proportional control parameter K_p seconds^{-1} and derivative control parameter K_d that result in a closed-loop system that satisfies the specifications listed in Example 6.2 and improves performance with respect to that which can be achieved with only discrete-time proportional control.

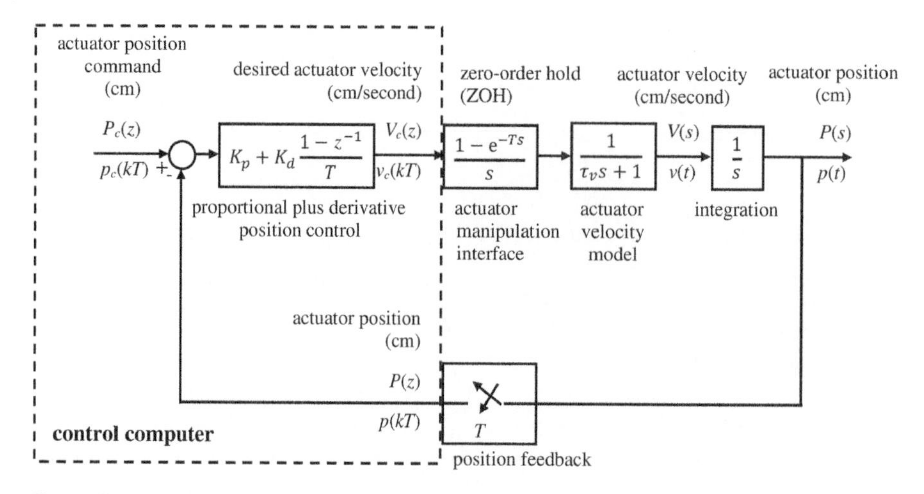

Figure 6.6 Actuator with proportional plus derivative position control.

Program 6.3 can be used to calculate the unit step response of actuator position for $T = 0.004$ seconds. By varying K_p and K_d it is found that $K_p = 20$ seconds^{-1} and $K_d = 1.2$ is one combination that results in step response that satisfies the requirements for no overshoot and a settling time of approximately $t_s = 0.2$ seconds. The results for various combinations of K_p and K_d are shown in Figure 6.7; more than one meets the requirements. For comparison, the response found for $K_p = 4$ seconds^{-1} in Example 6.2 is included; this value is approximately the highest value of K_p that does not result in overshoot with only proportional control ($K_d = 0$). The benefit of including derivative control in this application, avoiding overshoot, and permitting shorter settling times, is clearly demonstrated.

Program 6.3 Calculation of settling time for proportional plus derivative discrete-time control of actuator position

```
tau=0.06;  % actuator time constant (seconds)
t=0.004;   % period between velocity command adjustments (seconds)

gpz=c2d(tf(1,[tau, 1, 0]),t);  % discrete-time actuator transfer function

z=tf('z',t);

kp=10; kd=1.2; gcz=kp+kd*(1-z^-1)/t;  % PD control
gclz1=feedback(gcz*gpz,1);  % closed-loop transfer function for position
stepinfo(gclz1)
```

```
    ans =
            risetime: 0.2680
        settlingtime: 0.5800
         settlingmin: 0.9019
         settlingmax: 0.9993
```

```
        overshoot: 0
       undershoot: 0
             peak: 0.9993
         peaktime: 1.1960
```

```
kp=20; kd=1.2; gcz=kp+kd*(1-z^-1)/t;  % PD control
gclz2=feedback(gcz*gpz,1);  % closed-loop transfer function for position
stepinfo(gclz2)
```

```
ans =
         risetime: 0.1000
     settlingtime: 0.1920
      settlingmin: 0.9001
      settlingmax: 0.9998
        overshoot: 0
       undershoot: 0
             peak: 0.9998
         peaktime: 0.4760
```

```
kp=30; kd=1.2; gcz=kp+kd*(1-z^-1)/t;  % PD control
gclz3=feedback(gcz*gpz,1);  % closed-loop transfer function for position
stepinfo(gclz3)
```

```
ans =
         risetime: 0.0680
     settlingtime: 0.2040
      settlingmin: 0.9157
      settlingmax: 1.0336
        overshoot: 3.3574
       undershoot: 0
             peak: 1.0336
         peaktime: 0.1480
```

```
kp=4; kd=0; gcz=kp;  % proportional control
gclz0=feedback(gcz*gpz,1);  % closed-loop transfer function for position
stepinfo(gclz0)
```

```
ans =
         risetime: 0.4200
     settlingtime: 0.7320
      settlingmin: 0.9024
      settlingmax: 0.9985
        overshoot: 0
       undershoot: 0
             peak: 0.9985
         peaktime: 1.1160
```

```
step(gclz1,gclz2,gclz3,gclz0,1) %  position step responses - figure 6.7
xlabel('time kt'); ylabel('position p(kt) (cm)')
legend('kp=10 seconds^-^1, kd=1.2','kp=20 seconds^-^1, kd=1.2',...
    'kp=30 seconds^-^1, kd=1.2','kp=4 seconds^-^1, kd=0')
```

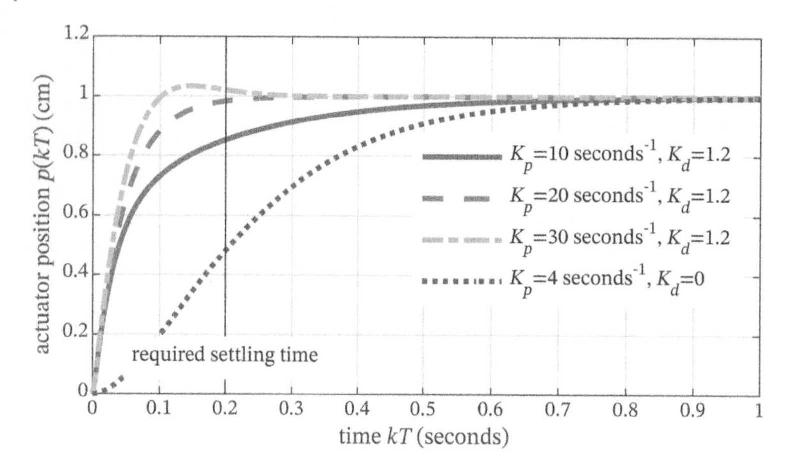

Figure 6.7 Step response with various values of control parameters K_p and K_d.

6.4 Direct Design of Decision-Making

Although searching strategies can facilitate convergence to an acceptable design, the required values of parameters in control transfer functions that represent decision-making often can be found directly using the characteristic equation obtained from the closed-loop transfer function. If the transfer function of a production system is of low order, then it often is possible to

- algebraically calculate the values of control parameters that result in desired closed-loop dynamic characteristics such as time constants and damping ratios
- add control components that cancel or replace unfavorable dynamic characteristics.

This approach is illustrated in several examples in this section. If the transfer function of a production system is not of low order[1] or contains delays, it may be possible to simplify the transfer function by ignoring insignificant factors and delays, thereby facilitating direct controller design using transfer function algebra. This is discussed at the end of the section.

Example 6.4 Direct Design for Settling Time of Discrete-Time Integral Control of Production Using Metal Forming

In Example 6.1 continuous-time integral control was designed for making decisions regarding how many presses are required in production using metal forming. The rate of production of parts that pass quality requirements was assumed to be continuously

1 Pole placement methods can be used to calculate values of control parameter that result in a characteristic equation with specified roots. The reader is referred to the Bibliography and many control-theory publications that address this approach for state-space models.

observed and adjustments in the number of presses were assumed to be continuous. In this example, the rate of production of parts that pass quality requirements is assumed to be observed twice per hour and the number of presses is adjusted twice per hour using discrete-time integral control. The following requirements for closed-loop dynamic behavior are to be met:

- non-oscillatory response to step disturbance $r_d(kT)$
- zero steady-state error for step disturbance $r_d(kT)$
- closed-loop settling time $t_s = 2$ hours.

The value of integral control parameter K_i that results in the desired closed-loop time constant can be calculated directly from the discrete-time closed-loop characteristic equation, and it can be shown that this choice of the integral control parameter satisfies the other requirements.

A discrete-time block diagram for this production system is shown in Figure 6.8. The simple discrete-time model of production using metal forming is

$$r_m(kT) = K_m m((k-1)T)$$

where $m(kT)$ is the number of presses, which is not required to be an integer. Production constant $K_m > 0$ relates the number of presses to rate of production $r_m(kT)$ parts/hour. There is a delay of T hours in the model because the effect of adjustment $m(kT)$ presses is first observed in $r_o((k+1)T)$ parts/hour rather than immediately in $r_o(kT)$ parts/hour.

The portion of the production rate that meets quality requirements is

$$r_o(kT) = r_m(kT) - r_d(kT)$$

where $r_d(kT)$ parts/hour is the portion of the production rate that is rejected. $r_d(t)$ is assumed to be constant over the period $(k\text{-}1)T \le t < kT$ hours.

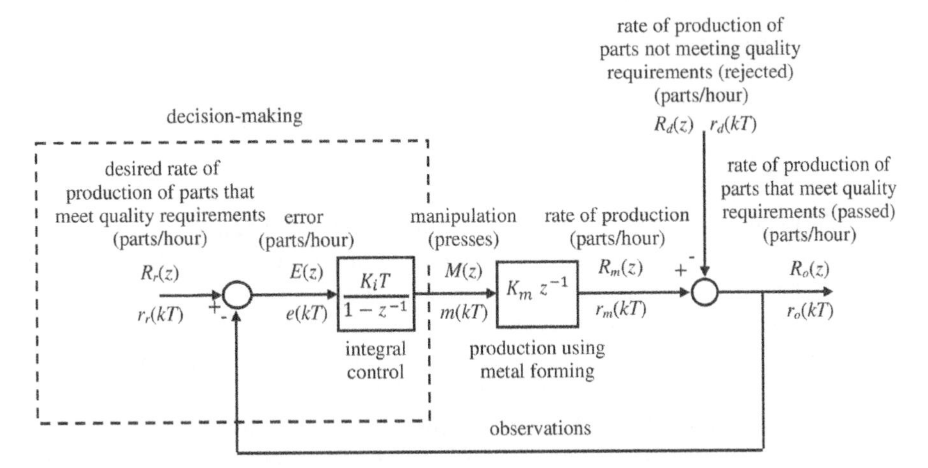

Figure 6.8 Block diagram for discrete-time integral control of production using metal forming.

The number of presses is adjusted using discrete-time integral control in the form of Equation 6.22:

$$e(kT) = r_r(kT) - r_o(kT)$$

$$m(kT) = m((k-1)T) + K_i Te(kT)$$

where $r_r(kT)$ parts/hour is the desired rate of production of parts that meet quality requirements, $e(kT)$ parts/hour is the error between this desired rate of production and the actual rate of production that meets quality requirements, and $K_i > 0$ (presses/hour)/(parts/hour) is the integral control parameter that must be designed.

Desired rate of production $r_r(kT)$ parts/hour is changed infrequently but disturbances $r_d(kT)$ parts/hour often perturb production and the response of error $e(kT)$ parts/hour and manipulation $m(kT)$ presses to this disturbance therefore is of significant interest. The closed-loop transfer functions of interest therefore are

$$\frac{E(z)}{R_d(z)} = \frac{1 - z^{-1}}{1 - (1 - K_i K_m T)z^{-1}}$$

$$\frac{M(z)}{R_d(s)} = \frac{K_i T}{1 - (1 - K_i K_m T)z^{-1}}$$

and the characteristic equation is

$$z - (1 - K_i K_m T) = 0$$

The single root of the characteristic equation is $1 - K_i K_m T$ and from Equation 4.31, the desired root is

$$1 - K_i K_m T = e^{-T/\tau}$$

where $0 < K_i K_m T < 1$, and τ is the desired time constant of the closed-loop response. Hence,

$$K_i = \frac{1 - e^{-T/\tau}}{K_m T}$$

The results in Figure 4.1a show that response can be expected to be approximately 98% complete at time 4τ. Hence, the desired closed-loop time constant can be chosen to be $\tau = t_s/4 = 0.5$ hours and the discrete-time response obtained can be verified to confirm that the settling time requirement has been met.

Program 6.4 can be used to obtain the result $K_i = 0.0063$ (presses/hour)/(parts/hour) given production constant $K_m = 200$ (parts/hour)/press, period between adjustments $T = 0.5$ hours, and desired closed-loop time constant $\tau = 0.5$ hours. The discrete-time response of error $e(kT)$ parts/hour and manipulation $m(kT)$ presses for a step disturbance of 300 rejected parts/hour is shown in Figure 6.9 for this value of K_i. The settling time is 2 hours, there is zero steady-state error, and the response is not oscillatory. The design requirements therefore are satisfied.

Program 6.4 Calculation of parameter K_i for discrete-time integral control of production using metal forming

```
ts=2; % desired settling time (hours)
T=0.5; % period between adjustments (hours)
Km=200; % forming process constant (parts/hour/press)
rd=300; % production rate rejected due to quality (parts/hour)

tau=0.5; % desired time constant (hours)
Ki=(1-exp(-T/tau))/Km/T % control parameter (presses/hour)/(parts/hour))
```

```
    Ki = 0.0063
```

```
z=tf('z',T,'TimeUnit','hours');
Gez=minreal((1-z^-1)/(1-(1-Ki*Km*T)*z^-1)); % transfer function for error
Gmz=minreal((Ki*T)/(1-(1-Ki*Km*T)*z^-1)); % transfer function for presses

damp(Gez)
```

Pole	Magnitude	Damping	Frequency (rad/hours)	Time Constant (hours)
3.68e-01	3.68e-01	1.00e+00	2.00e+00	5.00e-01

```
step(rd*Gez,4) % step response of error - Figure 6.9a
xlabel('time kT'); ylabel('error e(kT) (parts/hour)')

step(rd*Gmz,4)   step response of manipulation - Figure 6.9b
xlabel('time kT'); ylabel('manipulation m(kT) (presses)')

stepinfo(rd*Gmz)  % step response characteristics
```

```
    ans =
            RiseTime: 1
        SettlingTime: 2
         SettlingMin: 1.4725
         SettlingMax: 1.5000
           Overshoot: 0
          Undershoot: 0
                Peak: 1.5000
            PeakTime: 18
```

Example 6.5 Direct Design for Dead-Beat Response of Discrete-Time Integral Control of Production Using Metal Forming

For the production system described in Example 6.4 in which discrete-time integral control is used to decide how many metal forming presses are required, a value of parameter K_i can be found that results in complete response to step disturbances in one adjustment period T. This often is called dead-beat response which refers, in general, to bringing a closed-loop discrete-time system to its steady state in a finite number of periods T. This is achieved, if it is possible given the closed-loop topology chosen, by choosing control decision parameter values that make all of the

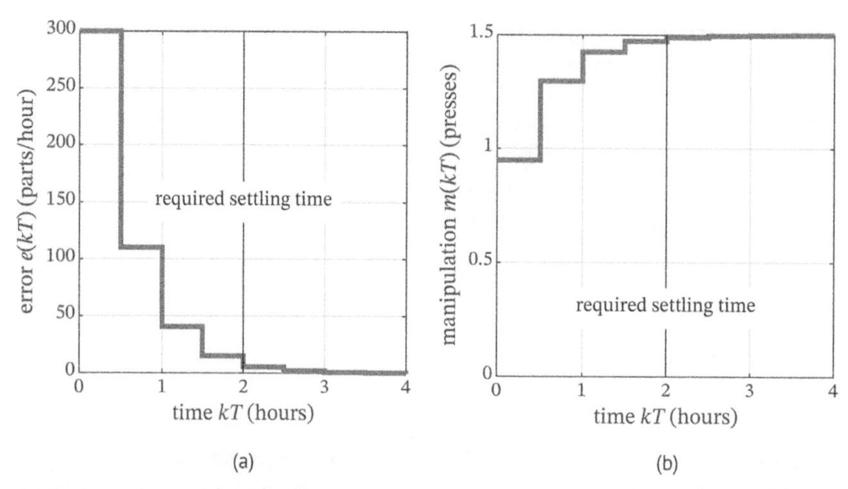

Figure 6.9 Response to step disturbance of 300 rejected parts/hour for discrete-time integral control with K_i = 0.0063 (presses/hour)/(parts/hour) and T = 0.5 hours. (a) Error between desired and actual rate of production of parts that meet quality equipments. (b) Number of presses.

roots of the characteristic equation of the closed-loop production system transfer function zero.

For the closed-loop production system with the block diagram shown in Figure 6.8, the closed-loop characteristic equation found in Example 6.4 is

$$z-(1-K_iK_mT)=0$$

The root of this characteristic equation is zero when

$$K_iK_mT=1$$

The closed-loop transfer functions that relate response of error $e(kT)$ parts/hour and manipulation $m(kT)$ presses to disturbance $r_d(kT)$ parts/hour then are

$$\frac{E(z)}{R_d(z)}=1-z^{-1}$$

$$\frac{M(z)}{R_d(s)}=K_iT$$

The period between adjustments in the number of presses can be increased to T = 2 hours and for production constant K_m = 200 parts/hour/press, integral control parameter K_i = 0.005 (presses/hour)/(parts/hour). These values of T and K_i can be used in Program 6.4 to calculate the discrete-time response of error $e(kT)$ for the step disturbance of 300 rejected parts/hour shown in Figure 6.10a and manipulation $m(kT)$ shown in Figure 6.10b. The manipulation is the required number of presses at time T = 0, and the observed error is zero after time T = 2 hours.

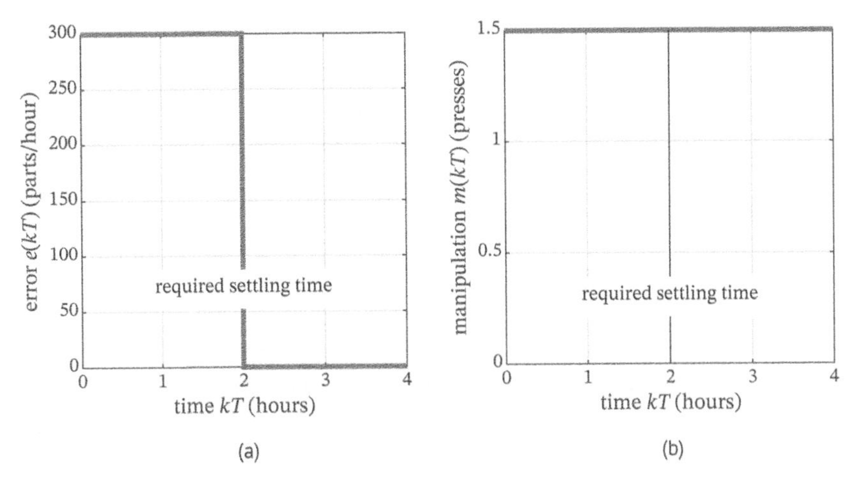

Figure 6.10 Response to step disturbance of 300 rejected parts/hour for discrete-time integral control with K_i = 0.005 (presses/hour)/(parts/hour) and T = 1 hour. (a) Error between desired and actual rate of production of parts that meet quality requirements. (b) Number of presses.

Example 6.6 Direct Design of Damping Ratio and Settling Time for Continuous-Time Proportional Plus Integral Regulation of Mixture Temperature

Figure 2.4 shows the heater and temperature regulation components that form a portion of a production process. A continuous-time model for the change in mixture temperature $\Delta h(t)$ °C that results from voltage $v(t)$ input to the heater was found in Example 2.3:

$$\tau \frac{d\Delta h(t)}{dt} + \Delta h(t) = K_h v(t)$$

$$\frac{\Delta H(s)}{V(s)} = \frac{K_h}{\tau_h s + 1}$$

where τ_h seconds is the mixture heating time constant and K_h °C/V is the mixture heating parameter.

It is desired to specify both the closed-loop damping ratio $\zeta_{cl} < 1$ and settling time t_s seconds. One option is to use continuous-time proportional plus integral control in the form of Equation 6.11, which has two parameters K_p V/°C and K_i (V/second)/°C that can be chosen to obtain the desired damping ratio and settling time:

$$M(s) = \left(K_p + \frac{K_i}{s}\right)\left(H_d(s) - H(s)\right)$$

The block diagram for proportional plus integral mixture temperature regulation is shown in Figure 6.11.

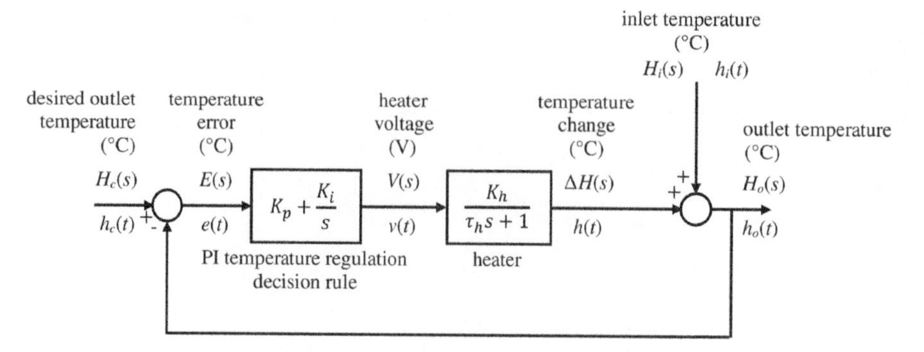

Figure 6.11 Proportional plus integral regulation of mixture temperature.

The closed-loop transfer function for the inlet mixture temperature input $h_i(t)$ is

$$\frac{H_o(s)}{H_i(s)} = \frac{1}{1+\left(K_p+\frac{K_i}{s}\right)\left(\frac{K_h}{\tau_h s+1}\right)} = \frac{s(\tau_h s+1)}{\tau_h s^2+(1+K_pK_h)s+K_iK_h}$$

and the characteristic equation is

$$s^2+\frac{1+K_pK_h}{\tau_h}s+\frac{K_iK_h}{\tau_h}=0$$

The desired characteristic equation in the form of Equation 4.14 is

$$\frac{1}{\omega_n^2}s^2+\frac{2\zeta_{cl}}{\omega_n}s+1=0$$

where ζ_{cl} is the desired closed-loop damping ratio and from Equation 4.21, the desired closed-loop natural frequency is

$$\omega_n=\frac{1}{\zeta_{cl}\tau_{cl}}$$

where τ_{cl} is the desired closed-loop time constant. Equating like terms yields

$$\zeta_{cl}=\frac{1+K_pK_h}{2\sqrt{\tau_h K_i K_h}}$$

$$\omega_n=\sqrt{\frac{K_iK_h}{\tau_h}}$$

and

$$\tau_{cl}=\frac{\tau_h}{2(1+K_pK_h)}$$

If damping ratio ζ_{cl} and settling time $t_s \approx 4\tau_{cl}$ seconds are specified, then the required K_p and K_i are

$$K_p = \frac{8\frac{\tau_h}{t_s} - 1}{K_h}$$

$$K_i = \frac{16\tau_h}{\zeta_{cl}^2 t_s^2 K_h}$$

Example 6.7 Direct Design of Time Constant Cancellation for Continuous-Time Integral-Lead Regulation of Mixture Temperature

The alternative form of proportional plus integral control in Equation 6.12,

$$M(s) = \frac{K_c}{s}(\tau_c s + 1)(H_d(s) - H(s))$$

can be designed for mixture temperature regulation in the block diagram in Figure 6.11 in two steps to achieve obtain settling time $t_s \approx 4\tau_{cl}$ seconds where τ_{cl} is the corresponding closed-loop time constant. In this case, the closed-loop transfer function for the inlet mixture temperature input $h_i(t)$ is

$$\frac{H_o(s)}{H_i(s)} = \frac{1}{1 + \frac{K_c}{s}(\tau_c s + 1)\left(\frac{K_h}{\tau_h s + 1}\right)}$$

As a first step, the lead control time constant is chosen so that

$$\tau_c = \tau_h$$

where τ_h seconds is the mixture time constant. The effect of the heater time constant has been canceled, and the closed-loop transfer function becomes

$$\frac{H_o(s)}{H_i(s)} = \frac{1}{1 + \frac{K_c}{s} K_h} = \frac{\frac{1}{K_c K_h} s}{\frac{1}{K_c K_h} s + 1}$$

The closed-loop time constant then is $\tau_{cl} = 1/(K_c K_h)$ seconds and the corresponding settling time is $t_s = 4/(K_c K_h)$ seconds. Given a desired settling time t_s, the required value of parameter K_c is

$$K_c = \frac{4}{t_s K_h}$$

6.4.1 Model Simplification by Eliminating Small Time Constants and Delays

Control system engineering software supports modeling and analysis of relatively complicated models of closed-loop production systems. However, the results of these analyses can be difficult to interpret when the models and associated transfer functions are of relatively high order. A simpler model often can provide clearer and more intuitive understanding of the performance that can be expected and can provide guidance in designing decision-making including selection of topology and values for parameters. It may be possible to simplify a model of a production system and its corresponding transfer function, obtaining a convenient, relatively low-order linear model that is of sufficient fidelity to enable it to be used in an initial direct design of decision-making.

Goals in design of decision-making often include achieving response to commands and disturbances that has a desired or maximum settling time t_s, which is the approximate period required for the closed-loop production system to return to a steady-state condition after it has been perturbed. Time constants and delays in a model then can be considered with respect to this settling time, and they often can be eliminated from the model when they are small compared to the desired settling time t_s.

As shown in Figure 4.1, the response associated with a time constant τ in the model of a stable closed-loop production system is approximately 98% complete at time $t = 4\tau$. Therefore, a factor with time constant τ in the numerator or denominator polynomial of a continuous-time transfer function in the form of Equation 4.2 often can be eliminated when $0 < \tau \ll t_s/4$ by assuming

$$\tau s + 1 \approx 1 \tag{6.29}$$

Similarly, a factor in the numerator or denominator of a discrete-time transfer function in the form of Equation 5.35 often can be eliminated when $0 < \tau \ll t_s/4$ by assuming

$$\frac{z - e^{-T/\tau}}{1 - e^{-T/\tau}} \approx 1 \qquad 6.30)$$

A factor with damping ratio $0 < \zeta < 1$ and natural frequency ω_n in the numerator or denominator of a continuous-time transfer function has an associated time constant of exponential decay $1/\zeta\omega_n$ and often can be eliminated when $\zeta\omega_n \gg 4/t_s$ by assuming[2]

$$\frac{1}{\omega_n^2} s^2 + \frac{2\zeta}{\omega_n} s + 1 \approx 1 \tag{6.31}$$

Similarly, a factor in the numerator or denominator of a discrete-time transfer function that has time constant of exponential decay $1/\zeta\omega_n$ often can be eliminated when $\zeta\omega_n \gg 4/t_s$ by assuming

---

2 If damping ratio $\zeta \ll 1$ in a factor of a transfer function, potentially large amplitudes of oscillation in responses may preclude elimination of that factor from a model even if $\zeta\omega_n \gg 4/t_s$.

$$\frac{z^2 - 2e^{-\zeta\omega_n T}\cos\left(\omega_n\sqrt{1-\zeta^2}T\right)z + e^{-2\zeta\omega_n T}}{1 - 2e^{-\zeta\omega_n T}\cos\left(\omega_n\sqrt{1-\zeta^2}T\right) + e^{-2\zeta\omega_n T}} \approx 1$$

Delay D in a continuous-time transfer function in the form of Equation 5.4 often can be eliminated when $D \ll t_s$ by assuming

$$e^{-Ds} \approx 1 \tag{6.32}$$

Similarly, a delay dT in a discrete-time transfer function in the form of Equation 5.35 often can be eliminated when $dT \ll t_s$ by assuming

$$z^{-d} \approx 1 \tag{6.33}$$

After decision-making has been designed using a simplified model, the resulting dynamic behavior should be evaluated using the original model and the design should be modified if necessary to meet requirements.

Example 6.8 Simplified Design of Closed-loop Production System with WIP and Backlog Regulation

It is desired to find values for decision-making parameters in the closed-loop production system illustrated in Figure 1.3 in which planning and control includes backlog and WIP regulation. Decisions are made at the planning level regarding production capacity, which is adjusted with the goal of eliminating backlog. Here, backlog is the difference between the amount of work that has been planned to be completed and the amount of work that actually has been completed by the production work system. At the scheduling level, decisions are made regarding release of orders to the production work system, with the goal of eliminating any difference between actual work in progress and planned work in progress. Hence, planning and control of this production work system include independent backlog regulation and WIP regulation decision-making, each with different goals. However, both backlog and WIP regulation decisions affect the release of orders to production. Furthermore, both capacity and work disturbances can occur in the production work system. These disturbances are not directly detected at either the planning level or the scheduling level; rather, the effects they have on backlog and WIP are responded to by backlog and WIP regulation, respectively, and these disturbances therefore can result in compensating adjustments in both planed capacity and order release rate.

The block diagram in Figure 6.12 can be used to better understand the fundamental behavior of this closed-loop production system. Proportional control parameter K_b for backlog decision-making needs to be chosen to achieve

- settling time $t_s \leq 40$ days
- step response that has little or no overshoot.

The period between adjustments in the production capacity plan is $T = 5$ days and there is a delay $dT = 5$ days in implementing capacity plans.

WIP regulation already has been designed to have damping ratio $\zeta = 0.85$, natural frequency $\omega_n = 1.18$ radians/day, and time constant of exponential decay $\tau_w = 1/(\zeta\omega_n) = 1$ day.

In this case, WIP regulation time constant τ_w is likely to be insignificant when desired settling time $t_s \gg 4\tau_w$ days. Therefore, it is convenient to apply Equation 6.31 and ignore the 2nd-order transfer function that describes WIP regulation. The resulting simplified model allows the direct design approach to be used to choose a value for proportional control parameter K_b. Delay dT is more significant and remains in the model.

The following variables and parameters are used in the model in Figure 6.12:

T	period between adjustments in production capacity plan (days)
$w_r(t)$	orders received (orders)
$w_i(t)$	orders released (orders)
$w_b(t)$	backlog (orders)
$r_f(kT)$	forecasted order input rate (orders/day)
K_b	backlog regulation parameter (days^{-1})
d	non-negative integer (dT is delay in implementing planned capacity adjustments)
$r_p(kT)$	production capacity plan (orders/day)
$r_d(t)$	capacity disturbances (orders/day)
$r_a(t)$	actual production capacity (orders/day)
$w_d(t)$	work disturbances (orders)
$w_p(t)$	WIP plan (orders)

The backlog is

$$w_b(t) = w_r(t) - w_i(t)$$

Backlog is observed and the production capacity plan is adjusted with period T days. Proportional control is used to make capacity plan adjustment decisions with the goal

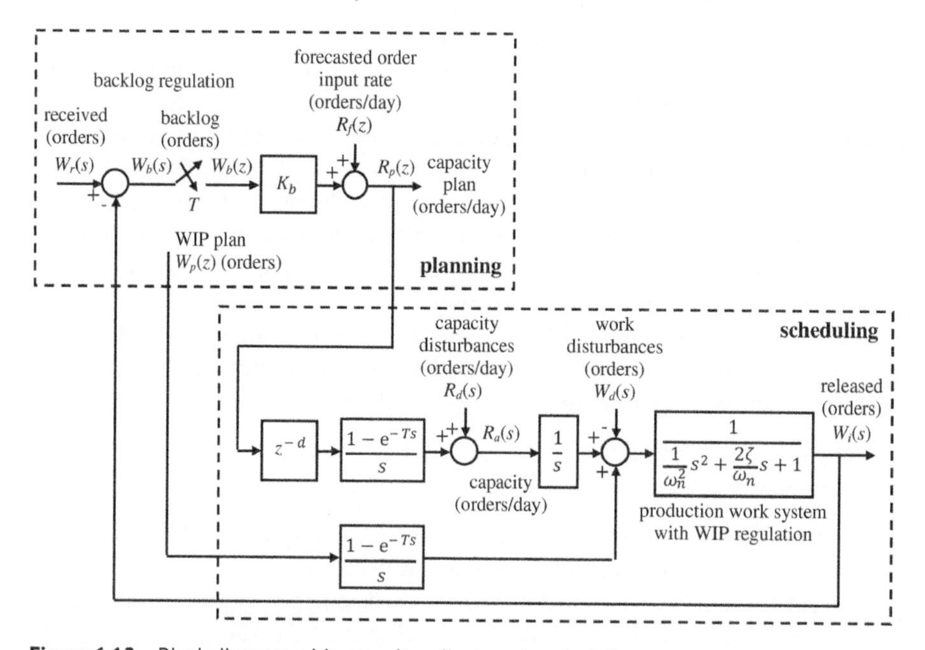

Figure 6.12 Block diagram with capacity adjustment period T.

of eliminating backlog. The relationship between the capacity plan, forecasted rate of order input, and backlog is

$$r_p(kT) = r_f(kT) + K_b w_b(kT)$$

where the value of proportional control parameter K_b needs to be chosen.

The transfer function that relates capacity plan to orders released is

$$\frac{W_i(z)}{R_p(z)} = \mathcal{Z}\left\{\left(\frac{1-e^{-Ts}}{s}\right)\frac{1}{s}\left(\frac{1}{\omega_n^2}s^2 + \frac{2\zeta}{\omega_n}s + 1\right)\right\}z^{-d}$$

and the following simplified transfer function can be used in design of decision-making:

$$\frac{W_i(z)}{R_p(z)} \approx \mathcal{Z}\left\{\left(\frac{1-e^{-Ts}}{s}\right)\frac{1}{s}\right\}z^{-d} = \frac{Tz^{-1}}{1-z^{-1}}z^{-d}$$

The simplified closed-loop characteristic equation for $d = 1$ is

$$z^2 - z + K_b T = 0$$

Choosing $K_b = 0.25/T$ results in two equal real roots $r = 0.5$, which can be expected to result in relatively rapid step response that has little or no overshoot. The corresponding time constant is

$$\tau_b = \frac{-T}{\ln(0.5)}$$

which is $\tau_b = 7.21$ days when $T = 5$ days. As shown in Figure 4.16, the closed-loop settling time with two equal roots can be predicted to be somewhat longer than $4\tau_b = 29$ days.

Program 6.5 can be used to calculate the settling time for a negative unit step disturbance in production capacity that occurs at time $t = 0$. While the simplified WIP

Program 6.5 Calculation of proportional control parameter for backlog regulation using simplified model

```
T=5;  % period between adjustments in capacity plan (days)
d=1;  % delay dT days in implementing adjustments in capacity plan

zeta=0.85;  % WIP regulation damping ratio
tauW=1;  % WIP regulation time constant of exponential decay (days)
Wn=1/zeta/tauW;  % WIP regulation natural frequency (radians/day)

% transfer function
Gz=c2d(tf(1, [1/Wn/Wn, 2*zeta/Wn, 1, 0],'TimeUnit','days'),T);

Kb=0.25/T  % backlog regulation parameter for two equal real roots
```

```
Kb = 0.0500
```

```
Gclz=-feedback(Gz,Kb*tf(1,[1, 0],T,'TimeUnit','days')); % backlog

step(-Gclz,100)  % response to negative capacity disturbance Figure 6.13
xlabel('time kT'); ylabel(' backlog w_b(kT) (orders)')

stepinfo(Gclz)
```

```
ans =
        RiseTime: 20
    SettlingTime: 35
     SettlingMin: 18.2342
     SettlingMax: 20.2865
       Overshoot: 1.4324
      Undershoot: 0
            Peak: 20.2865
        PeakTime: 45
```

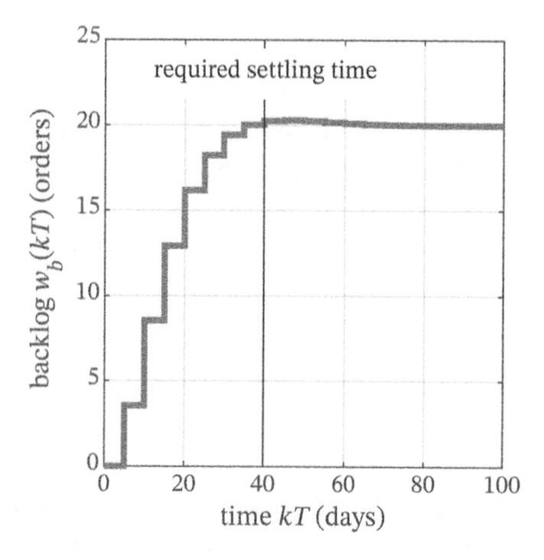

Figure 6.13 Response of backlog to unit step capacity decrease disturbance with decision-making designed using a simplified model.

regulation transfer function was used to choose the value for K_b, the actual WIP regulation transfer function is used to calculate the step response. For $T = 5$ days, $dT = 5$ days, and $K_b = 0.05$ days^{-1} the settling time is 35 days and the overshoot is 1.4%; these satisfy the design requirements. The step response is shown in Figure 6.13. The simplified transfer function allowed a convenient, direct design approach to be used to find a value for proportional control parameter K_b.

6.5 Design Using Frequency Response

Design of decision-making using time response as described in Section 6.3 requires searching the parameter space for suitable values, the difficulty of which increases with the number of decision-making parameters and the complexity of the closed-loop production system topology. Direct design of decision-making as described in Section 6.4 requires derivation of relationships between desired closed-loop dynamic characteristics and decision-making parameters for which values need to be chosen. Then these relationships are used to obtain the values of the parameters that result in specified closed-loop dynamic characteristics; this tends to increase in difficulty and become intractable as the number of decision-making parameters increases, as the order of the production system model increases, and when there is delay in the closed-loop production system.

As an alternative, the open-loop frequency response can be used to design decision-making. Calculation of closed-loop frequency response or time response is not required for making primary design decisions, which facilitates design because values of decision-making parameters and choice of control actions are not yet known. Design guidelines can be used to modify the open-loop frequency response by adding appropriate control components, anticipating that the result will be a closed-loop production system with dynamic characteristics that meet requirements such as settling time, damping, and steady-state error.

The design guidelines that are presented in the following sections can be used for design of both continuous-time and discrete-time decision-making. However, they are developed and justified for a benchmark continuous-time production system model where the components in Figure 6.2 have the open-loop transfer function

$$G_{ol}(s)=G_c(s)G_p(s)H_p(s)H_c(s)=\frac{K}{s(\tau s+1)} \tag{6.34}$$

The magnitude and phase of the open-loop frequency response of the benchmark model for radian frequency ω are

$$M_{ol}(\omega)=|G_c(j\omega)||G_p(j\omega)||H_p(j\omega)||H_c(j\omega)|=\frac{K}{\omega\sqrt{1+(\omega\tau)^2}} \tag{6.35}$$

$$\phi_{ol}(\omega)=\angle G_c(j\omega)+\angle G_p(j\omega)+\angle H_p(j\omega)+\angle H_c(j\omega)=-\frac{\pi}{2}-\tan^{-1}(\omega\tau) \tag{6.36}$$

where the phase is in radians.

A continuous-time or discrete-time production system with open-loop frequency response that resembles Equations 6.35 and 6.36 can be obtained by adding appropriate decision-making components G_c and H_c to given manipulated components G_p and H_p in Figure 6.2. There can be significant differences between the actual open-loop frequency response and Equations 6.35 and 6.36 because the open-loop production system transfer function can be discrete-time, can include delay, can have terms in the numerator, and can be higher order. Nevertheless, resulting closed-loop dynamic behavior tends to resemble what is expected from the following open-loop design guidelines, and the values of decision-making parameters obtained using the design guidelines subsequently can be adjusted to compensate for approximations implicit in applying the guidelines.

Design Guideline 1 High Open-Loop Frequency Response Magnitude at Relatively Low Frequencies

The open-loop magnitude should be $M_{ol}(\omega) >> 1$ at relatively low frequencies ω, resulting in closed-loop magnitude $M_{cl}(\omega) \approx 1$ at those frequencies. This is desirable because closed-loop production systems often are expected to respond effectively to relatively slowly changing inputs. As an example, for the benchmark production system model in Equation 6.34, the closed-loop transfer function is

$$G_{cl}(s) = \frac{B(s)}{R(s)} = \frac{K}{\tau s^2 + s + K} \tag{6.37}$$

In this case, the open-loop frequency response magnitude at frequency $\omega = 0$ is $M_{ol}(0) = \infty$ and the closed-loop frequency response is magnitude $M_{cl}(0) = 1$.

If the open-loop transfer function G_pH_p does not result in $M_{ol}(\omega) >> 1$ at relatively low frequencies, then decision-making transfer functions G_c and H_c can be modified to make this the case. Often, this is accomplished by adding integral control to decision-making as described in Sections 6.1.3 and 6.2.3.

Design Guideline 2 Low Open-Loop Frequency Response Magnitude at Relatively High Frequencies

The open-loop magnitude should be $M_{ol}(\omega) << 1$ at relatively high frequencies ω, resulting in closed-loop magnitude $M_{cl}(\omega) \approx 0$ at those frequencies. This is desirable because closed-loop production systems often are not physically capable of being manipulated at relatively high frequencies or are not expected to respond to manipulations at relatively high frequencies. Furthermore, as will be noted in Design Guidelines 3, 4 and 5, it is desirable for the open-loop magnitude $M_{ol}(\omega)$ to decrease with increasing frequency ω. As an example, for the benchmark production system model in Equation 6.34, the open-loop frequency response magnitude at frequency $\omega = \infty$ is $M_{ol}(\infty) = 0$ and the closed-loop frequency response magnitude is $M_{cl}(\infty) = 0$.

If open-loop transfer function G_pH_p does not result in $M_{ol}(\omega) << 1$ at relatively high frequencies, then decision-making transfer functions G_c and H_c can be modified to achieve this design goal. As with Design Guideline 1, this design goal often is accomplished by adding integral action to decision-making as described in Sections 6.1.3 and 6.2.3.

Design Guideline 3 Relationship Between Open-Loop Phase Margin and Closed-Loop Damping Ratio

The open-loop phase margin can be used to predict the closed-loop damping ratio. This is of significant utility because closed-loop damping ratio ζ_{cl} often is included as a design requirement; for example, $\zeta_{cl} \geq 0.707$ can be specified to avoid magnification in closed-loop frequency response. The corresponding open-loop phase margin can be identified, and decision-making components can be added to modify the open-loop frequency response to achieve the desired phase margin.

As described in Section 5.3.2, the phase margin is measured at frequency ω_{cp} where $M_{ol}(\omega_{cp}) = 1$. From Equation 6.35,

$$\omega_{cp} = \frac{1}{\tau}\sqrt{\frac{\sqrt{1+4(K\tau)^2}-1}{2}} \tag{6.38}$$

and from Equation 6.36, the phase margin in radians, as defined in Section 5.3.2, is

$$P_m = 180^\circ + \phi_{ol}(\omega_{cp}) = \left(\frac{\pi}{2} - \tan^{-1}\left(\sqrt{\frac{\sqrt{1+4(K\tau)^2}-1}{2}}\right)\right)\frac{180^\circ}{\pi} \tag{6.39}$$

Equating like terms in Equations 4.14 and 6.37, the closed-loop damping ratio is

$$\zeta_{cl} = \frac{1}{2\sqrt{K\tau}} \tag{6.40}$$

Substituting for $K\tau$ in Equation 6.39 yields the following relationship between open-loop phase margin PM in degrees and closed-loop damping ratio ζ_{cl} for the benchmark open-loop transfer function in Equation 6.34:

$$P_m = \left(\frac{\pi}{2} - \tan^{-1}\left(\sqrt{\frac{\sqrt{\left(1+\frac{1}{4\zeta_{cl}^4}\right)}-1}{2}}\right)\right)\frac{180^\circ}{\pi} \tag{6.41}$$

Closed-loop damping ratio ζ_{cl} is plotted versus open-loop phase margin P_m in degrees for this benchmark open-loop transfer function in Figure 6.14. The approximation

$$\zeta_{cl} \approx \frac{P_m}{100^\circ} \tag{6.42}$$

also is plotted along with phase margins 65.5° and 76.3° associated with $\zeta_{cl} = 0.707$ and $\zeta_{cl} = 1$, respectively. The approximation in Equation 6.42 often is called the "divide by 100° rule" and is most useful for the range $0^\circ \le P_m \le 70^\circ$ and $0 \le \zeta_{cl} \le 0.7$.

Equation 6.41 or 6.42 can be used to choose a desired open-loop phase margin based on a desired closed-loop damping ratio when the open-loop transfer function is approximately Equation 6.34. If this is not the case, decision-making transfer functions G_c and H_c can be modified to achieve the desired open-loop phase margin. This design goal often is accomplished by including lead control as described in Sections 6.1 and 6.2. Using this guideline often yields useful but approximate results because open-loop transfer functions can differ significantly from Equation 6.34. Therefore, decision-making parameters may need to be adjusted, and decision-making components G_c and H_c may need to be further modified to achieve a desired closed-loop damping ratio.

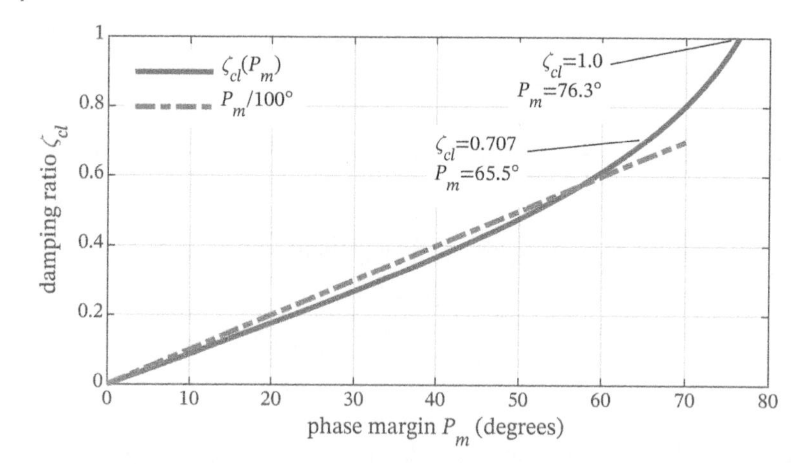

Figure 6.14 Benchmark relationship between closed-loop damping ratio and phase margin.

Design Guideline 4 Open-Loop Frequency Response Magnitude in the Vicinity of Frequency ω_{cp}

In the vicinity of unity magnitude crossing frequency ω_{cp} where $M_{ol}(\omega_{cp}) = 1$ and the phase margin is measured, it is desirable for the magnitude of the open-loop frequency response $M_{ol}(\omega)$ to decrease by a factor of 10 for a factor of 10 increase in frequency ω (this is a slope of −20 dB/decade on a logarithmic magnitude plot). This characteristic is consistent with Design Guidelines 1 and 2 and facilitates application of Design Guideline 3; phase margin tends to be less favorable when open-loop magnitude decreases more rapidly with frequency, and phase margin tends to be more difficult to adjust when open-loop magnitude decreases less rapidly with frequency. If open-loop transfer function G_pH_p does not have this characteristic, then decision-making transfer functions G_c and H_c can be modified to achieve this design goal. This often is accomplished by adding proportional, integral, or lead control decision-making.

Design Guideline 5 Relationship Between Closed-Loop Bandwidth, Open-Loop Phase Margin, and Frequency ω_{cp}

For closed-loop production systems with open-loop frequency response that either has the characteristics described in Design Guidelines 1, 2, and 4 or approximately those characteristics, the unity magnitude crossing frequency ω_{cp} often can be used to predict the closed-loop bandwidth ω_b, which tends to be $\omega_b \geq \omega_{cp}$. This is of significant utility because closed-loop bandwidth often is a design requirement.

As an example, for $\tau = 0$ and $K > 0$ in Equation 6.34, the open-loop and closed-loop transfer functions become

$$G_{ol}(s) = \frac{B(s)}{E(s)} = \frac{K}{s} \tag{6.43}$$

$$G_{cl}(s) = \frac{B(s)}{R(s)} = \frac{K}{s+K} \tag{6.44}$$

The open-loop magnitude is $M_{ol}(\omega_{cp}) = 1$ at frequency $\omega_b = \omega_{cp} = K$ where the closed-loop bandwidth is measured.

A straightforward initial design assumption of $\omega_b \approx \omega_{cp}$ often is acceptable because the general trend for the benchmark model in Equations 6.34 and 6.37 as well as other models is $\omega_b \geq \omega_{cp}$. Assuming $\omega_b \approx \omega_{cp}$ also is convenient because ω_{cp} is used in Design Guideline 3 to determine the phase margin and predict close-loop damping ratio.

Design Guideline 5 is most useful when the open-loop frequency response has the characteristics established using Design Guidelines 1, 2, and 4. Then, decision-making transfer functions G_c and H_c can be modified so that unity magnitude (0 dB) crossover frequency ω_{cp} becomes the specified bandwidth. This can be combined with application of Design Guideline 3 by modifying the phase so that the phase measured at frequency ω_{cp} corresponds to the desired phase margin. Using this guideline often produces useful but approximate initial results. Therefore, decision-making parameters may need to be adjusted, and decision-making components G_c and H_c may need to be further modified to achieve the desired closed-loop bandwidth.

6.5.1 Using the Frequency Response Guidelines to Design Decision-Making

Most closed-loop production systems have open-loop transfer functions that deviate from the benchmark model in Equation 6.34 that was used to develop Design Guidelines 1 through 5. For example, the presence of delay can significantly affect the phase of the frequency response and significantly decrease phase margins. Similarly, phase margins are reduced in discrete-time models. Fortunately, delays and discrete-time behavior are readily included when designing using frequency response. The design guidelines often lead to useful results that are a valuable starting point for design of decision-making for closed-loop production systems.

Design of the dynamic behavior of closed-loop production systems often includes requirements such as desired closed-loop bandwidth, desired damping ratio, desired maximum overshoot, and desired settling time. Aided by the design guidelines, these requirements often can be achieved by

- adding integration to increase the frequency response magnitude at low frequencies and decrease the frequency response magnitude at high frequencies; however, production systems transfer functions often already include integration and additional integration tends to significantly decrease phase margins and is avoided unless necessary
- determining the required phase margin using specifications such as desired closed-loop damping ratio
- adding phase to the open-loop frequency response if bandwidth and phase margin or damping ratio requirements cannot both be met

- adjusting the magnitude of the open-loop frequency response to adjust phase margin and predicted bandwidth
- adjusting parameters after decision-making has been designed using the design guidelines to make the closed-loop time and frequency response better meet requirements.

The following examples illustrate application of the guidelines to design of decision-making for various production systems using frequency response.

Example 6.9 Frequency Response Design of Continuous-Time Control of Production Using Metal Forming

In Example 6.1 the simple continuous-time model

$$r_m(t) = K_m m(t)$$

$$r_o(t) = r_m(t) - r_d(t)$$

where $m(t)$ is the number of presses, $r_d(t)$ and $r_o(t)$ parts/hour are the rates of production that are rejected and accepted, respectively, based on quality requirements, and production parameter K_m (parts/hour)/press is used to approximately represent production using metal forming. The open-loop block diagram in Figure 6.15 corresponds to Figure 6.2b, and the open-loop transfer function with added decision-making components is

$$G_{ol}(s) = \frac{R_o(s)}{E(s)} = G_c(s) K_m H_c(s)$$

As in Example 6.1, the desired rate of production $r_r(t)$ is assumed to be constant, but disturbances $r_d(t)$ are frequent. The following closed-loop dynamic behavior is desired:

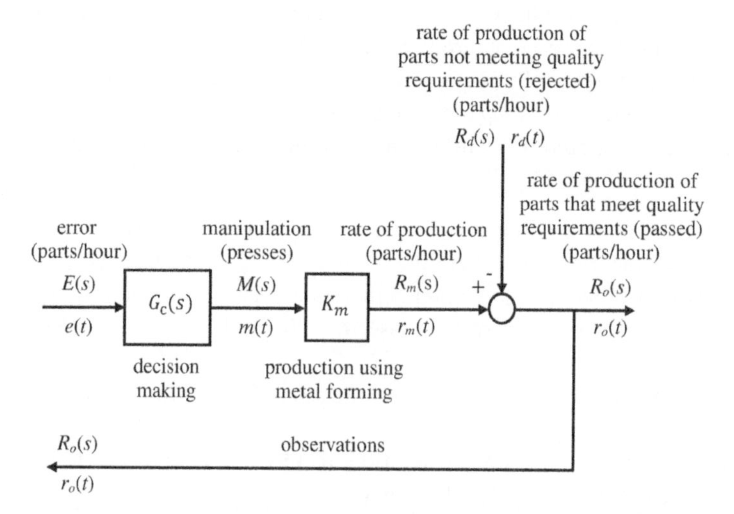

Figure 6.15 Open-loop block diagram for production using metal forming.

- little or no oscillation in response to step disturbances $r_d(t)$
- closed-loop bandwidth $f_b \geq 0.125$ cycles/hour (period $\delta_b \leq 8$ hours/cycle, frequency $\omega_b \geq \pi/4$ radians/hour), measured using frequency response for closed-loop transfer function $R_o(s)/R_r(s)$
- zero steady-state error for step disturbances $r_d(t)$.

Using Design Guideline 3, damping ratio $\zeta_{cl} = 1$ is predicted for phase margin $P_m = 76.3°$. Therefore, little or no oscillation in response to step disturbances is predicted for $P_m \geq 76.3°$. Also, using Design Guideline 5, bandwidth $f_b \geq 0.125$ cycles/hour is predicted when magnitude $M_{ol}(f) = 1$ at unity magnitude frequency crossing $f = 0.125$ cycles/hour.

The steps in design of the decision-making components for this example are

1. add integral control with parameter K_i (presses/hour)/(parts/hour) to obtain zero steady-steady error for step disturbances $r_d(t)$ and the frequency response magnitude characteristics described in Design Guidelines 1, 2 and 4
2. calculate the magnitude the open-loop frequency response at frequency $f = 0.125$ cycles/hour with integral control parameter $K_i = 1$ (presses/hour)/(parts/hour)
3. find the value of K_i that makes the magnitude of the open-loop frequency response $M_{ol}(f) = 1$
4. ensure that phase margin $P_m \geq 76.3°$, predicting little or no oscillation in response to step disturbance using Design Guideline 3
5. plot the closed-loop frequency response for this value of K_i; the closed-loop bandwidth should be in the vicinity of 0.125 cycles/hour ($\pi/4$ radians/hour) as predicted by Design Guideline 5
6. verify that the results meet requirements and adjust K_i if necessary.

Program 6.6 can be used to calculate the open-loop frequency response shown in Figure 6.16 for $K_m = 200$ (parts/hour)/press. When integral control $G_c(s) = K_i/s$ is

Program 6.6 Calculation of continuous-time integral control parameter for production using metal forming

```
Km=200;  % forming process constant ((parts/hour)/press)
rd=300;  % production rate rejected due to quality (parts/hour)

Gms=tf(Km,'TimeUnit','hours');  % transfer function

Gcs=tf(1,[1 0],'TimeUnit','hours');  % integral control with Ki=1

[M]=bode(Gcs*Gms,0.125*2*pi)  % magnitude at frequency 0.125 cycles/hour

    M = 254.6479

% Ki for magnitude=1 at 0.125 cycles/hour ((presses/hour)/(parts/hour))
Ki=1/M

    Ki = 0.0039

Gcs1=tf(Ki,[1 0],'TimeUnit','hours');  % modified integral control

W=logspace(-3, 1, 1000)*2*pi;  % frequency range
```

```
options=bodeoptions;
options.FreqUnits='cycles/hour'; options.MagScale='log';
options.MagUnits='abs';

bode(Gms,Gcs*Gms,Gcs1*Gms,W,options);  % open-loop - Figure 6.16
xlabel('frequency f')
legend('without integral control','K_i=1 (presses/hour)/(parts/hour)',...
    'K_i=0.00393 (presses/hour)/(parts/hour)')

bode(feedback(1,Gcs*Gms),W,options);  % closed-loop error - Figure 6.17
xlabel('frequency f')

bandwidth(feedback(Gcs1*Gms,1))/2/pi  % bandwidth (cycles/hour)
```

```
ans = 0.1247
```

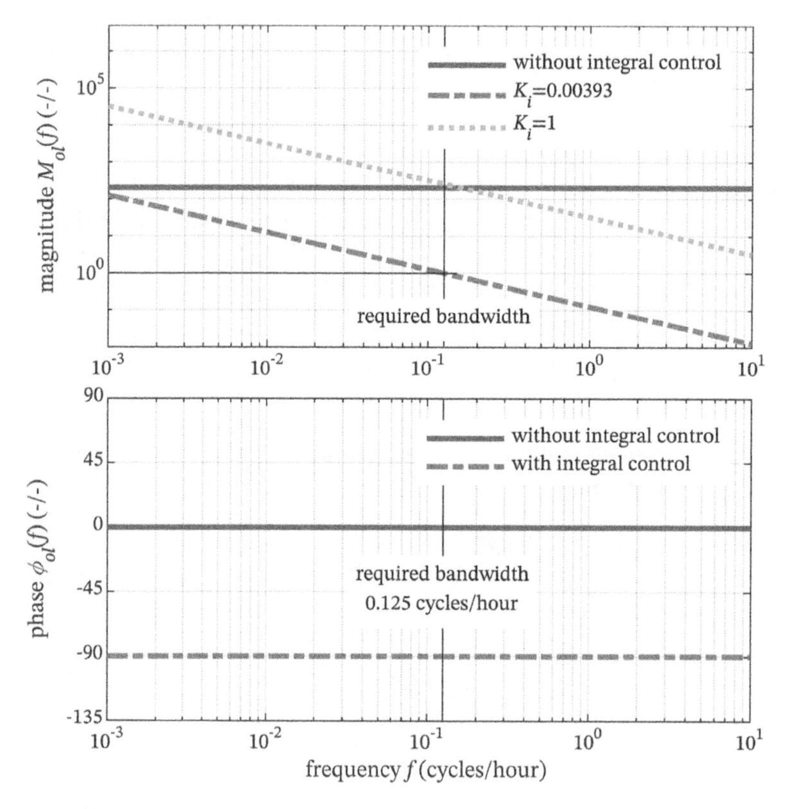

Figure 6.16 Open-loop frequency response with and without integral control with parameter K_i (presses/hour)/(parts/hour).

added, the open-loop frequency response with $K_i = 1$ (presses/hour)/(parts/hour) also is shown in Figure 6.16. The phase margin is always 90° regardless of K_i; this is predicted by Design Guideline 3 to result in response that is not oscillatory.

When $K_i = 0.00393$ (presses/hour)/(parts/hour), the magnitude of the modified open-loop frequency response has magnitude $M_{ol}(f) = 0$ at frequency 0.125 cycles/hour, which is predicted by Design Guideline 5 to satisfy the closed-loop bandwidth requirement $f_b \geq 0.125$ cycles/hour. The resulting open-loop frequency response also is shown in Figure 6.16.

The closed-loop transfer function for error $e(t)$ parts/hour in response to disturbance $r_d(t)$ parts/hour is

$$\frac{E(s)}{R_d(s)} = \frac{1}{1+\frac{K_i}{s}K_m} = \frac{s}{\frac{1}{K_i K_m}s+1}$$

and the corresponding closed-loop frequency response is shown in Figure 6.17 for $K_i = 0.00393$ (presses/hour)/(parts/hour). At frequency $\omega = 0$, the magnitude is zero, and the requirement for zero steady-state error for step disturbances $r_d(t)$ therefore is satisfied. As calculated using Program 6.6 the closed-loop system is 1st-order and therefore is not oscillatory. The closed-loop bandwidth is $f_b = 0.125$ cycles/hour, which satisfies the bandwidth requirement. Bandwidth was calculated in Program 6.6 using closed-loop transfer function

$$\frac{R_o(s)}{R_i(s)} = \frac{\frac{K_i}{s}K_m}{1+\frac{K_i}{s}K_m} = \frac{1}{\frac{1}{K_i K_m}s+1}$$

The magnitude of the error is relatively small when the rate of production of parts not meeting quality requirements fluctuates at frequencies significantly less than $f_b = 0.125$ cycles/hour (period significantly more than $\delta_b = 8$ hours/cycle).

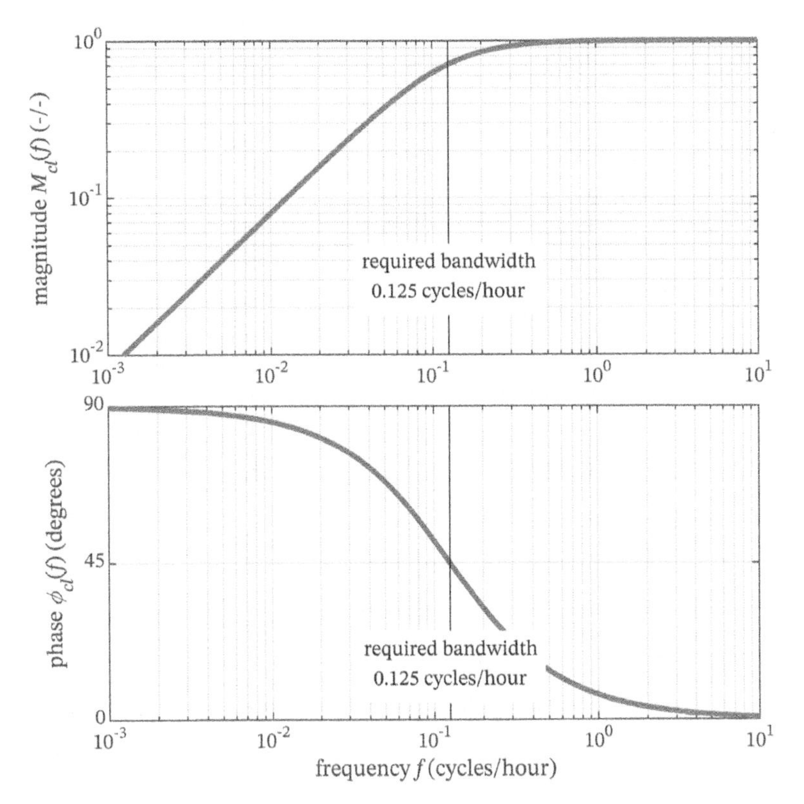

Figure 6.17 Closed-loop error frequency response with $K_i = 0.00393$ (presses/hour)/(parts/hour).

Example 6.10 Frequency Response Design of Proportional Control for Discrete-Time Production System with Delay Using Phase Margin

It is desired to regulate work in progress (WIP) in a production work system such as that illustrated in Figure 2.6. A block diagram for this closed-loop production system with discrete-time WIP regulation is shown in Figure 6.18 in which production capacity is adjusted with period T days with the goal of maintaining a planned amount of WIP. Proportional control decision-making is used to make adjustments in production capacity as a function of planned production capacity $r_p(kT)$ and the error between actual WIP $w_w(kT)$ hours and planned WIP $w_p(kT)$ hours. There is a delay of dT days in implementing capacity adjustment decisions, where d is a positive integer. The actual production capacity $r_a(t)$ is affected by capacity disturbances $r_d(t)$ hours/day such as equipment failures or worker absences that usually are negative. There are WIP disturbances $w_d(t)$ hours that can be positive or negative such as rush orders or order cancelations.

For the period $T = 1$ day between production capacity adjustments and delay $dT = 2$ days, it is desired to find a value of proportional control parameter K_p in decision-making equations

$$W_e(z) = W_w(z) - W_p(z)$$

$$R_c(z) = R_p(z) + K_p W_e(z)$$

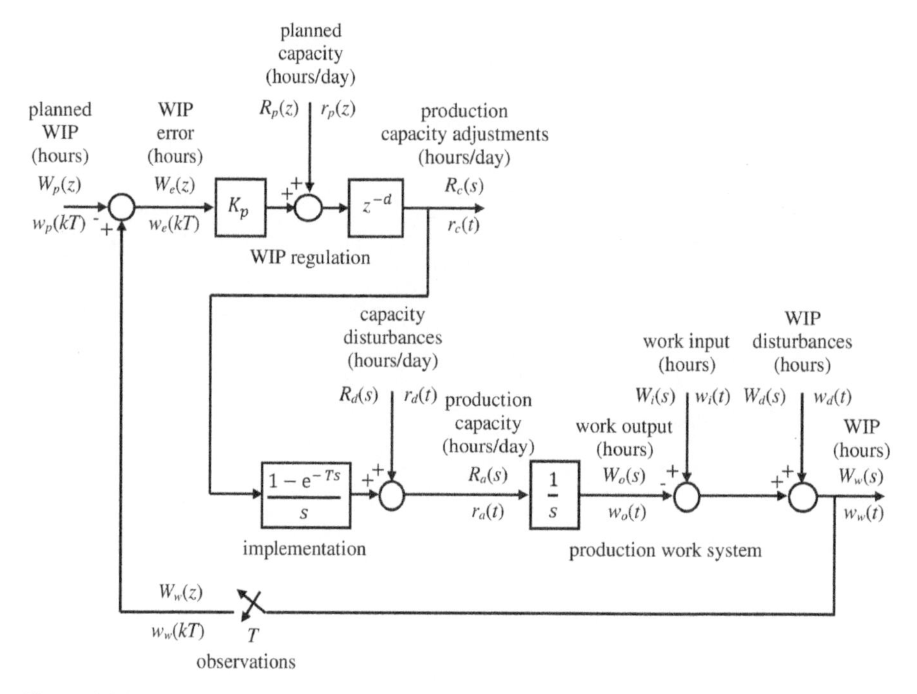

Figure 6.18 Production system with proportional discrete-time control.

that results in roots of the closed-loop characteristic equation with damping ratio 0.707 $\leq \zeta_{cl} \leq 1$. (There will be three roots and three associated damping ratios because the model with delay $dT = 2$ days is 3rd-order.) This is predicted to result in closed-loop dynamic behavior that has

- little or no magnification in its frequency response
- relatively high bandwidth
- relatively short settling times
- some overshoot in response to step changes in production system inputs.

The steps in design of the decision-making for this example are

1. Calculate the open-loop frequency response with proportional control parameter $K_p = 1$ days^{-1}
2. Determine the value of K_p that increases or decreases open-loop magnitude to obtain a phase margin 65.5°, which from Design Guideline 3 is predicted to result in a closed damping ratio ζ_{cl} in the vicinity of 0.707
3. Calculate the damping ratios of the closed-loop production system with this value of K_p
4. If necessary, modify the value of K_p to achieve $\zeta_{cl} \geq 0.707$
5. Also assess the settling time of the closed-loop step response and overshoot and modify the value of K_p if necessary.

The open-loop transfer function with $dT = 2$ days is

$$\frac{W_o(z)}{W_e(z)} = \frac{K_p T z^{-3}}{1 - z^{-1}}$$

and the open-loop frequency response with proportional control parameter $K_p = 1$ days^{-1} is shown in Figure 6.19, which can be calculated using Program 6.7. The magnitude margin is 0.618 and the phase margin is –60°. The closed-loop production systems would be unstable with this value of the proportional control parameter K_p, and the open-loop frequency response magnitude therefore needs to be reduced to obtain the desired phase margin and a stable closed-loop production system; this also will improve the magnitude margin.

This open-loop transfer function significantly differs from the open-loop transfer function in Equation 6.34 for which Design Guideline 3 was developed: there is no time constant, there is a delay dT days in implementing decisions, and the phase of the open-loop frequency response decreases with increasing frequency due to the discrete-time nature of decision-making. However, an approximate value of K_p can be obtained using Design Guideline 3, which predicts closed-loop damping ratio $\zeta_{cl} \approx 0.707$ when the phase margin is 65.5° as shown in Figure 6.14.

Program 6.7 can be used to calculate the value of K_p that results in a phase margin of 65.5°. The phase margin would be 65.5° if the unity magnitude crossing frequency was 0.0273 cycles/day. With $K_p = 1$ days^{-1}, the magnitude at this frequency is 5.85. If value of the proportional control parameter is decreased to $K_p = 1/5.85 = 0.171$ days^{-1}, the result will be magnitude $M_{ol}(f) = 1$ at this frequency.

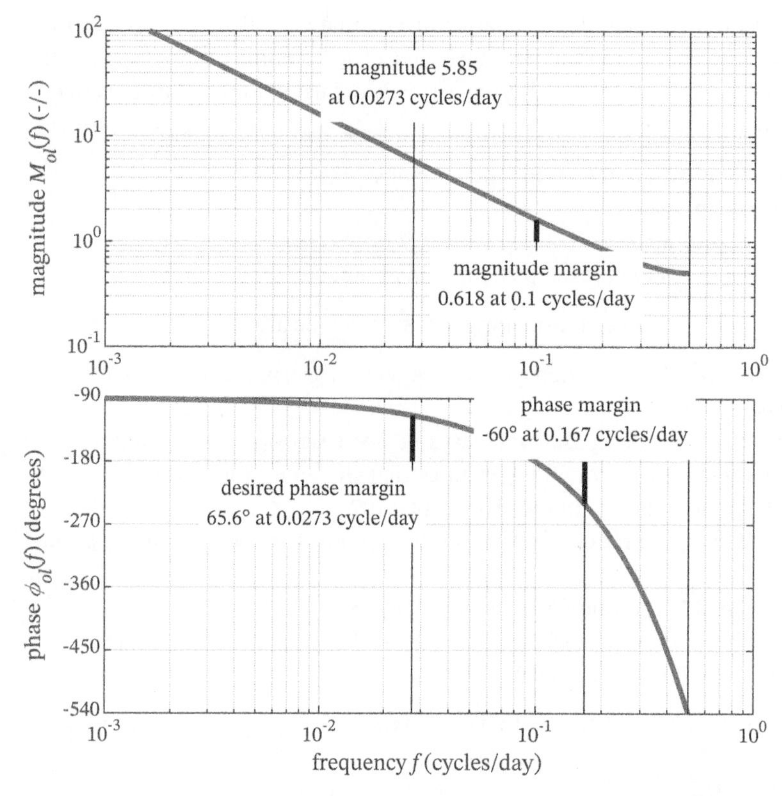

Figure 6.19 Open-loop frequency response with $T = 1$ day, delay $dT = 2$ days, and $K_p = 1$ days^{-1}.

Program 6.7 Calculate proportional control parameter using phase margin for discrete-time production system with delay

```
T=1;  % period (days)
d=2;  % delay dT=2 days in implementing capacity adjustments

% production work system transfer function
Golz=tf(T,[1 -1],T,'OutputDelay',d,'TimeUnit','days');

options=bodeoptions;
options.FreqUnits='cycles/day'; options.MagScale='log';
options.MagUnits='abs';

Kp=1; % initial value of Kp
margin(Kp*Golz,options)  % margins with Kp=1 - Figure 6.19

[Mol,Pol,Wol]=bode(Kp*Golz,logspace(-2,0,2000));
II=find(Pol<-180+65.5);  % find phase margin 65.5° (phase -114.5°)
M=Mol(II(1)),W=Wol(II(1))/2/pi  % find magnitude and frequency for 65.5°
```

```
M = 5.8467
W = 0.0273
```

```
Kp=1/M  % make phase margin 65.5°
```

```
    Kp = 0.1710
```

```
bode(Kp*Golz,options)  % modified frequency response - Figure 6.20
hold on; margin(Kp*Golz); hold off  % show gain and phase margins
Gclz=feedback(Kp*Golz,1); %  closed-loop transfer function

bode(Gclz,options)  % closed-loop frequency response - Figure 6.21
damp(Gclz)  % closed-loop damping ratios
```

Pole	Magnitude	Damping	Time Constant (days)
6.78e-01 + 1.49e-01i	6.94e-01	8.60e-01	2.74e+00
6.78e-01 - 1.49e-01i	6.94e-01	8.60e-01	2.74e+00
-3.55e-01	3.55e-01	3.13e-01	9.66e-01

```
stepinfo(Gclz)
```

```
    ans = struct with fields:
            RiseTime: 6
        SettlingTime: 12
         SettlingMin: 0.9097
         SettlingMax: 1.0052
           Overshoot: 0.5165
          Undershoot: 0
                Peak: 1.0052
            PeakTime: 16
```

The open-loop frequency response with $K_p = 0.171$ days^{-1}, shown in Figure 6.20, has the desired 65.5° phase margin. The corresponding closed-loop frequency response is shown in Figure 6.21. There is little or no magnification in the frequency response. The closed-loop damping ratio associated with the pair of complex conjugate roots calculated using Program 6.7 is $\zeta_{cl} = 0.860$; this satisfies the damping ratio requirement $0.707 \le \zeta_{cl} \le 1$. Settling time $t_s = 12$ days and 0.5 percent overshoot also are calculated in Program 6.7. These satisfy the requirements.

Example 6.11 Frequency Response Design of Discrete-Time Integral-Lead Control for Mixture Heating with Delay

The mixture temperature regulation system shown in Figure 3.5 has a delay of $D =$ 12 seconds between changes in heater voltage and changes in measured mixture temperature. In Example 6.6 it was shown how continuous-time proportional plus integral control could be designed directly using the transfer function for mixture heating without delay. In this example, temperature is regulated using a control computer, delay makes direct design difficult, and design of discrete-time integral control using frequency response will be investigated. Discrete-time lead control then will be added because

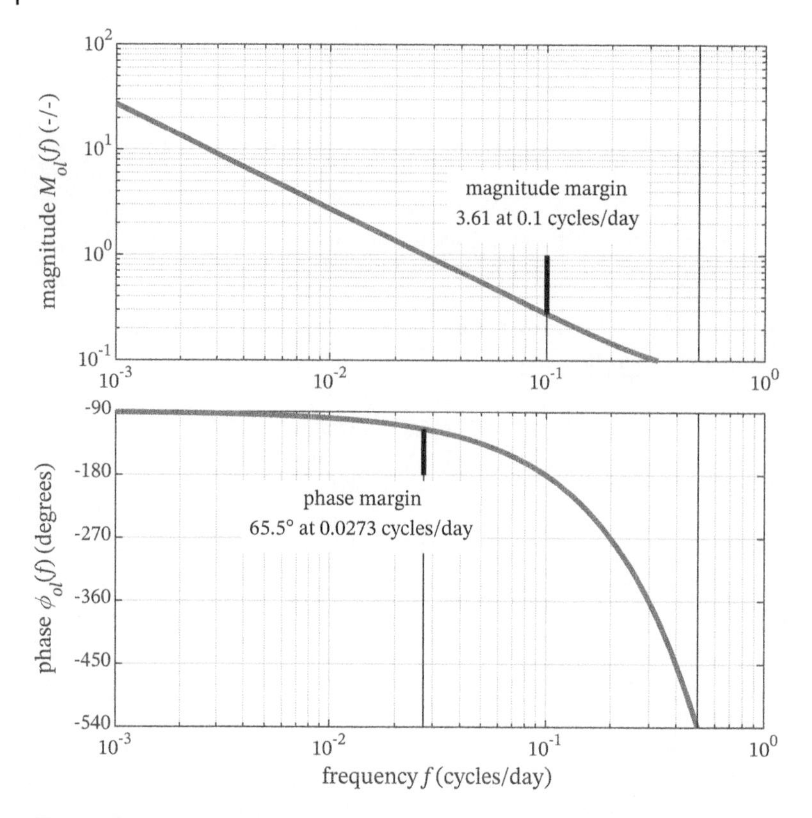

Figure 6.20 Open-loop frequency response with $T = 1$ day, delay $dT = 2$ days, and $K_p = 0.171$ days^{-1}.

integral control alone cannot satisfy dynamic behavior requirements. A block diagram for this system with discrete-time mixture temperature regulation is shown in Figure 6.22.

To ensure that the mixture is delivered to production at the proper temperature, discrete-time decision-making needs to be designed to achieve the following closed-loop dynamic behavior:

- closed-loop bandwidth greater than $f_b = 0.003$ cycles/second (period $\delta_b = 333$ seconds/cycle, $\omega_b = 0.019$ radians/second)
- phase margin PM $\geq 76.3°$ so little or no oscillation is expected in mixture temperature
- no steady-state error between desired and actual mixture temperature when inlet temperature is constant.

The steps in design of the decision-making components for this example using frequency response are

1. choose period T seconds between adjustments in heater voltage
2. calculate the open-loop mixture temperature frequency response
3. add integral control to satisfy Design Guidelines 1, 2 and 4 and obtain a closed-loop production system that has no steady-state error when inputs are constant
4. add lead control to increase phase margin so that bandwidth can be increased

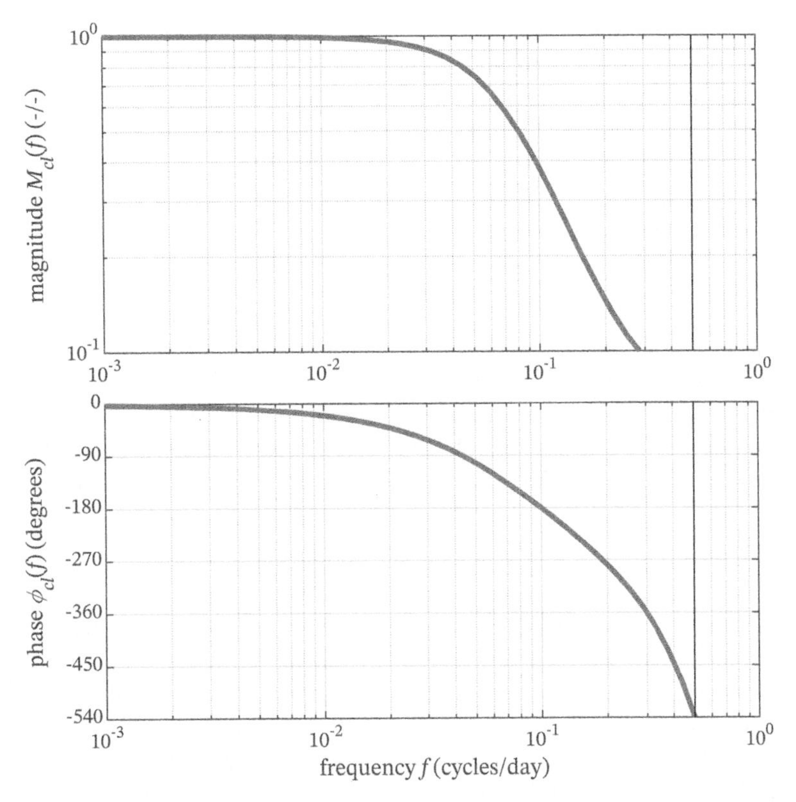

Figure 6.21 Closed-loop frequency response with T = 1 day, delay dT = 2 days, and K_p = 0.171 days^{-1}.

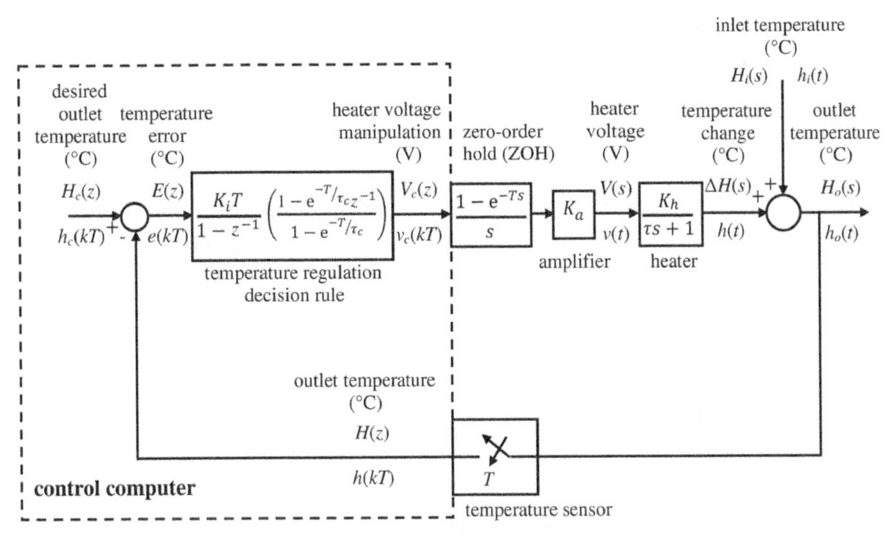

Figure 6.22 Block diagram for mixture heater with delay and discrete-time integral-lead temperature regulation.

5. design the lead-control time constant to make the phase $-180° + 76.3° = -103.7°$ at frequency $f_b = 0.003$ cycles/second, which is the desired closed-loop bandwidth
6. adjust the magnitude of the frequency response by adjusting integral control parameter K_i seconds^{-1} to make the open-loop phase margin 76.3° at frequency $f_{cp} = f_b = 0.003$ cycles/second so that the closed-loop damping ratio predicted by Design Guideline 3 is $\zeta_{cl} \approx 1$ and the closed-loop bandwidth predicted by Design Guideline 5 is $f_b = 0.003$ cycles/second
7. if necessary, modify the integral and lead control components to achieve the required closed-loop bandwidth and damping ratio.

The period between adjustments in heater voltage can be chosen to be $T = 10$ seconds, which is relatively short with respect to period $\delta_b = 333$ seconds/cycle associated with the desired closed-loop bandwidth. The magnitude of the open-loop frequency response of mixture heating is not high at lower frequencies, and integral control therefore is added as a decision-making component to satisfy Design Guideline 1:

$$G_i(z) = \frac{K_i T}{1 - z^{-1}}$$

where K_i (V/second)/°C is the integral control parameter. Integral control will tend to eliminate steady-state error between desired mixture temperature $h_c(kT)$ and actual mixture temperature $h_o(kT)$. The open-loop frequency response of the mixture heating combined with discrete-time integral control shown in Figure 6.23 is calculated in Program 6.8 for mixture heating time constant $\tau_h = 49.8$ seconds, mixture heating parameter $K_h = 1.6$°C/V, and integral control parameter $K_i = 1$ (V/second)/°C.

The phase of the open-loop frequency response at the desired bandwidth $f_b = 0.003$ cycles/second is $-146.2°$ and the potential phase margin therefore is 33.8°, which is significantly less than the required 76.3°. Therefore, the phase of the frequency response needs to be increased in the vicinity of frequency $f_b = 0.003$ cycles per second; this can be accomplished by adding lead control with transfer function:

$$G_{lead}(z) = \frac{1 - e^{-T/\tau_c} z^{-1}}{1 - e^{-T/\tau_c}}$$

The additional phase desired is 42.5°. The desired phase of the lead-control frequency response at frequency $\omega_b = 2\pi f_b = 0.019$ radians/second therefore is $\phi_{lead}(\omega_b) = 0.741$ radians (42.5°). From Equation 5.40, the phase of discrete-time lead control is

$$\phi_{lead}(\omega_b) = \angle\left(1 - e^{-T/\tau_c}\left(\cos(T\omega_b) - j\sin(T\omega_b)\right)\right)$$

$$\phi_{lead}(\omega_b) = \tan^{-1}\left(\frac{e^{-T/\tau_c}\sin(T\omega_b)}{1 - e^{-T/\tau_c}\cos(T\omega_b)}\right)$$

and the lead time consant can be found from

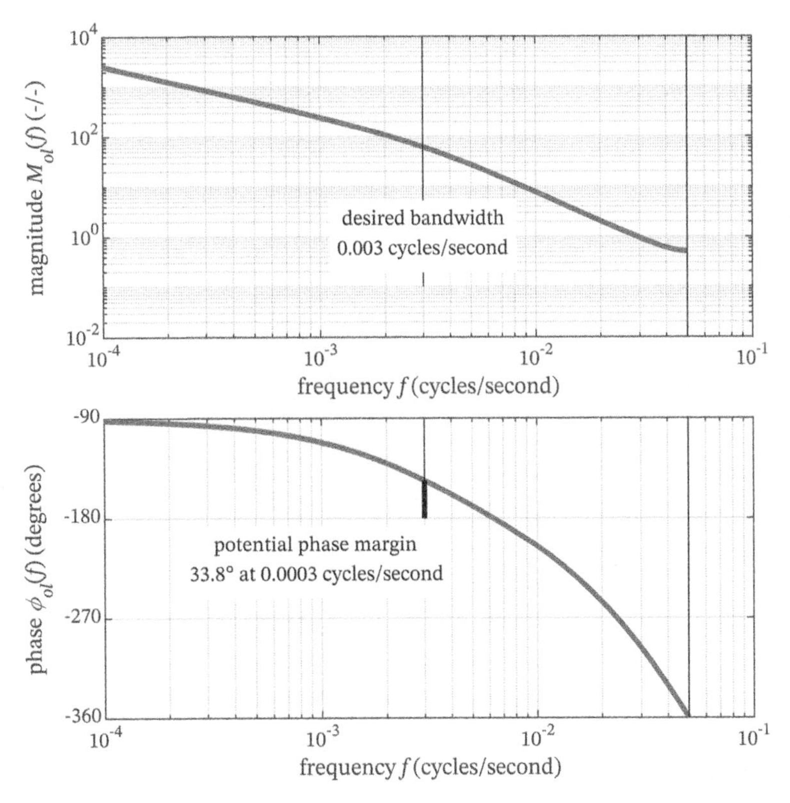

Figure 6.23 Open-loop frequency response of mixture heating and integral control with $K_i = 1$ (V/second)/°C.

Program 6.8 Calculation of discrete-time integral-lead control for mixture heating with delay

```
tauh=49.8; % mixture heating time constant (seconds)
Kh=1.6; % mixture heating parameter (°C/V)
D=12; % delay (seconds)

T=10; % period between adjustments in heater voltage (seconds)
GHoVz=c2d(tf(Kh,[tauh, 1],'OutputDelay',D),T); % mixture heating

PM=76.3; % desired phase margin (degrees)
fb=0.003; % desired bandwidth (cycles/second)

Gciz=tf([T, 0],[1, -1],T) % integral control with Ki=1
```

```
Gciz =

  10 z
  -----
  z - 1

Sample time: 10 seconds
Discrete-time transfer function.
```

```
options=bodeoptions;
options.FreqUnits='Hz'; options.MagScale='log'; options.MagUnits='abs';

% heating and integral control with Ki=1 - Figure 6.23
bode(Gciz*GHoVz,options)
xlabel('frequency f');

% magnitude and phase at desired bandwidth
[Mfb,Pfb]=bode(Gciz*GHoVz,fb*2*pi)
```

```
    Mfb = 61.9056
    Pfb = -146.1539
```

```
Pd=-180+PM-Pfb % additional phase needed
```

```
    Pd = 42.4539
```

```
% lead control transfer function that has phase Pd at frequency fb
a=tand(Pd)/(cos(fb*2*pi*T)*tand(Pd)+sin(fb*2*pi*T));
tauc=-T/log(a) % lead control time constant (seconds)
```

```
    tauc = 58.3020
```

```
Gcleadz=tf([1, -a],[1-a, 0],T) % lead control transfer function
```

```
    Gcleadz =
       z - 0.8424
       ----------
        0.1576 z
    Sample time: 10 seconds
    Discrete-time transfer function.
```

```
[Mol1,Pol1,Wol1]=bode(Gciz*Gcleadz*GHoVz,logspace(-3,-1,5000));
II=find(Pol1<-180+PM); M=Mol1(II(1)) % magnitude for phase -180°+PM°
```

```
    M = 91.8161
```

```
Ki=1/M % make phase margin PM
```

```
    Ki = 0.0109
```

```
% lead and integral-lead control frequency response - Figure 6.24
bode(Gcleadz, Gciz*Gcleadz,options)
legend('lead control with {\it\tau_c}=58.3 seconds',…
       'integral-lead control with {\itK_i}=1 (V/second)/°C')

Gclz=minreal(feedback(Ki*Gciz*Gcleadz*GHoVz,1)); % closed-loop
bode(Gclz,options) % closed-loop frequency response - Figure 6.25

bandwidth(Gclz)/2/pi % closed-loop bandwidth (cycles/second)
```

```
ans = 0.0046
```

```
damp(Gclz) % closed-loop damping ratios
```

```
             Pole              Magnitude      Damping      Time Constant
                                             (seconds)
    8.71e-01                   8.71e-01      1.00e+00      7.26e+01
    5.35e-01 + 5.62e-02i       5.38e-01      9.86e-01      1.61e+01
    5.35e-01 - 5.62e-02i       5.38e-01      9.86e-01      1.61e+01
   -1.24e-01                   1.24e-01      5.54e-01      4.79e+00
```

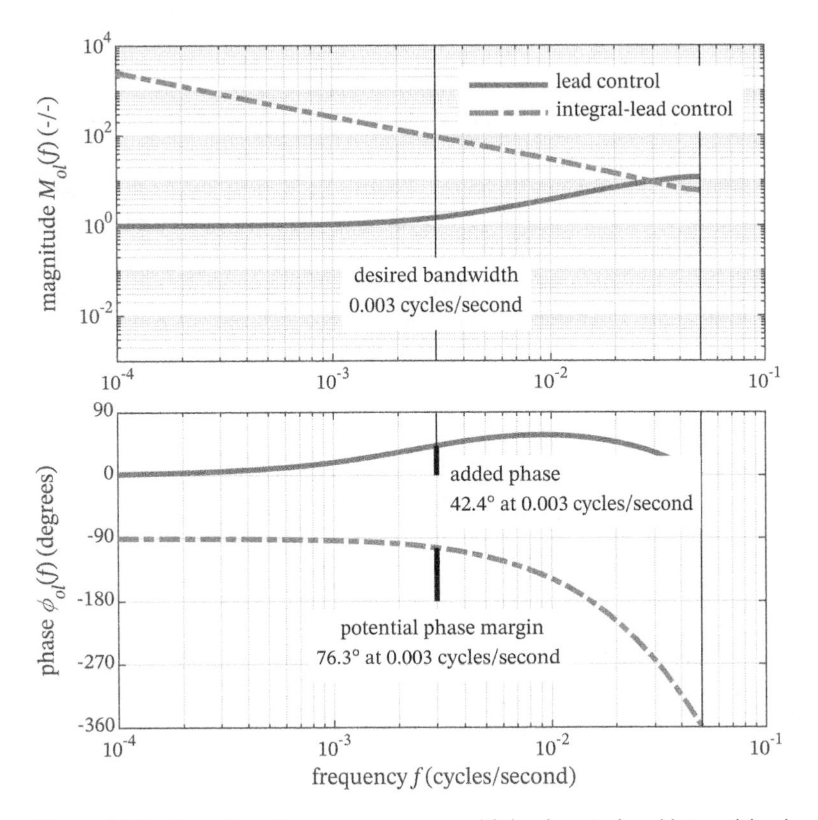

Figure 6.24 Open-loop frequency response with lead control and integral-lead control with K_i = 0.011 (V/second)/°C and τ_c = 58.3 seconds.

$$e^{-T/\tau_c} = \frac{\tan(\phi_{lead}(\omega_b))}{\cos(\omega_b T)\tan(\phi_{lead}(\omega_b)) + \sin(T\omega_b)}$$

$\tau_c = 58.3$ seconds is found using Program 6.8 for $T = 10$ seconds and $\omega_b = 0.006\pi$ radians/second ($f_b = 0.003$ cycles/second). The lead control frequency response shown in Figure 6.24 is the combined open-loop frequency response of mixture heating, integral control, and lead control:

$$\frac{H(z)}{E(z)} = \frac{K_i T}{\left(1-z^{-1}\right)} \frac{\left(1-e^{-T/\tau_c} z^{-1}\right)}{\left(1-e^{-T/\tau_c}\right)} \mathcal{Z}\left\{\left(\frac{1-e^{-Ts}}{s}\right) \frac{K_h}{\tau_h s+1} e^{-Ds}\right\}$$

With $K_i = 1$ (V/second)°C and $\tau_c = 58.3$ seconds. The resulting potential phase margin is 76.3° as desired.

To make the phase margin 76.3° at $f_b = 0.003$ cycles/second, the magnitude of the open-loop frequency response must be adjusted by adjusting integral control parameter K_i. The magnitude with integral control parameter $K_i = 1$ (V/second)°C is 91.8 at frequency 0.003 cycles/second, and reducing the magnitude using $K_i = 1/91.8 = 0.011$ (V/second)/°C makes the unity magnitude crossing frequency 0.003 cycles/second, the desired bandwidth f_b.

The resulting closed-loop frequency response is shown in Figure 6.25 for $T = 10$ seconds, $K_i = 0.011$ seconds^{-1} and $\tau_c = 58.3$ seconds. As calculated using Program 6.8, the actual closed-loop bandwidth is $f_b = 0.0046$ cycles/second, which satisfies the bandwidth requirement. The closed-loop damping ratios associated with complex conjugate roots are approximately $\zeta = 1$, satisfying the requirement for little or no oscillation.

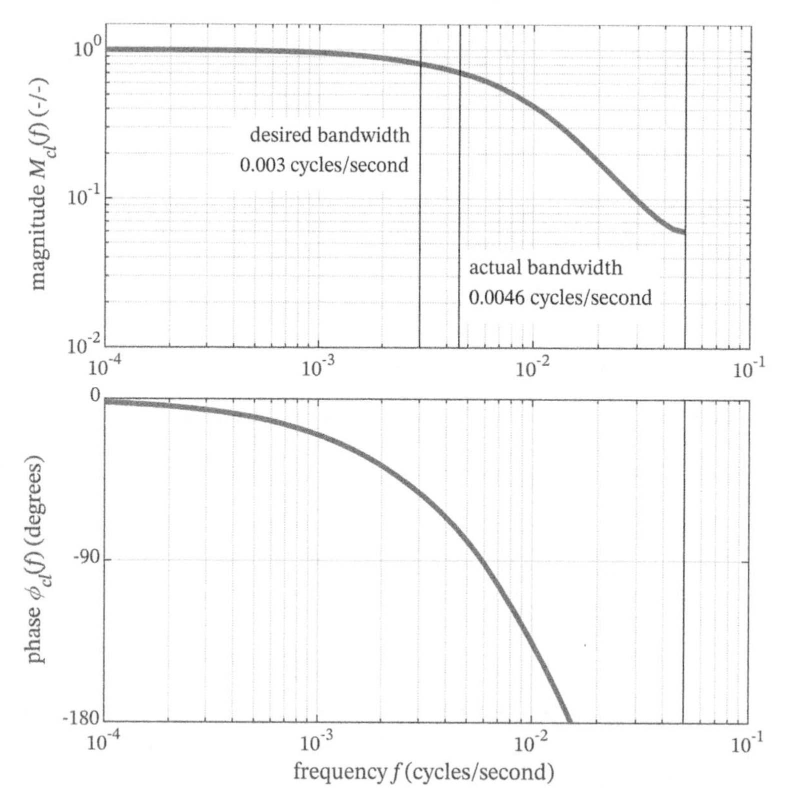

Figure 6.25 Closed-loop mixture heating frequency response with integral-lead control and $T = 10$ seconds, $K_i = 0.011$ (V/second)/°C, and $\tau_c = 58.3$ seconds.

6.6 Closed-Loop Decision-Making Topologies

There are many ways in which closed-loop decision-making can be structured in production systems. Besides allocating decision-making components to the forward path and feedback path components as shown in Figure 6.2, decision-making can be implemented with multiple feedback loops, and models can be integrated that predict dynamic behavior; adding a feedforward path, combining closed-loop and open-loop control, is an example of the latter. In this section, some of the many possible decision-making topologies will be described and examples will be presented to illustrate their design. This is by no means exhaustive, either with respect to possible topologies or associated design methodologies; there are many permutations and combinations, and whether a particular topology or design methodology is appropriate is highly dependent on the specific nature of the production system application and the required closed-loop behavior.

6.6.1 PID Control

Proportional (P), integral (I), and derivative (D) control along with proportional plus derivative (PD) and proportional plus integral (PI) control combinations already have been described in Sections 6.1 and 6.2, as well as in Example 3.17. Proportional plus integral plus derivative (PID) control combines these in closed-loop decision-making, and PID control often is implemented in industrial control software and in production process control equipment.

Combining Equations 6.5 and 6.8 to obtain continuous-time PID control yields

$$m(t)=K_p e(t)+K_i\int_0^t e(t)\,dt+K_d\frac{de(t)}{dt} \tag{6.45}$$

and the corresponding transform is

$$M(s)=\left(K_p+\frac{K_i}{s}+K_d s\right)E(s) \tag{6.46}$$

This topology is shown in the block diagram Figure 6.26a.

Similarly, combining Equations 6.17 and 6.22 to obtain discrete-time PID yields

$$m(kT)=K_p e(kT)+K_i T\sum_{n=0}^{k}e(nT)+K_d\frac{e(kT)-e((k-1)T)}{T} \tag{6.47}$$

The transform is

$$M(z)=\left(K_p+\frac{K_i T}{1-z^{-1}}+\frac{K_d\left(1-z^{-1}\right)}{T}\right)E(z) \tag{6.48}$$

This topology is shown in the block diagram Figure 6.26b. Alternatively, discrete-time PID control can be implemented using the difference equation

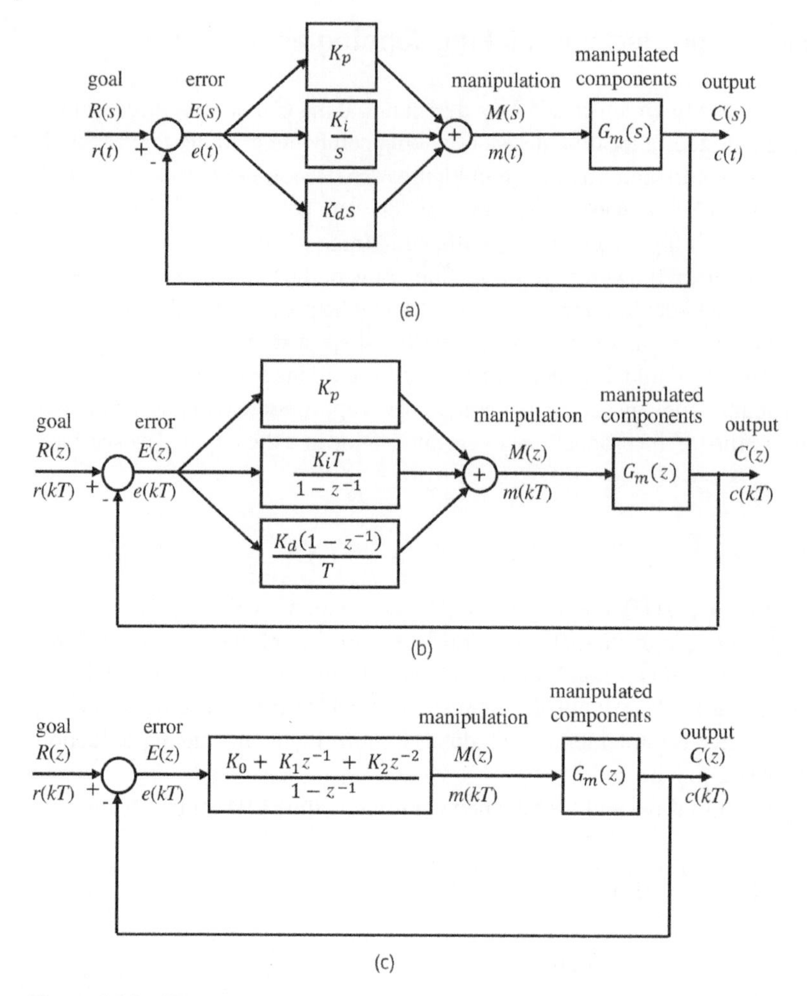

Figure 6.26 PID control topologies. (a) Continous-time PID control. (b) Discrete-time PID control. (c) Alternative discrete-time PID control.

$$m(kT) = m((k-1)T) + K_0 e(kT) + K_1 e((k-1)T) + K_2 e((k-2)T) \tag{6.49}$$

which has the transform

$$M(z) = \frac{(K_0 + K_1 z^{-1} + K_2 z^{-2})}{1 - z^{-1}} E(z) \tag{6.50}$$

where

$$K_0 = K_p + K_i T + \frac{K_d}{T} \tag{6.51}$$

$$K_1 = -K_p - \frac{2K_d}{T} \tag{6.52}$$

$$K_2 = \frac{K_d}{T} \tag{6.53}$$

This topology is shown in the block diagram in Figure 6.26c. If there is no integral control, $K_i = 0$ and Equations 6.49 and 6.50 can be simplified:

$$m(kT) = \left(K_p + \frac{K_d}{T}\right)e(kT) - K_p e((k-1)T) \tag{6.54}$$

$$M(z) = \frac{\left(K_p + \frac{K_d}{T}\right) - K_p z^{-1}}{1 - z^{-1}} E(z) \tag{6.55}$$

Example 6.12 Equivalent Discrete-Time PID Control

It is desired to replace continuous-time decision-making with discrete-time decision-making using a control computer in a closed-loop production system with the topology in Figure 6.2a. The continuous-time decision rule is

$$m(t) = 6.7e(t) + 0.2\int_0^t e(t)dt$$

where manipulation $m(t)$ is continuously adjusted as a function of error $e(t)$. The corresponding continuous-time transfer function is

$$\frac{M(s)}{E(s)} = K_p + \frac{K_i}{s}$$

where $K_p = 6.7$ and $K_i = 0.2$ seconds^{-1}. Manipulation $m(kT)$ is to be adjusted 10 times per second in discrete-time decision-making, and the period between adjustments therefore is $T = 0.1$ seconds.

In this case derivative control parameter $K_d = 0$, and the equivalent discrete-time transfer function in Equation 6.48 becomes

$$\frac{M(z)}{E(z)} = \left(K_p + \frac{K_i T}{1 - z^{-1}}\right)$$

From Equations 6.49, 6.51, and 6.52, the difference equation that would be implemented in the control computer is

$$m(kT) = m((k-1)T) + 6.72e(kT) - 6.70e((k-1)T)$$

6.6.2 Decision-Making Components in the Feedback Path

Placement of decision-making components in the feedback path shown in Figure 6.2 instead of in the forward path can be advantageous when goals R can change suddenly. Large manipulations can be generated when derivative control and lead control are

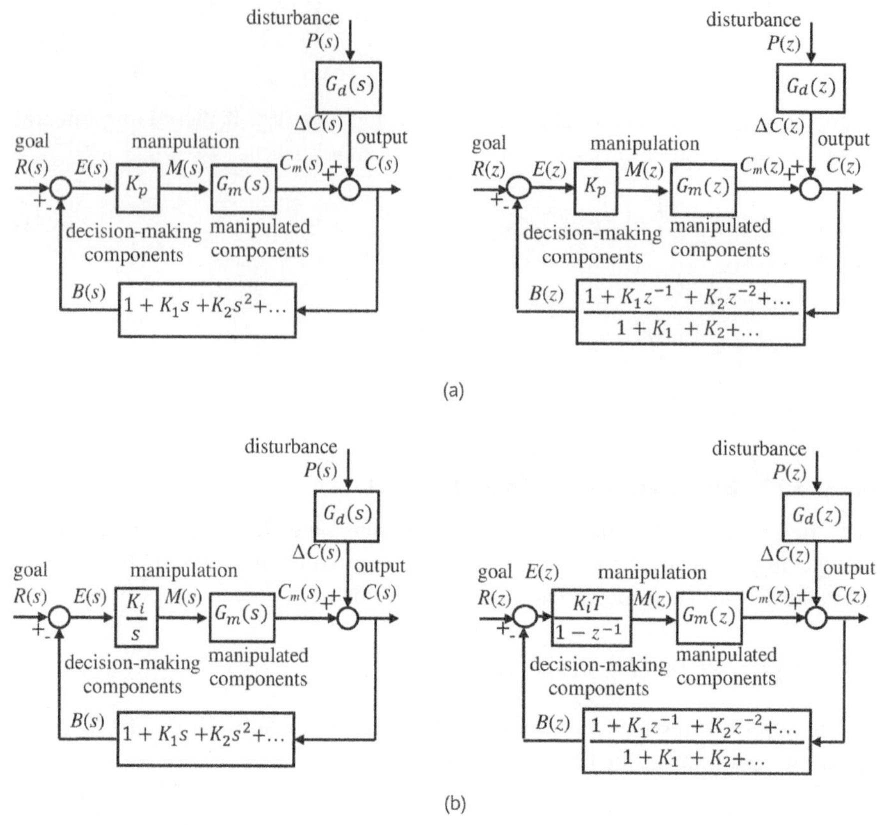

Figure 6.27 Decision-making divided between the forward and feedback paths. (a) Integral control not needed. (b) Integral control needed.

present in the forward path, and these manipulations can result in responses that, while not oscillatory, have significant overshoot. Placing derivative and lead control in the feedback path can be preferable because, in many cases, the production system output does not change suddenly. The need to consider this alternative topology depends on the requirements for closed-loop behavior, the presence of derivative or lead control in decision-making, whether step changes in inputs can be kept small or deliberately avoided in production system operation, and whether disturbances can cause sudden, significant changes in the production system output.

If integral control is not needed, one way that decision-making can be divided between the forward path and the feedback path is shown in Figure 6.27a. On the other hand, if integral control is needed, it often is preferable to divide decision-making between the forward path and the feedback path as shown in Figure 6.27b. Values can be found for parameters K_p or K_i in the forward path and K_0, K_1, K_2, ... in the feedback path using the methods such as those described in Sections 6.3, 6.4, and 6.5.

Example 6.13 Discrete-Time Lead Control in the Feedback Path

In a closed-loop discrete-time production system modeled as shown in the block diagrams in Figure 6.28, lead time is indirectly regulated by requiring that work actually

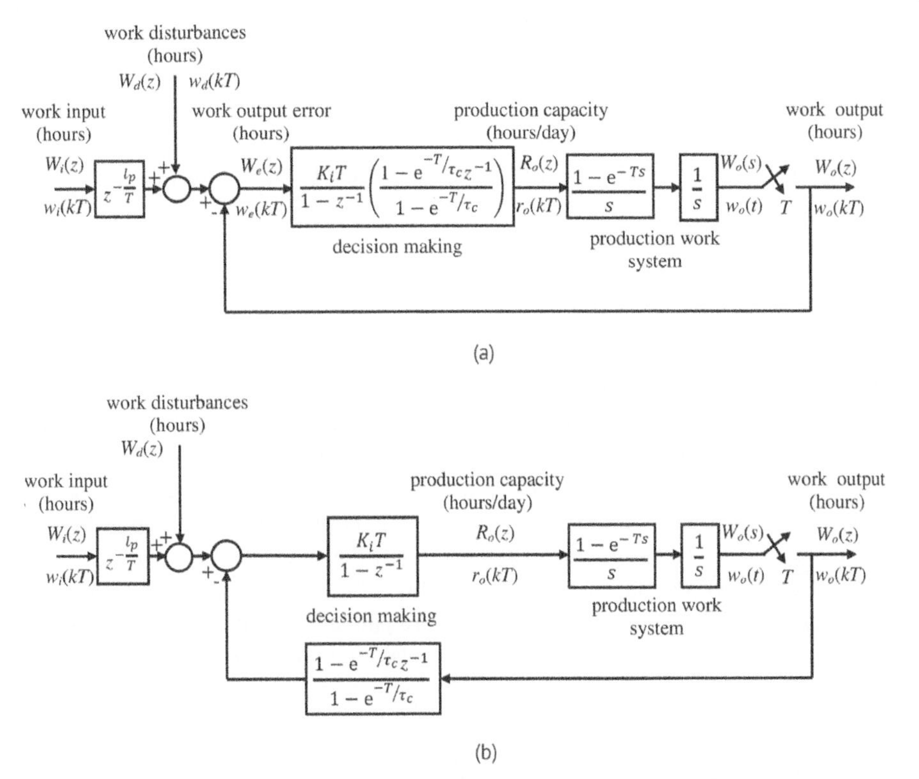

Figure 6.28 Block diagrams for lead-time regulation in a production system by adjusting production capacity based on the difference between expected work output and actual work output. (a) Discrete-time integral and lead control in the forward path. (b) Discrete-time integral control in the forward path and lead control in the feedback path.

done accurately follow the work that is expected to be done, which includes the work input delayed by planned lead time l_p days and any work disturbances, such as rush orders and order cancelations. $w_o(kT)$ hours is the work that has actually been done, $w_i(kT)$ hours is the work input, $w_d(kT)$ represents work disturbances, and $w_e(kT)$ is the work output error, which is the difference between the work that is expected to be done and the work that actually has been done. $r_o(kT)$ is the production capacity, which is adjusted with period T days. It is assumed that l_p is an integer multiple of T.

Decision-making for using integral-lead control in the forward path has the following transfer functions as shown in Figure 6.28a:

$$G_c(z) = \frac{K_iT}{1-z^{-1}}\left(\frac{1-e^{-T/\tau_c}z^{-1}}{1-e^{-T/\tau_c}}\right)$$

$$H_c(z) = 1$$

The direct design approach described in Section 6.4 can be used to obtain dead-beat response by choosing the values of integral and lead control parameters K_i and τ_c that make the two roots of characteristic equation

$$z^2 - \left(2 - \frac{K_i T^2}{1 - e^{-T/T_c}}\right) z + \left(1 - \frac{K_i T^2 e^{-T/T_c}}{1 - e^{-T/T_c}}\right) = 0$$

equal to zero. The required integral-lead control parameter values are

$$b = e^{-T/T_c} = 0.5$$

$$K_i = \frac{1}{T^2}$$

Program 6.9 can be used to calculate the change in work output and production capacity in response to a rush order disturbance of 10 hours of work when the designed decision-making parameters are $T = 3$ days, $K_i = 0.111$ days^{-2}, and $b = 0.5$. Response is complete in $2T = 6$ days as shown in Figure 6.29a. The characteristic equation has no complex roots but there is a 100% overshoot in work output. The initial change in production capacity is 6.667 hours/day when lead control is in the forward path.

If these characteristics are not acceptable, then one option is to modify decision-making by keeping integral control in the forward path and moving lead control to the feedback path as shown in Figure 6.28b. The decision-making transfer functions then are

$$G_c(z) = \frac{K_i T}{1 - z^{-1}}$$

$$H_c(z) = \frac{1 - e^{-T/T_c} z^{-1}}{1 - e^{-T/T_c}}$$

Program 6.9 Calculation of response with lead control in forward and feedback paths

```
T=3; % period between capacity adjustments (days)

Ki=1/T^2;  % integral control parameter (days^-2)
b=Kp/(Kp+Ki*T);  % lead control parameter

Gciz=tf([Ki*T, 0],[1, -1],T,'TimeUnit','days');  % integral control
Gcleadz=tf([1, -b],[1-b, 0],T,'TimeUnit','days');

Gmz=c2d(tf(1,[1, 0],'TimeUnit','days'),T);  % production work system

% closed-loop transfer functions with lead control in the forward path
Gcloz=minreal(feedback(Gciz*Gcleadz*Gmz,1))  % work output
```

```
Gcloz =

  2 z - 1
  -------
    z^2

Sample time: 3 days
Discrete-time transfer function.
```

```
Gclmz=minreal(feedback(Gciz*Gcleadz,Gmz))  % production capacity
```

```
Gclmz =

  0.6667 z^2 - z + 0.3333
  -----------------------
            z^2

Sample time: 3 days
Discrete-time transfer function.
```

```
% closed-loop transfer functions with lead control in the feedback path
Gcloz1=minreal(feedback(Gciz*Gmz,Gcleadz))   % work output
```

```
Gcloz1 =

  1
  -
  z

Sample time: 3 days
Discrete-time transfer function.
```

```
Gclmz1=minreal(feedback(Gciz,Gcleadz*Gmz))  % capacity
```

```
Gclmz1 =

  0.3333 z - 0.3333
  -----------------
          z

Sample time: 3 days
Discrete-time transfer function.
```

```
step(Wrush*Gcloz,40)  % work output, forward path - Figure 6.29a
xlabel('time kT'); ylabel('work output w_o(t) (hours)')

step(Wrush*Gclmz,40)  % capacity, forward path - Figure 6.29a
xlabel('time kT'); ylabel('capacity r_o(t) (hours/day)')

step(Wrush*Gcloz1,40)  % work output, feedback path - Figure 6.29b
xlabel('time kT '); ylabel('work output w_o(t) (hours)')

step(Wrush*Gclmz1,40)  % capacity, feedback path - Figure 6.29b
xlabel('time kT '); ylabel('capacity r_o(t) (hours/day)')
```

With lead control in the feedback path there is no overshoot in the response shown in Figure 6.29b and the response is complete in $T = 3$ days. Furthermore, the peak change in production capacity is reduced to 3.333 hours/day.

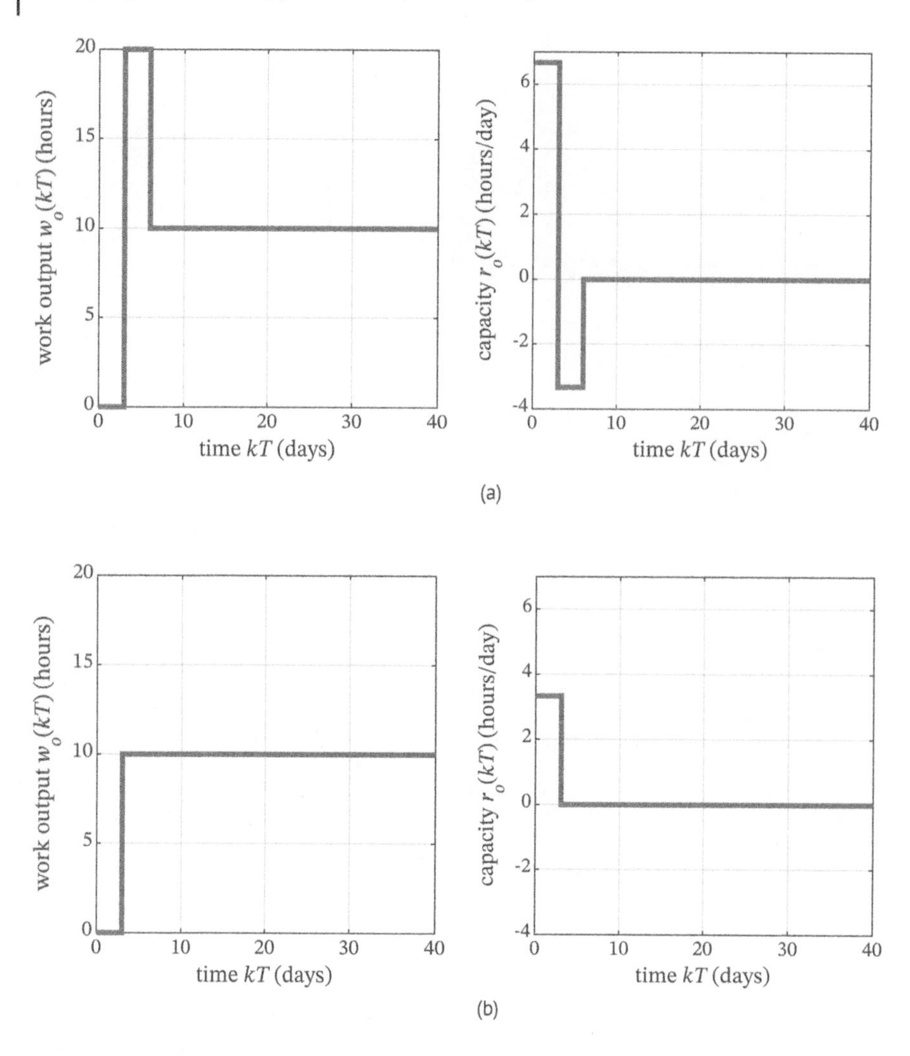

Figure 6.29 Response lead-time regulation to a 10-hour rush order work disturbance. (a) Lead control in the forward path. (b) Lead control in feedback path in the feedback path.

6.6.3 Cascade Control

Although the primary goal may be to control a specific output variable of a production system, it often is possible to measure and control additional output variables and to improve system performance by designing decision-making for each of them. For example, as illustrated in the block diagram in Figure 6.30, continuous-time decision-making component $G_{c1}(s)$ can be designed for production system output $C(s)$ and a separate decision-making component $G_{c2}(s)$ can be designed for internal variable $V(s)$. These components are in a series or cascade configuration, with manipulations generated by $G_{c1}(s)$ serving as goals for $G_{c2}(s)$. The production system then is manipulated by $G_{c2}(s)$. This example has two nested feedback loops. Another example of a cascade

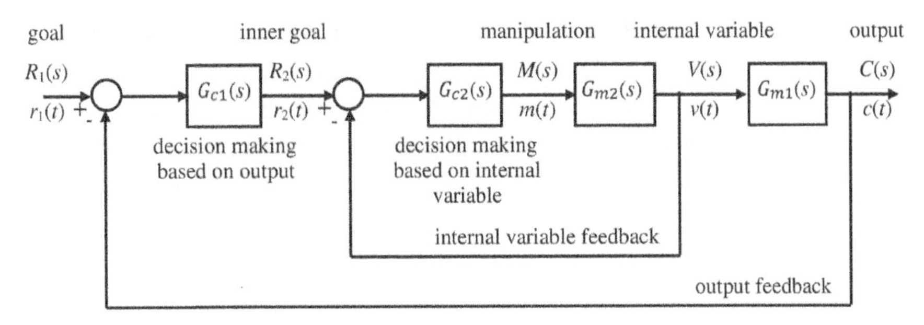

Figure 6.30 Example of a continuous-time cascade control topology.

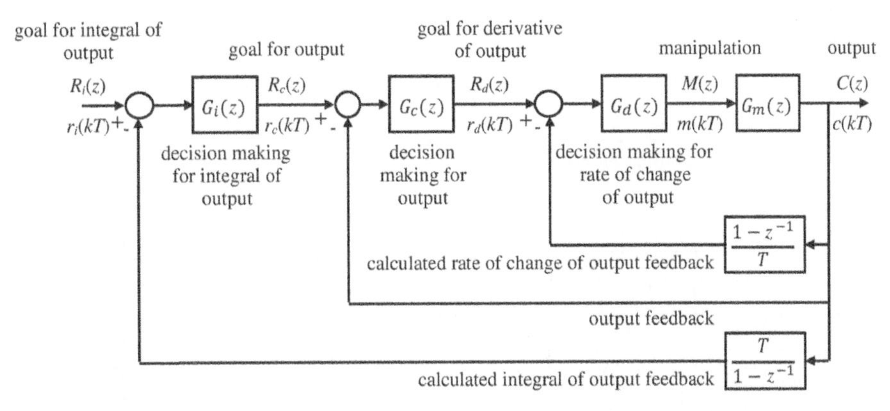

Figure 6.31 Example of discrete-time cascade control with measurement of the output of the production system and calculation of both the integral of the output and the rate of change of the output.

control topology is the discrete-time block diagram in Figure 6.31 in which feedback of the rate of change[3] and the integral of the production system output are calculated. Again, decision-making is implemented in nested loops.

Decision-making components in a cascade topology can be designed together, but it often is more convenient to design the inner-most feedback loop and its decision-making first, ignoring outer feedback loops. If well-behaved dynamic behavior is achieved in design of the inner-most loop that responds relatively quickly compared to what is required of the next, surrounding loop, then the dynamics of the designed inner-most loop often can be ignored when designing the next, surrounding loop. This procedure can be repeated until the outer-most loop and its decision-making have been designed, simplifying the task of designing the dynamic behavior of the production system.

3 When rate of change is calculated using differencing of feedback measured with resolution R_c, the resolution of the calculated rate of change is $R_d = R_c/T$; hence, smaller sampling periods T result in a need for a smaller resolution R_c (higher precision) for a specified resolution R_d.

Example 6.14 Cascade Discrete-Time Control of Force in a Pressing Operation

In the pressing operation illustrated in Figure 6.32, force $f(t)$ N is exerted on a part during its production. A block diagram for this production system is shown in Figure 6.33 in which a force control $G_f(z)$ and actuator velocity control $G_v(z)$ are included in a cascaded topology. Only actuator position $p(t)$ cm is measured, and force can be calculated from position using

$$f(kT) = K_f p(kT)$$

where T seconds is the period between measurements and K_f N/cm is the stiffness of the material being pressed. It is known that $p(t) = 0$ when the actuator is in contact with but not compressing the material. Furthermore, velocity feedback $v_d(kT)$ cm/second also can be calculated using position feedback, eliminating the need for a velocity sensor:

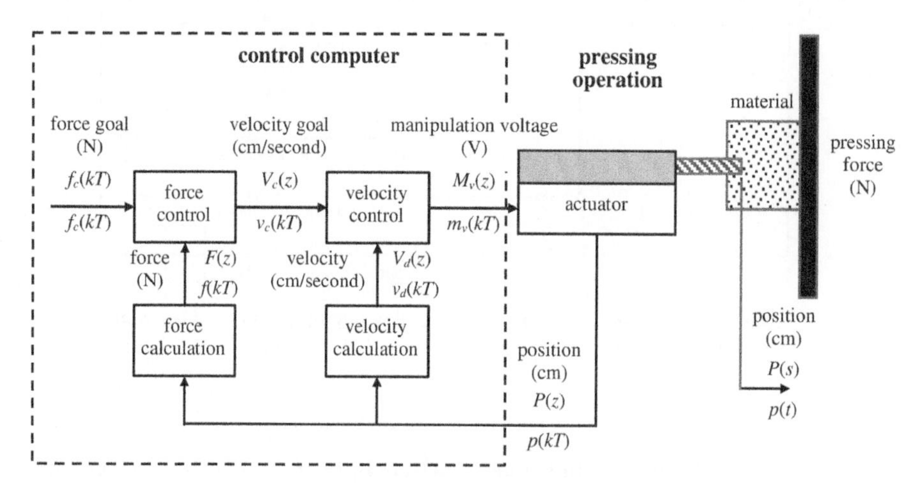

Figure 6.32 Cascade computer control of pressing force and actuator velocity.

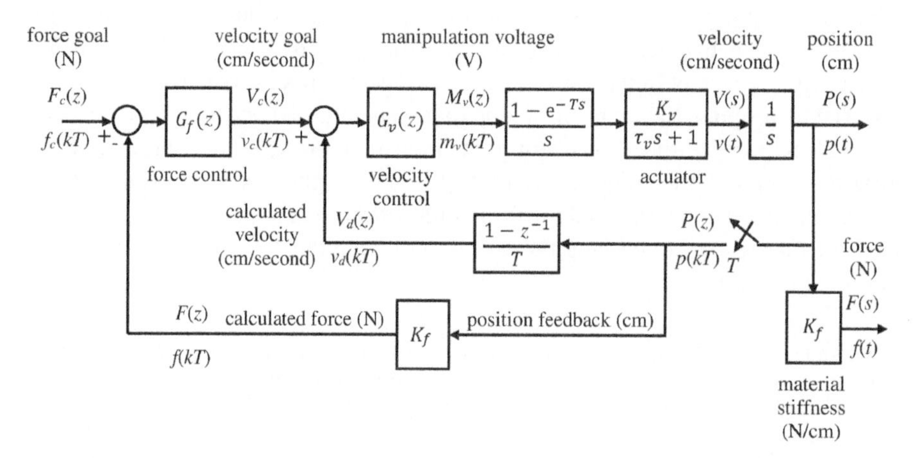

Figure 6.33 Block diagram for cascade discrete-time control of force and actuator velocity.

$$v_d(kT) = \frac{p(kT) - p((k-1)T)}{T}$$

The transfer function of the actuator is

$$G_a(z) = \frac{P(z)}{M_v(z)} = \mathcal{Z}\left\{\frac{\left(1 - e^{-Ts}\right)}{s}\left(\frac{K_v}{\tau_v s + 1}\right)\frac{1}{s}\right\}$$

where τ_v is a time constant in the model used to characterize how quickly actuator velocity responds to changes in manipulation voltage $m_v(kT)$, and K_v (cm/second)/V is a constant that relates actuator velocity to manipulation voltage.

If the desired settling time of the outer force control loop is approximately 0.2 seconds and the settling time of the inner velocity control loop is be chosen to be significantly less, 0.032 seconds, for example, then it is likely that the outer force control loop can be ignored when designing decision-making transfer function $G_v(z)$ for control of actuator velocity. Then, decision-making transfer function $G_f(z)$ can be designed for the outer force control loop as a second step. This two-step design sequence can be significantly more straightforward than designing $G_f(z)$ and $G_v(z)$ at the same time.

Discrete-time integral-lead control in the form of Equation 6.27 can be used to introduce integration into the inner velocity loop and cancel actuator time constant τ_v as described in Example 6.7:

$$G_v(z) = \frac{M_v(z)}{V_c(z) - V_d(z)} = \left(\frac{K_i T}{1 - z^{-1}}\right)\left(\frac{1 - e^{-T/\tau_v} z^{-1}}{1 - e^{-T/\tau_v}}\right)$$

where $v_c(kT)$ is the velocity goal calculated by the force control. Velocity control parameter K_i (V/second)/(cm/second) can be adjusted to make the inner velocity loop settling time approximately 0.032 seconds. The period between adjustments in both velocity command and voltage can be chosen to be $T = 0.004$ seconds, which is significantly less than this settling time.

Program 6.10 can be used to calculate the step response of the inner velocity loop for actuator parameters $K_v = 0.125$ (cm/second)/V, $\tau_v = 0.07$ seconds, and $K_f = 22{,}250$ N/cm. Among the methods available for choosing a value of integral control parameter K_i are those described in Sections 6.3, 6.4, and 6.5. A result obtained by searching is $K_i = 650$ (V/second)/(cm/second). The response of the inner velocity loop to a step change in velocity goal is shown in Figure 6.34a. The response is not oscillatory, and the settling time is 0.032 seconds as desired.

Force control can be proportional:

$$G_f(z) = \frac{V_c(z)}{F_c(z) - F(z)} = K_p$$

where $f_c(kT)$ is the force goal. Proportional force control parameter K_p then can be adjusted to make the outer force control loop settling time approximately 0.2 seconds. A result obtained by searching is $K_p = 0.00075$ (cm/second)/N. The response of the

Program 6.10 Calculate force and position for pressing operation with cascade control

```
T=0.004;  % period between velocity adjustments (seconds)

Kv=0.125;  % actuator velocity parameter ((cm/second)/v)
tauv=0.07;  % actuator velocity time constant (seconds)
Kf=22250;  % force constant (N/cm)

Gaz=c2d(tf(Kv,[tauv, 1, 0]),T);  % actuator position transfer function

Gdz=tf([1, -1],[T, 0],T);  % discrete differentiation

Ki=650;  % velocity control parameter found by searching

% integral-lead velocity control
Gvz=tf(Ki*T*[1, -exp(-T/tauv)],(1-exp(-T/tauv))*[1, -1],T);

% inner closed-loop transfer function for velocity
Gvclz=minreal(feedback(Gvz*Gaz*Gdz,1));

step(Gvclz,0.4)  % inner closed-loop step response - Figure 6.34a
ylabel('Actuator velocity v_d(kT) (cm/second)')

stepinfo(Gvclz)  % settling time of inner velocity loop
```

```
ans =
        RiseTime: 0.0160
    SettlingTime: 0.0320
     SettlingMin: 0.9029
     SettlingMax: 1.0000
       Overshoot: 0
      Undershoot: 0
            Peak: 1.0000
        PeakTime: 0.2360
```

```
% inner closed-loop transfer function for position
Gpclz=minreal(feedback(Gvz*Gaz,Gdz));

Kp=0.00075;  % force control parameter found by searching (cm/N)

% outer closed-loop transfer function for force
Gfclz=Kf*feedback(minreal(Kp*Gpclz),Kf);

step(Gfclz,0.4)  % outer closed-loop step response - Figure 6.34b
ylabel('Pressing force f(kT) (N)')

stepinfo(Gfclz)  % settling time of outer force loop
```

```
ans =
        RiseTime: 0.1080
    SettlingTime: 0.2000
     SettlingMin: 0.9061
     SettlingMax: 0.9999
       Overshoot: 0
      Undershoot: 0
            Peak: 0.9999
        PeakTime: 0.4360
```

```
% pressing force example
t=[0:T:1];
fc=8900*t;   % increasing force goal to 8900 N over 1 second
t=[t, 1+T:T:1.8];
fc=[fc, 8900*ones(1,50) zeros(1,150)];   % 0.2 seconds at 8900 N then 0 N

f=lsim(Gfclz,fc,t);   % calculate response to force goal input

stairs(t,fc)  % plot force goal - Figure 6.35
hold on; stairs(t,f); hold off  % plot pressing force response
xlabel('time t'); ylabel('force (N)')
legend('force goal f_c(kT)','pressing force f(kT)')
```

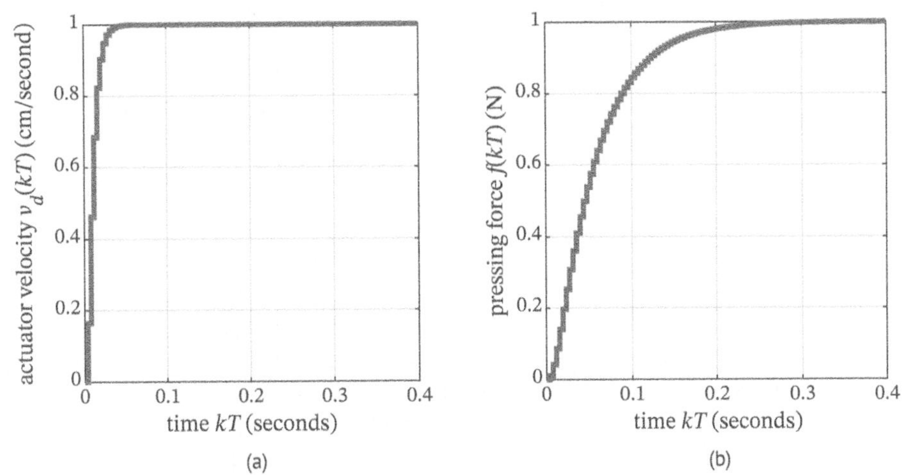

Figure 6.34 Cascaded discrete-time control of pressing force control. (a) Step response of inner velocity loop. (b) Step response of outer force loop.

outer force control loop, including the already designed inner velocity loop, to a step change in the force goal is shown in Figure 6.34b. The response is not oscillatory, and the settling time is 0.2 seconds as desired.

An example of the dynamic behavior of the pressing operation with cascade force and velocity control that has been designed is shown in Figure 6.35. The force goal increases linearly from 0 to 8900 N over a period of 1 second. The force goal then is constant at 8900 N for 0.2 seconds, the force control settling time, and then returns as a step to zero.

6.6.4 Feedforward Control

Decision-making in a closed-loop production system can be modified as shown in the block diagram in Figure 6.36a to include an additional forward path that anticipates the manipulations required to produce the desired output. Feedback control using decision-making component G_c is combined with feedforward control using decision-making component G_{ff}, where the path with feedforward control G_{ff} can be viewed as open-loop control as emphasized in Figure 6.36b. Two decision-making components contribute to manipulation M:

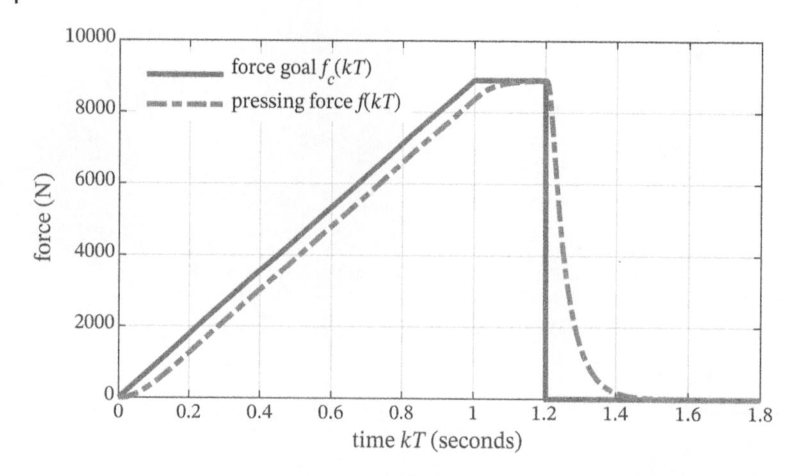

Figure 6.35 Response of cascaded discrete-time control to an 8900 N force goal profile.

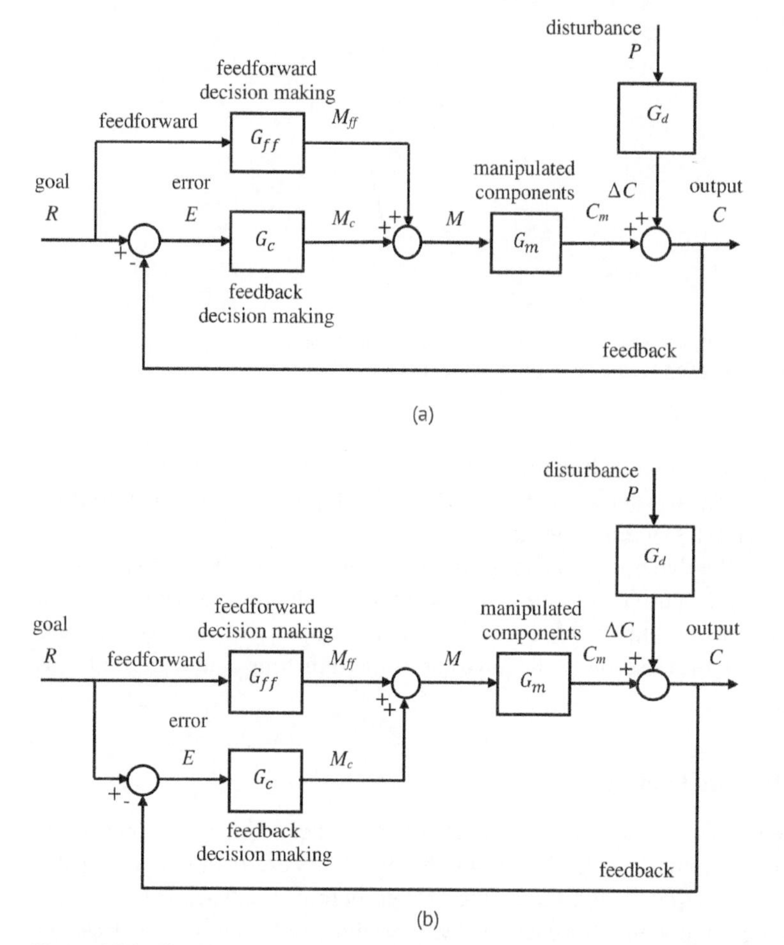

Figure 6.36 Feedforward control. (a) Feedforward path added to closed-loop production system. (b) Alternative perspective with open-loop control augmented by closed-loop control.

$$M = M_c + M_{ff} = G_c E + G_{ff} R \tag{6.56}$$

Ideally, the feedforward control transfer function is

$$G_{ff} = \frac{1}{G_m} \tag{6.57}$$

which would result in output C exactly following goal R and error $E = 0$. However, transfer function G_{ff} found using Equation 6.57 may be impractical because the result is a complicated feedforward component, the result is an unrealizable feedforward component that requires unknown values of the future of command inputs R, or the result is a feedforward control component that is unstable or produces otherwise undesirable behavior of M_{ff}. The feedforward control transfer function G_{ff} as calculated using Equation 6.57, therefore, may have to be simplified to make its implementation feasible. Regardless, the feedback loop with decision-making component G_c is needed to correct for disturbances and inaccuracies in production system transfer function G_m. With the addition of feedforward control, the relationship between goal R and output C is

$$\frac{C}{R} = \frac{\left(G_f + G_c\right)G_m}{1 + G_c G_m} \tag{6.58}$$

Note that feedforward control transfer function G_{ff} does not appear as the denominator of this transfer function. Hence, the characteristic equation and the associated fundamental dynamic behavior of production system are not changed by the addition of feedforward control.

Example 6.15 Discrete-Time Regulation of Lead Time With Feedforward Decision-Making

Lead time is indirectly regulated in the production system described in Example 6.13 by requiring that work actually done accurately follows the work that is expected to be done, which is the work input, delayed by planned lead time l_p days plus any work disturbances such as rush orders and order cancellations. As shown in Figure 6.37, feedforward decision-making can be added to this production system to eliminate the need for integral control and, instead, use measurements of work input to anticipate the needed production capacity. Capacity is adjusted with period T days for the purpose of eliminating error between work done and work that is expected to be done.

The transfer function of the production work system in the block diagram in Figure 6.37 is

$$G_m(z) = \frac{W_o(z)}{R_o(z)} = \frac{Tz^{-1}}{1 - z^{-1}} z^{-d}$$

where there is delay dT in implementing capacity adjustment decisions. The ideal feedforward transfer function, from Equation 6.57, is

$$\frac{1}{G_m(z)} = \frac{1 - z^{-1}}{T} z^{d+1}$$

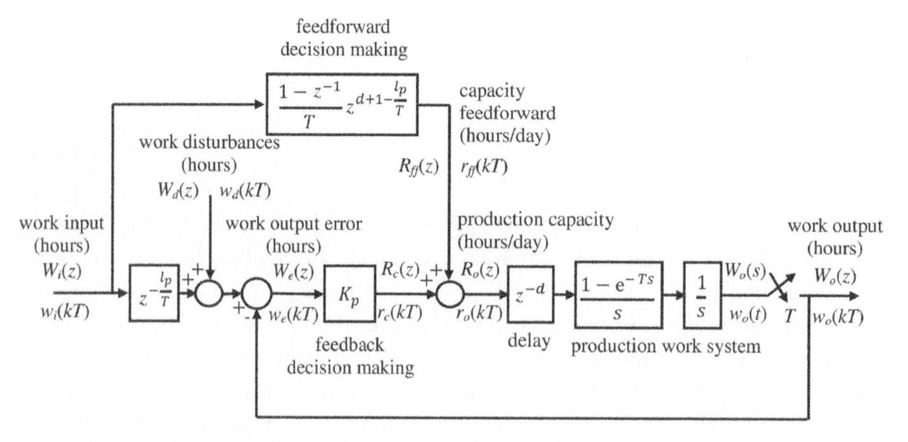

Figure 6.37 Discrete-time regulation of lead time with feedforward decision-making, where lead time l_p is an integer multiple of period T and $l_p \geqslant (d + 1)T$.

This requires knowledge of work input $w_i(kT)$ at time $(d + 1)T$ days in the future. Fortunately, the approach to lead-time regulation shown in Figure 6.37 has delay l_p days in work input $w_i(kT)$. Assuming l_p is an integer multiple of T and $l_p \geq (d + 1)T$, the feedforward path can be implemented without requiring unavailable, future knowledge of work input $w_i(kT)$

$$G_{ff}(z) = \frac{R_{ff}(z)}{W_i(z)} = \frac{1 - z^{-1}}{T} z^{d+1-\frac{l_p}{T}}$$

This allows the production capacity needed to satisfy work input $w_i(kT)$ to be anticipated considering the delay of dT days in implementation. The result is work output error $w_e(kT) = 0$ when there are no disturbances $w_d(kT) = 0$.

Proportional control parameter K_p can be chosen using the methods such as those described in Sections 6.3, 6.4, and 6.5. The response to work disturbances $w_d(kT)$ is not affected by the addition of feedforward decision-making. Delay dT and choice of K_p, therefore, determine the dynamic behavior of the production system in response to disturbances.

Example 6.16 Discrete-Time Feedforward Control of Force in a Pressing Operation

Cascade discrete-time control of pressing force was designed in Example 6.14 for the pressing operating illustrated in Figure 6.32. It can be observed in Figure 6.35 that there is a significant error in force when the force command is changing; this often is referred to as following error, and the following error is constant in the steady-state condition when the rate of change of the force command is constant. Addition of feedforward control of force to this system as shown in the block diagram in Figure 6.38 can eliminate much of the following error in the force produced in the pressing operation.

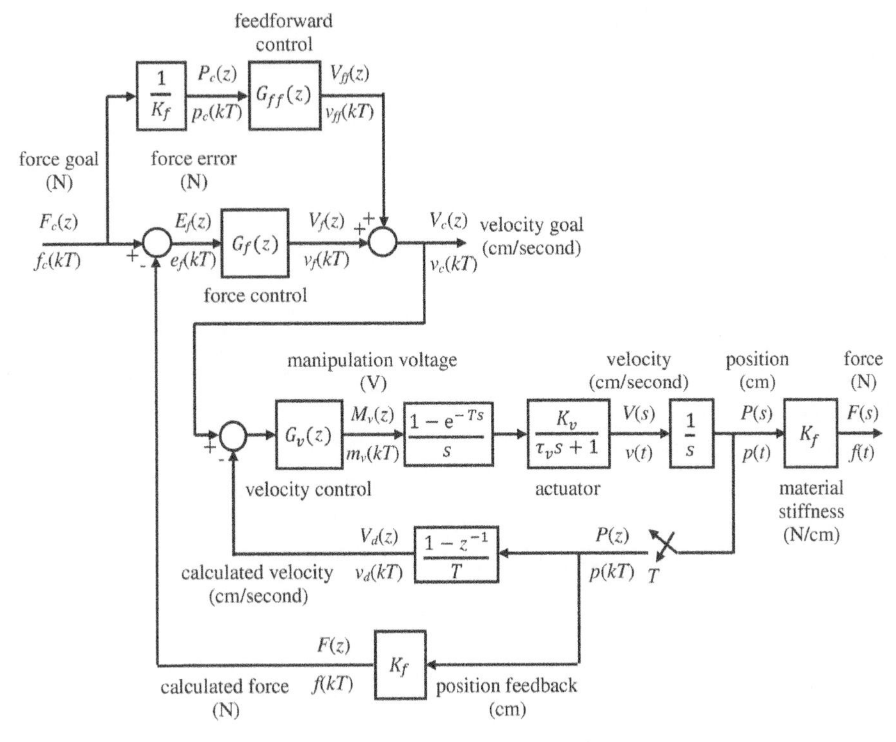

Figure 6.38 Feedforward control added to cascade discrete-time control of pressing force.

When feedforward control is added, velocity goal $v_c(kT)$ has feedback control component $v_f(kT)$ and feedforward control component $v_{ff}(kT)$:

$$v_c(kT) = v_f(kT) + v_{ff}(kT)$$

When Equation 6.57 is used to calculate the ideal feedforward transfer function $G_{ff}(z)$, the numerator of the inner closed-loop velocity transfer function becomes the denominator of the feedforward transfer function. As calculated using Program 6.11, production system parameters $K_v = 0.125$ cm/seconds/V, $\tau_v = 0.07$ seconds and $K_f = 22250$ N/cm and $T = 0.004$ seconds result in a factor $(z + 0.9811)$ in the denominator of ideal feedforward transfer function $G_{ff}(z)$; hence, its characteristic equation has root $r = -0.9811$. The feedforward control component therefore is stable because $|r| < 1$, but its behavior is undesirable because $r < 0$ and output $v_{ff}(kT)$ tends to resemble Figure 4.4c. Furthermore, this ideal feedforward transfer function has a 3rd-order numerator, which makes it a relatively complicated decision-making component.

As an alternative, the feedforward controller transfer function can be simplified to

$$G_{ff}(z) = \frac{V_{ff}(z)}{P_c(z)} = \frac{z-1}{T}$$

Program 6.11 Calculation of pressing force with discrete-time feedforward control

```
T=0.004;  % period between velocity adjustments (seconds)

Kv=0.125;  % actuator velocity constant (cm/seconds/v)
tauv=0.07;  % actuator velocity time constant (seconds)
Kf=22250;  % force constant (N/cm)

Ki=650;  % velocity control parameter ((V/second)/cm)
Kp=0.00075;  % force control parameter (cm/N)

Gaz=c2d(tf(Kv,[tauv, 1, 0]),T);  % actuator position transfer function

Gdz=tf([1, -1],[T, 0],T);  % digital differentiation

% integral-lead velocity control
Gvz=tf(Ki*T*[1, -exp(-T/tauv)],(1-exp(-T/tauv))*[1, -1],T);

Gvclz=minreal(feedback(Gvz*Gaz*Gdz,1));  % velocity transfer function
Gpclz=minreal(feedback(Gvz*Gaz,Gdz));  % position transfer function

Gfclz=Kf*feedback(minreal(Kp*Gpclz),Kf); % force transfer function

zpk(1/Gpclz/Kf)  % ideal feedforward transfer function
```

```
ans =
  0.068492 (z-1) (z-0.5352) (z-0.3007)
  ------------------------------------
             z (z+0.9811)
Sample time: 0.004 seconds
Discrete-time zero/pole/gain model.
```

```
Gffz=tf([1, -1],T,T)/Kf  % simplified feedforward transfer function
```

```
Gffz =
  z - 1
  -----
   89
Sample time: 0.004 seconds
Discrete-time transfer function.
```

```
Gffclz=Gfclz*(1+Gffz/Kp);  % closed-loop including feedforward

% ramp to 8900 N, 0.2 seconds at 8900 N, ramp to 0 N, remain at 0 N
fc=[8900*(0:T:1) 8900*ones(1,50) 8900*(1:-0.025:0) zeros(1,150-41)];
t=[0:T:1.8];

f=lsim(Gfclz,fc,t);  % calculate response without feed forward control
ff=lsim(Gffclz,fc,t);  % calculate response with feedforward control

ef=fc-f';  % error without feedforward control
eff=fc-ff';  % error with feedforward control

stairs(t,ef)  % plot error - Figure 6.39
hold on; stairs(t,eff); hold off
legend('without feedforward control','with feedforward control',...
       'Location','southwest')
xlabel('time kT'); ylabel('pressing force f(kT) (N)')
```

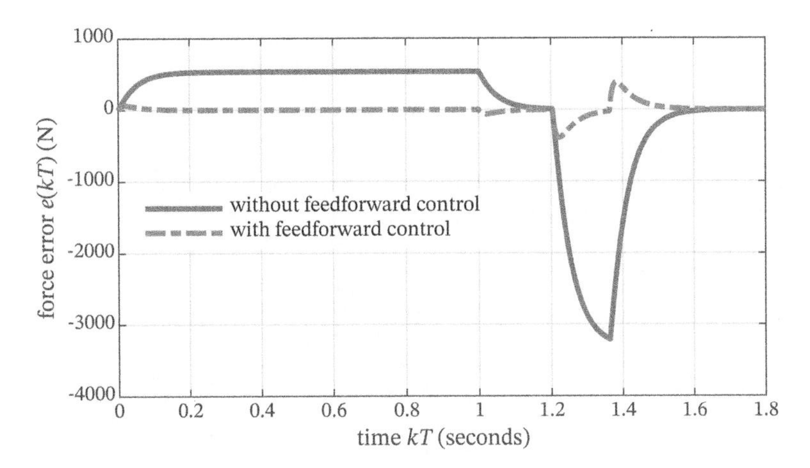

Figure 6.39 Response of cascade discrete-time control to a 8900 N force goal profile with and without feedforward control.

The corresponding difference equation is

$$v_{ff}(kT) = \frac{p_c((k+1)T) - p_c(kT)}{T}$$

This feedforward control requires knowledge of force goal profile $f_c(kT)$ one period T in the future. However, it can be assumed in this example that force goal profile is precalculated and stored in memory of a control computer. If this is not the case and the force goal profile is not known in advance, then the feedforward control transfer function can be further modified:

$$G_{ff}(z) = \frac{V_{ff}(z)}{P_c(z)} = \frac{1 - z^{-1}}{T}$$

and the corresponding difference equation is

$$v_{ff}(kT) = \frac{p_c(kT) - p_c((k-1)T)}{T}$$

Program 6.11 can be used to calculate the force error with and without feedforward control. The force error $e_f(kT)$ is shown in Figure 6.39 where, unlike the case in Figure 6.35 without feedforward control, force error $e_f(kT)$ decreases to approximately zero when the force goal is increasing or decreasing at a constant rate of change. Because feedforward control tends to generate large manipulations when there are abrupt changes in the force goal, profile $f_c(kT)$ used in this example has been modified with respect to that used in Example 6.14 to eliminate the step change at $kT = 1.2$ seconds. The force goal now returns to zero in 0.16 seconds.

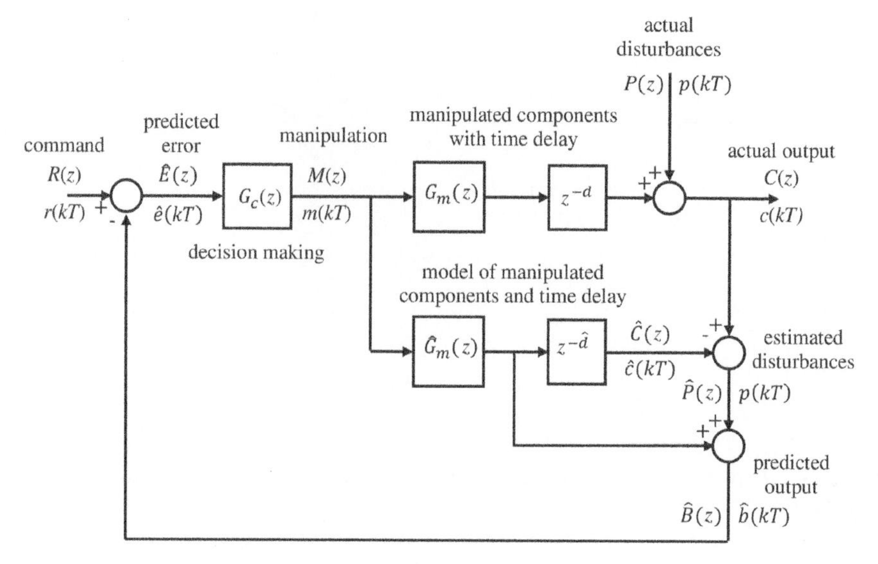

Figure 6.40 Discrete-time closed-loop decision-making in which time delay is circumvented using a Smith Predictor topology.

6.6.5 Circumventing Time Delay Using a Smith Predictor Topology

The presence of significant time delays in a closed-loop production system tends to result in relatively long closed-loop setting times, particularly when it is necessary to include integration in decision-making. It can be possible to circumvent time delay by incorporating a model of the production system into the closed-loop control. An example is the Smith Predictor topology[4] shown in the discrete-time block diagram in Figure 6.40. This example illustrates the use of an internal model to obtain estimates of production system variables that cannot be readily measured. In Figure 6.40, the output of a production system with disturbances is compared with the output of a model of the system with no disturbances, with the difference being an estimate of disturbances that is obtained without physical sensing. Furthermore, decision-making then can use feedback from a portion of the model of manipulated components of the production system from which time delay has been separated.

The relationship between the actual production system output $C(z)$ and command input $R(z)$ and disturbance input $P(z)$ in Figure 6.40 is

$$C(z)=\frac{G_c(z)G_p(z)z^{-d}}{1+G_c(z)\left(\hat{G}_m(z)\left(1-z^{-\hat{d}}\right)+G_m(z)z^{-d}\right)}R(z)$$
$$+\frac{1+G_c(z)\hat{G}_m(z)\left(1-z^{-\hat{d}}\right)}{1+G_c(z)\left(\hat{G}_m(z)\left(1-z^{-\hat{d}}\right)+G_m(z)z^{-d}\right)}P(z) \tag{6.59}$$

4 The Smith Predictor topology is well known both as a means for circumventing time delay in control systems and for being sensitive to modeling errors. The reader is encouraged to consult the many publications that describe this approach, improvements on the basic Smith Predictor topology, and other approaches that employ internal models in closed-loop control systems.

If the model of the manipulated components $\hat{G}_m(z) \approx G_m(z)$ and integer delay $\hat{d} \approx d$, then

$$C(z) \approx \frac{G_c(z)G_m(z)}{1+G_c(z)G_m(z)} z^{-d} R(z) + \frac{1+G_c(z)G_m(z)\left(1-z^{-d}\right)}{1+G_c(z)G_m(z)} P(z) \tag{6.60}$$

Thus, if the model of the production system is sufficiently accurate, the Smith Predictor topology effectively places the time delay of d sampling periods outside the closed loop. The characteristic equation then is

$$1+G_c(z)G_m(z) \approx 0, \tag{6.61}$$

which is not a function of delay. The detrimental effects of delay in a feedback loop therefore are circumvented, improving opportunities for obtaining better closed-loop dynamic behavior in design of decision-making including shorter settling times.

The Smith Predictor topology is well known for being sensitive to modeling errors. A minor difference between the dynamic behavior of the actual production system and the dynamic behavior of the production system model can result in an undesirable or unstable closed-loop system. Therefore, the dynamics of the manipulated components of the production system, including delay, must be accurately known in general if application of the Smith Predictor topology is to be successful.

Example 6.17 Circumventing Delay in Discrete-Time Regulation of Lead Time Using Work Output Error and a Smith Predictor Topology

A block diagram for regulation of lead time in a production system is shown in Figure 6.41 in which there is delay D days in implementing production capacity adjustment decisions. This delay limits achievable performance in regulation of lead time and it can be circumvented using the Smith Predictor decision-making topology. As described in Example 6.13, the planned lead time is used to calculate the amount of work that a production system is expected to complete as a function of time. Any difference between the expected or desired amount of work completed and the actual amount of work completed then is used to adjust the capacity of the production system to eliminate this difference so that work is completed on time. In Example 6.13 there was no delay in adjusting production capacity, and decision-making was designed for dead-beat response in which integral control was in the forward path and lead control was in the feedback path.

In this example, delay D days is inside the closed-loop portion of the production system and significantly limits dynamic behavior such as settling time for which decision-making can be successfully designed. Fortunately, the delay in implementing capacity adjustments is likely to be well known because it is established by logistic practices in production. Furthermore, the remainder of the production system model also is well known because it simply is an accumulation of work. In this case, the Smith Predictor topology may be a good candidate for circumventing delay D in adjusting production capacity.

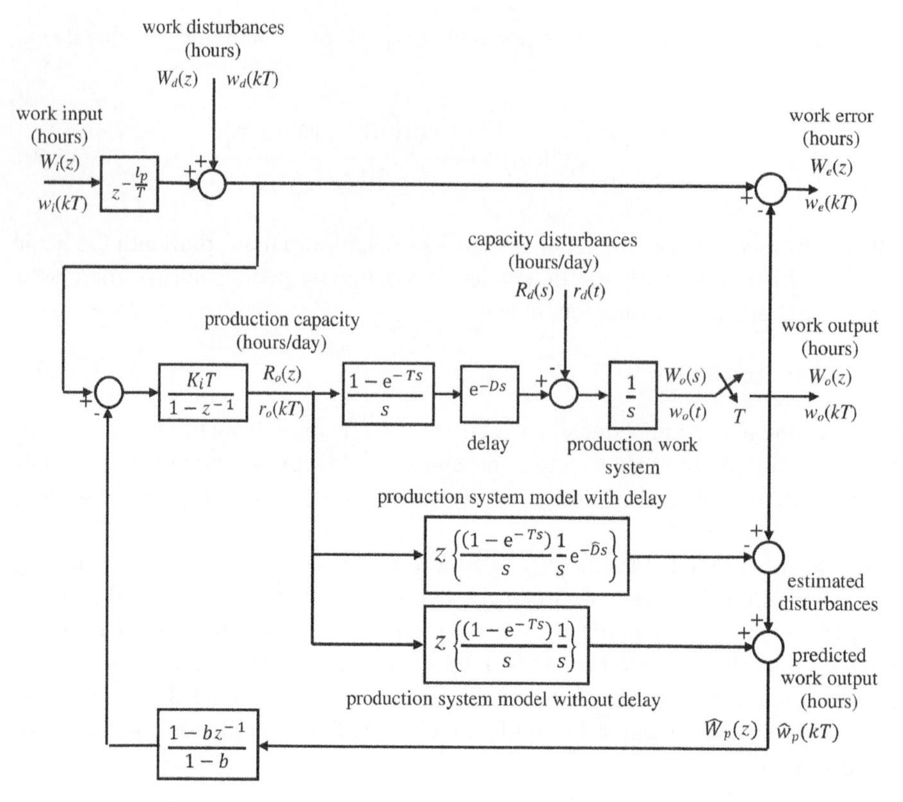

Figure 6.41 Smith Predictor topology for regulation of lead time using work output.

A block diagram for regulation of lead time using the Smith predictor topology is shown in Figure 6.41. The discrete-time transfer function of the production system without delay is

$$G_p(z) = \frac{T}{1 - z^{-1}}$$

where T days is the period between adjustments in production capacity. Decision-making with integral-lead control in the form of Equation 6.27 is used, with component

$$G_c(z) = \frac{K_i T}{1 - z^{-1}}$$

in the forward path and component

$$H_c(z) = \frac{1 - bz^{-1}}{1 - b}$$

in the feedback path. The Smith predictor topology effectively places the delay outside the closed-loop portion of the control system when modeled delay $\hat{D}$ days is equal to the actual delay D days. When $D = T$ days, the characteristic equation in Equation 6.61 becomes

$$z^2 - \left(2 - \frac{K_i T^2}{1-b}\right)z + \left(1 - \frac{K_i T^2 b}{1-b}\right) = 0$$

and decision-making components $G_c(z)$ and $H_c(z)$ can be designed without considering the delay in implementing capacity adjustments.

Program 6.12 demonstrates how the response of work output to a step work disturbance of 10 hours, a rush order for example, can be calculated when period $T = 3$ days,

Program 6.12 Calculation of response of discrete-time integral-lead control with Smith Predictor topology

```
T=3;  % period between production capacity adjustments (days)

D=3;  % actual delay in implementing capacity adjustments (days)
Gmz=c2d(tf(1,[1, 0],'TimeUnit','days','InputDelay',D),T);  % actual

Dhat=D;  % modeled delay in implementing capacity adjustments (days)

% work system models with and without delay
GmhatDz=c2d(tf(1,[1, 0],'TimeUnit','days','InputDelay',Dhat),T);
Gmhatz=c2d(tf(1,[1, 0],'TimeUnit','days'),T);

% integral-lead control for closed-loop roots=0
b=0.5; Ki=1/T^2
```

```
Ki = 0.1111
```

```
Gciz=tf([Ki*T 0],[1 -1],T,'TimeUnit','days');
GcleadzSP0=tf([1, -b],[1-b, 0],T,'TimeUnit','days');

Gcloz=minreal(feedback(feedback(Gciz,...
    (Gmhatz-GmhatDz)*GcleadzSP0)*Gmz,GcleadzSP0));  % work output
Gclmz=minreal(feedback(feedback(Gciz,...
    (Gmhatz-GmhatDz)*GcleadzSP0),GcleadzSP0*Gmz));  % capacity
damp(Gcloz)
```

Pole	Magnitude	Damping	Frequency (rad/days)	Time Constant (days)
0.00e+00	0.00e+00	NaN	NaN	NaN
0.00e+00	0.00e+00	NaN	NaN	NaN

```
Wrush=10;  % plot response to 10-day rush order work disturbance

step(Wrush*Gcloz,40)  % work output - Figure 6.42a
xlabel('time kT'); ylabel('work output w_o(kT) (hours)')

step(Wrush*Gclmz,40)  % production capacity - Figure 6.42a
xlabel('time kT'); ylabel('production capacity r_o(kT) (hours/day)')

% integral-lead control for closed-loop time constant tau=T days
tau=T; a=exp(-T/tau); b1=(1-a^2)/(1-a)/2, Ki1=(1-a)^2/T^2
```

```
b1 = 0.6839
Ki1 = 0.0444
```

```
Gciz1=tf([Ki1*T, 0],[1, -1],T,'TimeUnit','days');
Gcleadz1=tf([1, -b1],[1-b1, 0],T,'TimeUnit','days');

Gcloz1=minreal(feedback(minreal(feedback(Gciz1,...
    minreal(Gmhatz-GmhatDz)*Gcleadz1))*Gmz,Gcleadz1)); % work output
Gclmz1=minreal(feedback(feedback(Gciz1,...
    (Gmhatz-GmhatDz)*Gcleadz1),Gcleadz1*Gmz)); % capacity
damp(Gcloz1)
```

```
          Pole                Magnitude     Damping     Time Constant
                                                           (days)
 3.68e-01 + 1.39e-08i         3.68e-01      1.00e+00      3.00e+00
 3.68e-01 - 1.39e-08i         3.68e-01      1.00e+00      3.00e+00
```

```
step(Wrush* Gcloz1,40)  % work output - Figure 6.42b
xlabel('time kT'); ylabel('work output w_o(kT) (hours)')

step(Wrush* Gclmz1,40)  % productin capacity - Figure 6.42b
xlabel('time kT'); ylabel('production capacity r_o(kT) (hours/day)')
```

delay $D = 3$ days, and the delay in the production work system model is accurately known. The response of work output to the step work disturbance shown in Figure 6.42a is obtained when decision-making is designed for dead-beat response using the direct design approach described in Section 6.4:

$$b = 0.5$$

$$K_i = \frac{1}{T^2}$$

Integral control parameter $K_i = 0.111$ days^{-1} and the characteristic equation has two equal roots $r_1 = r_2 = 0$. The production capacity generated by the combined integral-lead control and internal model also are shown in Figure 6.42a. The settling time obtained with this design and the Smith Predictor topology is $2T = 6$ days because delay D is still present. It can be shown that for this design the closed-loop system is unstable when there is an unanticipated additional delay of one day in implementing capacity adjustments in the production work system.

An alternative design is to make the roots of the characteristic equation $r_1 = r_2 = e^{-T/\tau}$ where closed-loop time constant $\tau = T = 3$ days. Then

$$b = 2\frac{1 - e^{-2T/\tau}}{1 - e^{-T/\tau}}$$

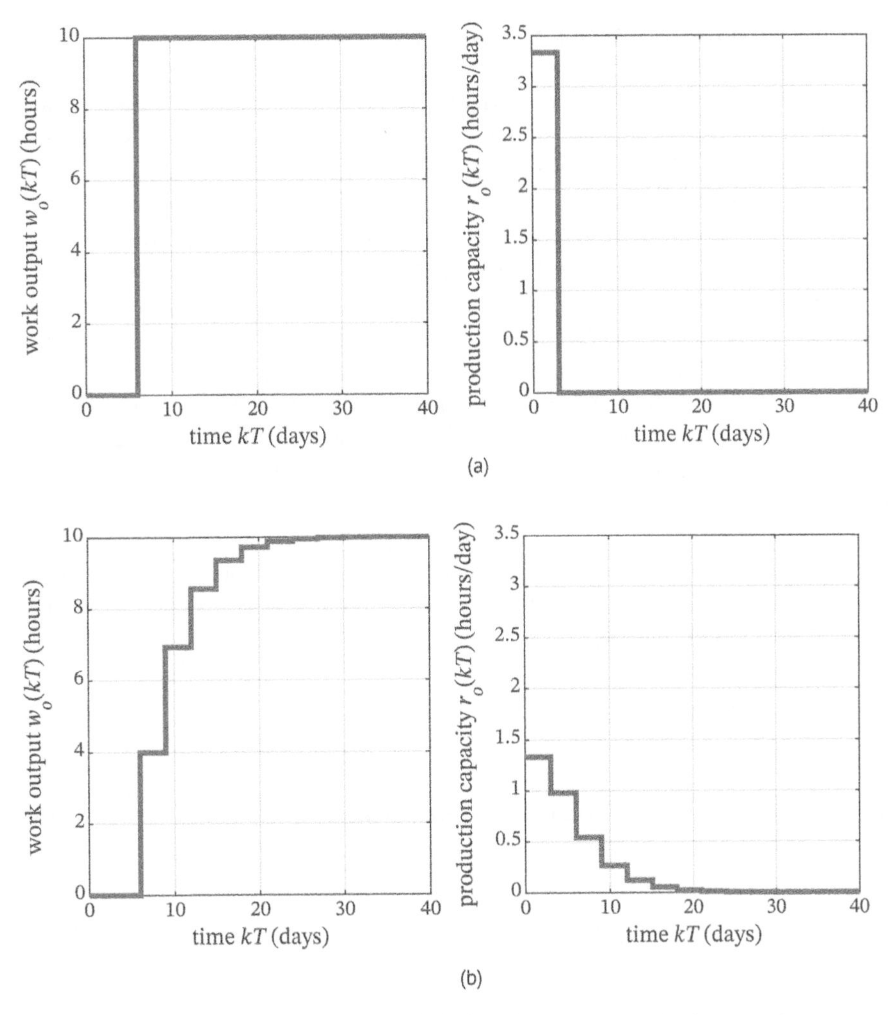

Figure 6.42 Response to step work disturbance input with Smith Predictor topology. (a) $b = 0.5$, $K_i = 0.111$ days^{-1}. (b) $b = 0.684$, $K_i = 0.0444$ days^{-1}.

$$K_i = \frac{\left(1 - e^{-T/\tau}\right)^2}{T^2}$$

Lead control parameter b = 0.684 and integral control parameter value $K_i = 0.0444$ days^{-1}.

The responses of work output and production capacity to the 10-hour step work disturbance with this alternative design are shown in Figure 6.42b. Production capacity adjustments are smaller and persist over a longer period compared to responses in Figure 6.42a. It can be shown that for this alternative design the closed-loop system remains stable when there is an unanticipated additional delay of one day in implementing capacity adjustments in the production work system, but response becomes oscillatory and settling time increases.

6.7 Sensitivity to Parameter Variations

The linear models that represent the dynamic behavior of production systems using differential and difference equations often are approximations. For various reasons, many aspects and details remain unmodeled: they are not well understood, they would make dynamic analysis intractable, etc. Often, production system behavior is only approximately represented by parameters in these models. The values of these parameters are uncertain and are assumed to be constant when in fact they may be variable, perhaps within some range, depending on operating conditions. It is useful to be able to quantify how measures of dynamic behavior vary as these parameters vary. This can be referred to as dynamic sensitivity with respect to parameters of interest. Dynamic sensitivity can be calculated as a function of fundamental dynamic characteristics and performance measures such as damping ratio, settling time, percent overshoot, bandwidth, magnitude margin, phase margin, and roots of the characteristic equation.

One way to evaluate sensitivity of measure m to production system parameter p is to determine the range $m_{min} \leq m \leq m_{max}$ that corresponds to range $p_{min} \leq p \leq p_{max}$. Alternatively, the normalized sensitivity of measure m to production system parameter p can be determined using

$$S_p^m = \frac{\partial m / m}{\partial p / p} = \left(\partial m / \partial p \right) \left(p / m \right) \tag{6.62}$$

Because this equation often is inconvenient to evaluate analytically, normalized sensitivity can be determined approximately using

$$S_p^m \approx \frac{\Delta m / m}{\Delta p / p} = \left(\Delta m / \Delta p \right) \left(p / m \right) \tag{6.63}$$

where Δp is a small change in parameter p and Δm is the change in measure m that results.

Evaluation of sensitivity is important in understanding the variation in dynamic behavior that can be expected in operation of a production system. Sensitivity can be different for different closed-loop decision-making topologies and if sensitivity is unacceptable for a chosen decision-making topology, a different topology can be selected. Additionally, models can be improved, production systems can be redesigned to be less variable, or decision-making can be designed to vary with production system operating conditions.

Example 6.18 Sensitivity of Closed-Loop Time Constant to Variation in Production Using Metal Forming Parameter K_m

In Example 6.4, decision-making was designed for deciding how many metal forming presses are required to produce parts that pass quality requirements. The rate of production of parts that pass quality requirements was assumed to be observed twice per hour and the number of presses, which was not required to be an integer, was adjusted twice per hour using discrete-time integral control. In the discrete-time block diagram for the production system in Figure 6.8, the simple discrete-time model of production using metal forming is

$$r_m(kT) = K_m m((k-1)T)$$

where $m(kT)$ is the number of presses and production parameter $K_m > 0$ (parts/hour)/press relates the number of presses to rate of production $r_m(kT)$. The number of presses is adjusted using discrete-time integral control:

$$e(kT) = r_r(kT) - r_o(kT)$$

$$m(kT) = m((k-1)T) + K_i T e(kT)$$

where $r_r(kT)$ parts/hour is the desired rate of production of parts that meet quality requirements, $e(kT)$ parts/hour is the error between this desired rate of production and the actual rate of production that meets quality requirements. Integral controller parameter K_i was designed in Example 6.4 using

$$K_i = \frac{1 - e^{-T/\tau}}{K_m T}$$

which resulted in closed-loop dynamic behavior with desired time constant τ hours. The sensitivity of this time constant to variation in parameter K_m is of interest because the model of production using metal forming is greatly simplified and many physical phenomena that affect quality are not modeled.

The relationship between closed-loop time constant τ and parameters K_m, K_i, and T is

$$\tau = \frac{-T}{\ln(1 - K_i K_m T)}$$

For $K_m = 200$ (parts/hour)/press, $T = 0.5$ hours, and $\tau = 0.5$ hours in Example 6.4, $K_i = 0.0063$ (presses/hour)/(parts/hour) was found. If this K_i is used, but K_m actually can have range $150 \le K_m \le 250$ (parts/hour)/press then, as calculated using Program 6.13,

Program 6.13 Calculation of sensitivity of closed-loop time contant to production parameter K_m

```
T=0.5;  % period between adjustments (hours)
Km=200;  % production constant ((parts/hour)/press)

tau=0.5;  % desired time constant (hours)
Ki=(1-exp-T/tau))/Km/T  % integral control (presses/(parts/hour))
```

```
    Ki = 0.0063
```

```
KmMin=150;  % 50≤Km≤250 (parts/hour)/press
KmMax=250;
tau1=-T/log(1-Ki*KmMin*T)  % time constant range (hours)
```

```
    tau1 = 0.7781

tau2=-T/log(1-Ki*KmMax*T)

    tau2 = 0.3202

% normalized sensitivity for time constant tau=0.5 hours
deltaKm=0.001*Km  % 0.1% change in Km

    deltaKm = 0.2000

deltaTau=-T/log(1-Ki*(Km+deltaKm)*T)-tau  % change in time constant

    deltaTau = -8.5840e-04

S=(deltaTau/tau)/(deltaKm/Km)  % normalized sensitivity

    S = -1.7168

% normalized sensitivity for alternative time constant tau=1 hour
tauA=1;
KiA=(1-exp(-T/tauA))/Km/T

    KiA = 0.0039

deltaTauA=-T/log(1-KiA*(Km+deltaKm)*T)-tauA  % change in time constant

    deltaTauA = -0.0013

S1=(deltaTauA/tauA)/(deltaKm/Km)  % new normalized sensitivity
S1 = -1.2962
```

the corresponding range of the resulting closed-loop time constant is $0.32 \leq \tau \leq 0.78$ hours.

The normalized sensitivity of closed-loop time constant τ to uncertainty in parameter K_m can be calculated using Equation 6.63:

$$S_{K_m}^{\tau} \approx \frac{\Delta\tau / \tau}{\Delta K_m / K_m}$$

In this case, the normalized sensitivity calculated in Program 6.13 using $\Delta K_m = 0.2$ (parts/hour)/press is $S_{Km}{}^{\tau} = -1.7$.

If the integral control parameter is reduced to $K_i = 0.0039$ (presses/hour)/(parts/hour), the normalized sensitivity calculated in Program 6.13 using $\Delta K_m = 0.2$ (parts/hour)/press is $S_{Km}{}^{\tau} = -1.3$. Hence, the sensitivity of the closed-loop time constant to variations in production parameter K_m has decreased with decreasing integral control parameter K_i.

6.8 Summary

In the closed-loop cycle illustrated in Figure 6.1, the operation of a production system and its components is monitored and collected data are analyzed to detect errors and deviations from goals and targets. Decisions are made and actions are taken to implement decisions with the objective of manipulating the production system to effectively correct errors and deviations. There are many possible topologies for closed-loop decision-making in production systems, and some of the most common and important alternatives have been described in this chapter for both continuous-time and discrete-time production systems. These included the proportional, integral, and derivative control, which is perhaps the most well-known, allocation of decision-making components to the forward and feedback paths in the closed loop, and cascade, feedforward, and Smith predictor topologies. Calculation of sensitivity was discussed at the end of the chapter, and the results of sensitivity analysis can be important in verifying the suitability of chosen decision-making topologies and parameters in the presence of modeling inaccuracies including uncertainty and variation in model parameters.

Examples have been presented that illustrate design of decision-making for closed-loop production systems to meet dynamic behavior requirements. The key steps include setting dynamic behavior requirements, choosing the closed-loop decision-making topology, choosing decision-making parameter values, assessing the design, and modifying the design, if necessary, to meet requirements. In Chapter 7, several additional examples will be presented that are of higher complexity and further illustrate analysis and design of the dynamic behavior of production systems from both the time and frequency perspectives.

7

Application Examples

The examples in previous chapters have been focused on illustrating the basic theoretical developments presented. In this chapter, several more complex application examples are presented that further illustrate modeling, analyzing, and designing the dynamic behavior of production systems. These application examples show how control theoretical modeling and analyses lead to better understanding of the dynamic behavior of production systems and how the dynamic behavior of such systems can be designed from both time and frequency perspectives.

In the first application example, relationships are explored between adjustment period and time to recover from disturbances and stability in replanning for lead time regulation in a production system. The goal is to quantitatively assess the potential of digitalization to reduce delays in replanning and decrease the time to recover from disturbances. The second application example is focused on modeling and analyzing the dynamic behavior of decision-making and adjustment of steel coil deliveries in a production network with inventory information sharing. This example illustrates the use of matrices of transfer functions in understanding dynamic interactions in a network of two steel companies and two galvanizing lines when demand for galvanizing steel coils fluctuates at various frequencies. The third application example illustrates modeling and analysis of the dynamic behavior of a network of production work systems and explores how this behavior is affected by the structure of order flows between the work systems and sharing of order flow information.

The final examples focus on applications in which there are multiple components that make decisions with different periods. The first of these application examples requires modeling of a production system in which cross-trained employee capacity and permanent employee capacity are adjusted to meet varying demand. The period between adjustments in cross-trained employee is relatively short, while the period between adjustments in permanent employee capacity is relatively long, and these adjustments are focused on different ranges of frequency of variations in demand. This application example illustrates the use of frequency response methods in both the design of decision-making for this system and in analysis of its dynamic behavior. The final application example requires modeling a production planning and control system in which backlog and work in progress (WIP) are regulated. There is a relatively long period between adjustments of work-system capacity in the outer backlog regulation loop, and there is a

Control Theory Applications for Dynamic Production Systems: Time and Frequency Methods for Analysis and Design, First Edition. Neil A. Duffie.

relatively short period between adjustments of order release rate in the inner work-system WIP regulation loop. Both backlog and WIP regulation need to be designed.

7.1 Potential Impact of Digitalization on Improving Recovery Time in Replanning by Reducing Delays

In this application example, opportunities are explored for digitalization to improve performance of replanning in a production system with lead time regulation [1]. Digitalization in a production system refers to adoption or increase in use of digital technologies. These technologies can be used to reduce delays by automating data collection, decision-making, and decision implementation. Reduction of delays can lead, in turn, to reduction in the time required to recover from disturbances, thereby improving due date reliability.

In the production system considered, incoming work is assumed to be scheduled using production planning and control software. At regularly scheduled planning meetings, the state of the production system is assessed and adjustments in work input are discussed with the goal of adjusting lead times to improve due-date reliability. This replanning cycle is illustrated in Figure 7.1. Instability is a known problem in replanning to regulate lead time. The effects of previous adjustments need to be considered when making replanning decisions because fluctuations in lead time can occur when the period between adjustments is too short compared to the delays in making and implementing planning decisions that are present in a replanning cycle.[1] To avoid

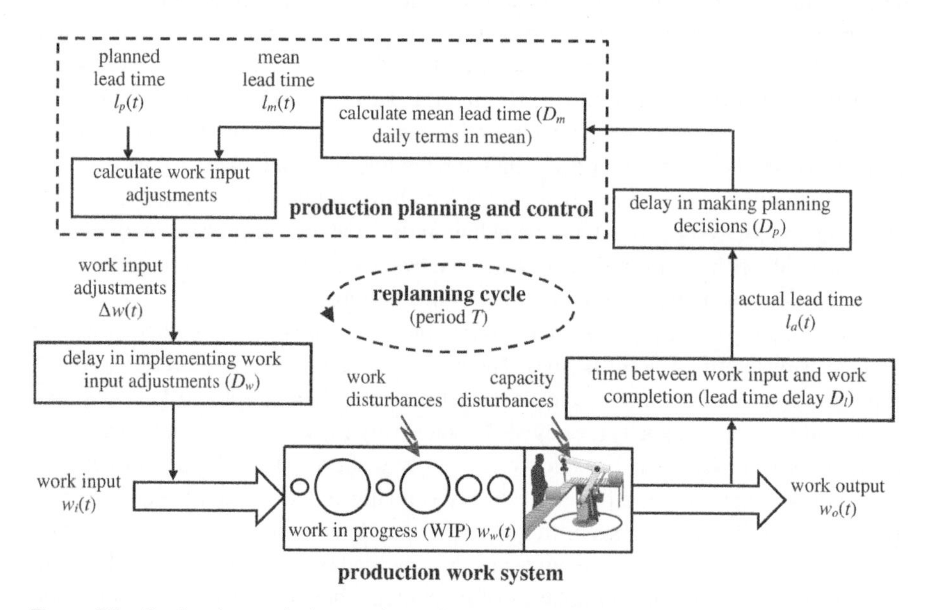

Figure 7.1 Replanning cycle for regulating lead time in a production system.

1 This sometimes is referred to as the lead time syndrome of production control. When a company's due-date reliability is low, production planners and managers can have the mistaken impression that "more often is better" when making adjustments that are intended to improve due date reliability, Unfortunately, the result can be instability and a further decrease due-date reliability.

instability, the period between replanning adjustments therefore must be relatively long when delays are relatively long, and the time required for the system to recover from disturbances then also is relatively long because this recovery only is achieved as a result of adjustments implemented in the replanning cycle. Performance can be improved by reducing the period between adjustments if digital technologies can be used to reduce or eliminate delays in the replanning cycle.

A dynamic model of a replanning cycle such as that shown in Figure 7.1 can be useful in understanding the relationship between dynamic behavior and the choice of adjustment period given the delays that are present in the cycle. Then, the potential benefit of reducing specific delays can be evaluated and means for reducing delay can be considered. A block diagram for such a model is shown in Figure 7.2, and the variables and parameters in the model are defined as follows:

T	period between work adjustments (days)
$w_i(t)$	work input (hours)
$w_o(t)$	work output (hours)
$w_w(t)$	work in progress (hours)
$l_a(t)$	lead time (days)
$l_m(kT)$	mean lead time (days)
$l_p(kT)$	planned lead time (days)
$\Delta w(kT)$	work adjustments (hours)
r_p	planned work system capacity (hours/day)
r_a	actual work system capacity (hours/day)
K_w	work adjustment parameter (-/-)
D_l	delay between arrival of work at the production system and the determination of the lead time based on completion of the work (days)

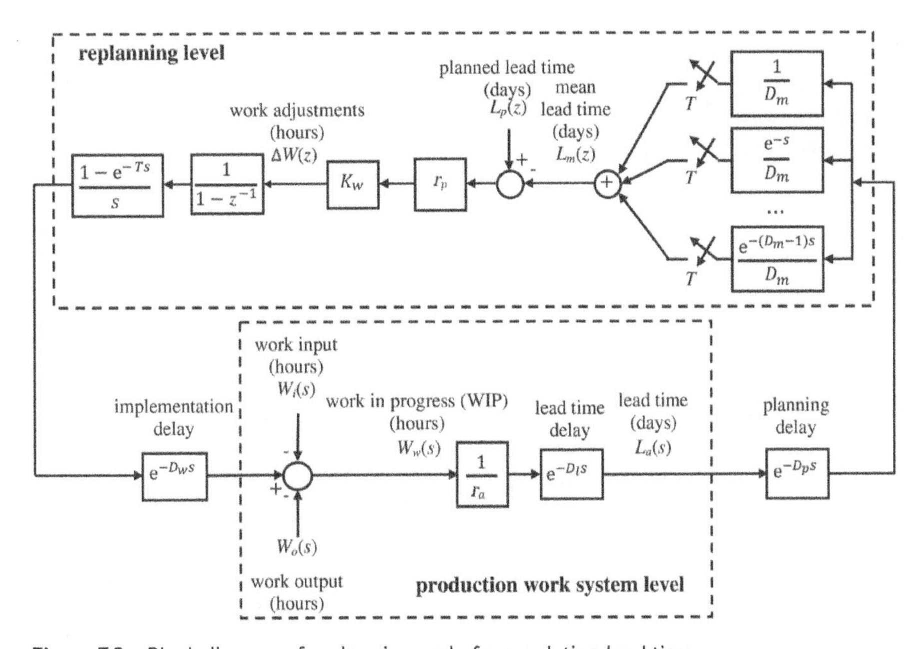

Figure 7.2 Block diagram of replanning cycle for regulating lead time.

D_p delay in decision-making (days)
D_w delay in work adjustment (days)
D_m number of daily terms in mean lead time calculation (days, positive integer)

The work in progress is

$$w_w(t) = w_i(t) - w_o(t) + \sum_{n=0}^{\text{floor}\left(\frac{kT - D_w}{T}\right)} \Delta w(nT)$$

where the amount of work that is input to the production system is adjusted by $\Delta w(nT)$ hours at the beginning of each adjustment period. The mean lead time is

$$l_m(kT) = \frac{1}{d_m} \sum_{m=0}^{D_m - 1} \frac{w_i(kT - D_l - D_p - m) - w_o(kT - D_l - D_p - m) + \sum_{n=0}^{\text{floor}\left(\frac{kT - D_l - D_p - D_w - m}{T}\right)} \Delta w(nT)}{r_a}$$

Delay D_l is included because lead time is determined after work is completed rather at the time work is input, and it is assumed to be constant even though lead time is a variable in the model. The numerical results presented later in this example illustrate the impact of different values of lead time delay D_l.

The mean lead time is used at the replanning level to adjust work entering the production work system with the goal of making the calculated mean lead time the same as the planned lead time:

$$\Delta w(kT) = K_w r_p \left(l_p(kT) - l_m(kT)\right)$$

where work adjustment parameter K_w is chosen to produce favorable dynamic behavior.

The work adjustments then are

$$\Delta w(z) = \frac{K_w \left(r_p L_p(z) - \frac{1}{D_m} \sum_{m=0}^{D_m - 1} \left(\mathcal{Z}\{w_i(t - D_l - D_p - m)\} - \mathcal{Z}\{w_o(t - D_l - D_p - m)\} \right) \right)}{1 + \frac{1}{1 - z^{-1}} K_w \frac{r_p}{r_a} \frac{1}{d_m} \sum_{m=0}^{D_m - 1} z^{-\text{floor}\left(\frac{D_l + D_p + D_w + m}{T}\right)} z^{-1}}$$

The characteristic equation associated with the replanning cycle then is

$$1 - z^{-1} + K_w \frac{r_p}{r_a} \frac{1}{D_m} \sum_{m=0}^{D_m - 1} z^{-\text{floor}\left(\frac{D_l + D_p + D_w + m}{T}\right)} z^{-1} = 0$$

The characteristic equation is the same for all inputs.

The roots of this characteristic equation provide insight into the fundamental dynamic properties of the replanning cycle, and the roots are functions of adjustment period T, work adjustment parameter K_w, the number of daily terms included in the mean lead time calculation D_m, and delays D_l, D_p, and D_w. The minimum (left-most on the number line) damping ratio ζ_{min} in the set of damping ratios associated with all of the roots of the characteristic equation can be used to predict whether the production system will be stable ($\zeta_{min} > 0$) or unstable ($\zeta_{min} \leq 0$). For stable production systems, the maximum time constant τ_{max} in the set of time constants associated with all of the roots of the characteristic equation is closely related to the time required to recover from disturbances. A factor of $4\tau_{max}$, henceforth referred to as the recovery time, is closely related to the 98% settling time, which is the time required for response to disturbances to decay to 2% of the initial response. If the production system is not stable ($\zeta_{min} \leq 0$), then the recovery time $4\tau_{max} = \infty$. These two measures, minimum damping ζ_{min} and recovery time $4\tau_{max}$, can be used to predict the impact of choice of adjustment period T on the dynamic behavior of the production system and assess the improvement in dynamic behavior that potentially can be obtained by applying digital technologies to reduce delays D_l, D_p, and D_w.

When $T > D_l + D_p + D_w + (D_m - 1)$ the characteristic equation becomes

$$1-\left(1-K_w\frac{r_p}{r_a}\right)z^{-1}=0$$

If $K_w = 1$, the lead time can be expected to be fully adjusted in one period T if capacities $r_a = r_p$; however, a choice $0 < K_w < 1$ may be better in practice to avoid overadjustment when there are significant variations in actual capacity r_a or significant variations or uncertainties in delays D_l, D_p, and D_w. The single time constant associated with the single root of the characteristic equation then is

$$\tau=\frac{-T}{\ln\left(1-K_w\frac{r_p}{r_a}\right)}$$

Recovery time $4\tau_{max} = 4\tau$ predicted in this manner increases linearly with adjustment period T, is not an integer multiple of T, and is longer for smaller K_w.

Consider a replanning cycle of the type shown in Figure 7.1 where $r_a = r_p$ and the delays and the number of daily terms in the mean lead time calculation are

lead time delay (D_l)	7.4 days
delay in making planning decisions (D_p)	9 days
delay in implementing work input adjustments (D_w)	4 to 8 days
daily terms in mean lead time calculation (D_m)	20 days

The delay in implementing work adjustments D_w can vary for logistical reasons. Program 7.1 can be used to calculate minimum damping ratio ζ_{min} and recovery time $4\tau_{max}$ for these delays and a range of adjustment periods T. $K_w = 0.8$ is chosen for lead time regulation, which can be expected to result in $4\tau_{max} \approx 2.5T$ when $T > D_l + D_p + D_w + D_m - 1$. This trend is clear in Figure 7.3, which shows the variation in recovery time $4\tau_{max}$ with integer adjustment period T days for both $D_w = 4$ and $D_w = 8$ days. Adjustment period $T = 43$ days can be selected, which results in the fastest recovery time of $4\tau_{max} = 107$ days for the worst-case delay in implementing adjustments of $D_w = 8$ days. Dynamic behavior deteriorates for $T < 43$ days when $D_w = 8$ days, and for $T < 39$ days when $D_w = 4$ days. The latter results in recovery time $4\tau_{max} = 97$ days, an indication of the benefit of reducing delays.

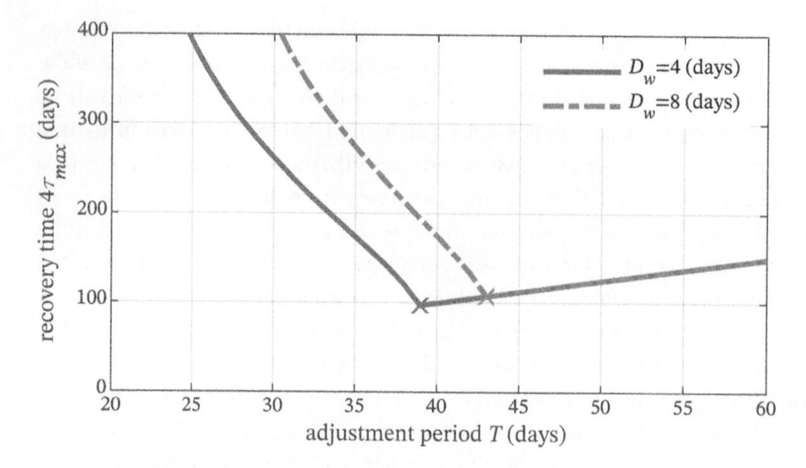

Figure 7.3 Recovery time $4\tau_{max}$ versus adjustment period T.

Program 7.1 Calculation of result of reducing delays in replanning cycle

```
Kw=0.8;  % work adjustment parameter (-/-)
rp=5;  % planned capacity (hours/day)
ra=rp;  % actual capacity (hours/day)

T=(1:60);  % adjustment periods investigated (days)
dampMin=zeros(6,length(T)); tauMax=zeros(6,length(T));  % preallocate

Dl=7.4;  % delay in measuring lead time (days)
Dp=9;  % delay in planning (days)
Dw=4;  % minimum delay in implementing decisions (days)
Dm=20;  % number of daily terms in mean lead time

for ii=1:6  % step by step reduction in delays
  switch ii
    case 2, Dw=8;  % delay in implementing decisions
    case 3, Dp=0;  % eliminate delay in planning
    case 4, Dl=0;  % eliminate delay in measuring lead time
    case 5, Dw=0;  % eliminate delay in implementing decisions
    case 6, Dm=5;  % reduce days in mean lead time calculation
  end

  D=Dl+Dp+Dw;  % all delays except number of daily terms in mean

  for n=1:length(T)  % cycle through T
    z=tf('z',T(n),'TimeUnit','days');  % define z for period T

    % transfer function relating deltaW to delayed mean lead time
    G=tf(0,1,T(n));
    for m=0:Dm-1 % dm components of tlm
      G=minreal(G+(1/Dm/ra/(1-z^-1))*z^-(1+floor((D+m)/T(n))));
    end

    H=minreal(1/(1+Kw*rp*G));  % closed-loop tranfer function
    [Wn,zeta]=damp(H);  % damping ratios and natural frequencies
    [dampMin(ii,n),nDamp]=min(zeta);  % minimum damping ratio
    [tauMax(ii,n),nTau]=max(1./(zeta.*Wn));  % maximum time constant
    if dampMin(ii,n)<=0  % infinite recovery time if not stable
      tauMax(ii,n)=Inf;
    end
```

```
    end
    [minTauMax(ii),nTau]=min(tauMax(ii,:));  % lowest recovery time
    TminTauMax(ii)=T(nTau);  % associated adjustment period T
end

plot(T,4*tauMax(1,:),T,4*tauMax(2,:));  % recovery time - Figure 7.3
xlim([0 60]); ylim([0 400]); grid
xlabel('adjustment period T (days)'); ylabel('recovery time 4\tau_{max}
    (days)')
legend('D_w=4 (days)','D_w=8 (days)')

[TminTauMax', 4*minTauMax']  % recovery times - Table 7.1
```

Table 7.1 Improvement of recovery time by application of digital technologies to reduce delays in replanning.

delay (days)	digital technology	best adjustment period T (days)	best recovery time $4\tau_{max}$ (days)
$D_p = 9$ days $D_l = 7.4$ days $D_w = 8$ days $D_m = 20$ days	base case	43	107
$D_p = 0$ days $D_l = 7.4$ days $D_w = 8$ days $D_m = 20$ days	automate lead time data gathering and automate work adjustment decision-making to eliminate planning delay	34	86
$D_p = 0$ days $D_l = 0$ days $D_w = 8$ days $D_m = 20$ days	use embedded model that predicts lead time based on current WIP, capacity, setup times, etc. to eliminate delay associated with measuring lead time	27	67
$D_p = 0$ days $D_l = 0$ days $D_w = 0$ days $D_m = 20$ days	automate implementation of work adjustments to eliminate delay	19	47
$D_p = 0$ days $D_l = 0$ days $D_w = 0$ days $D_m = 5$ days	reduce number of days in mean lead time calculation	5	12

Results obtained when delays D_l, D_p, and D_w. are successively eliminated and the number of daily terms in the mean lead time calculation D_m is reduced from 20 days to 5 days are summarized in Table 7.1, which also lists digital technologies that possibly could be used to eliminate or reduce delays. While the complexity of a replanning system may increase with increasing levels of digitalization, the benefits obtained are predicted to be significant. Digital technologies therefore have significant potential for improving the time to recover from disturbances in replanning.

7.2 Adjustment of Steel Coil Deliveries in a Production Network with Inventory Information Sharing

In this application example it is shown that the merit of information sharing in a production network that produces steel coils can be assessed quantitatively by analyzing frequency response for likely periods of variation of coil production rate [2]. This can help assess whether the increase in complexity that results from information sharing and decision-making based on information sharing is acceptable, and also help decide how often decisions and adjustments need to be made based on shared information.

Consider two companies, Company A and Company B, that produce coils of steel that require galvanizing. The companies deliver these coils to two galvanizing lines, Galvanizing Line 1 and Galvanizing Line 2, as illustrated in Figure 7.4. Company A delivers the majority of its coils by truck to Galvanizing Line 1, which is nearby, but it also delivers a portion of its coils by train to Galvanizing Line 2, which is farther away. Similarly, Company B delivers the majority of its coils by truck to Galvanizing Line 2, which is nearby, but it also delivers a portion of its coils by train to Galvanizing Line 1, which is farther away. Variations in the companies' output rates of coils that require galvanizing result in variations in coil inventory that is waiting to be processed at the two galvanizing lines.

Because galvanizing line coil inventory affects lead time in supplying galvanized steel coils to customers, it is of interest to investigate whether the companies should independently increase the portion of their coil output that they send to the galvanizing line with less inventory, and correspondingly decrease the portion of their coil output that they send to the galvanizing line with more inventory. This would be expected to decrease difference between the inventories of the two galvanizing lines, reduce the difference in lead times and avoid long lead times. To facilitate this, the galvanizing lines need to share inventory information with the companies.

The block diagram in Figure 7.5 shows how information sharing and coil delivery adjustments can be modeled for this production network. Both companies receive coil

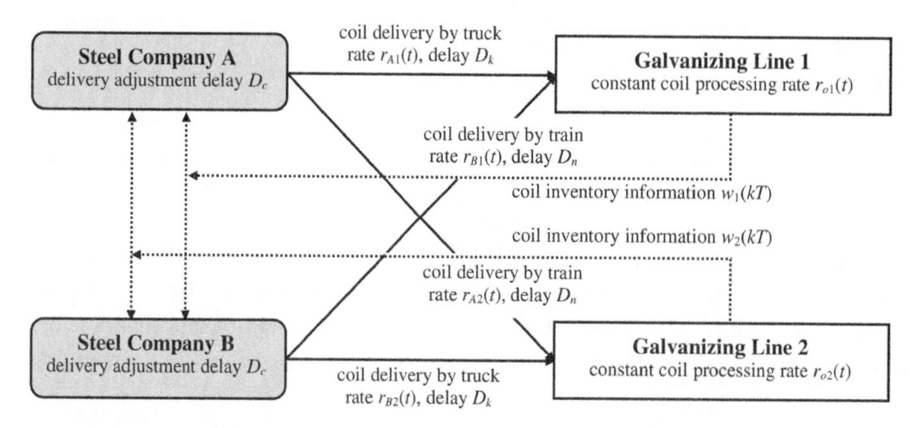

Figure 7.4 Galvanizing lines share coil inventory information with steel companies.

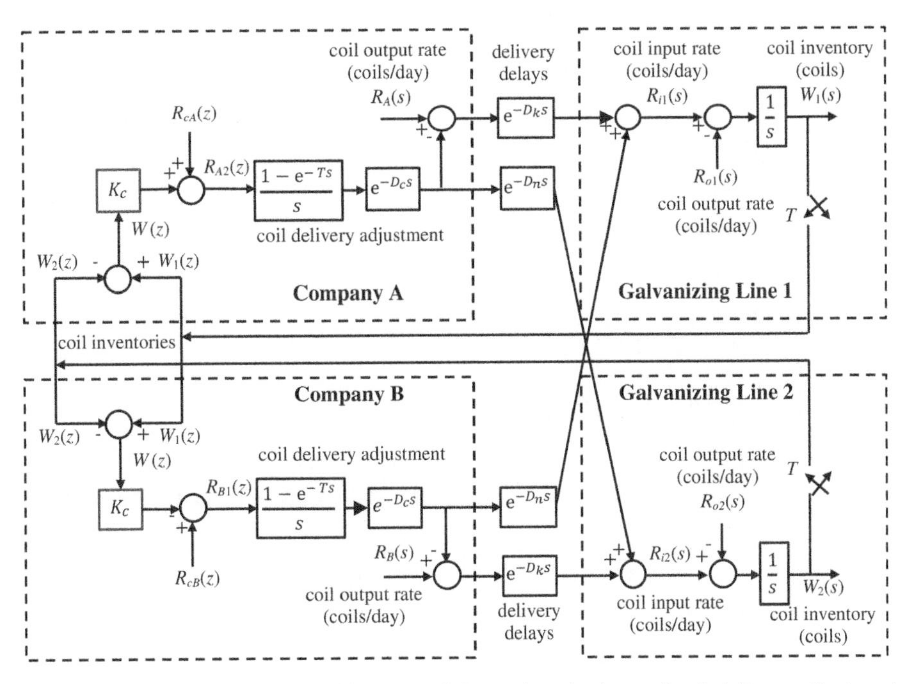

Figure 7.5 Block diagram for coil inventory information sharing and coil delivery adjustment.

inventory information from both galvanizing lines, and the flow of coils is modeled rather than the delivery and processing of individual coils. The coil delivery decision-making process, which is performed independently by each company, is modeled by assuming that the companies make regular adjustments in coil deliveries with a period of T days between adjustments. The variables and parameters in the model are defined as follows:

$r_A(t)$ coil output rate of Company A (orders/day)
$r_{A2}(kT)$ portion of the coil output rate of Company A that is delivered to more distant Galvanizing Line 2 (orders/day)
$r_B(t)$ coil output rate of Company B (orders/day)
$r_{B1}(kT)$ portion of the coil output rate of Company B that is delivered to more distant Galvanizing Line 1 (orders/day)
$r_{cA}(kT)$ nominal portion of the coil output rate of Company A that always is delivered by train to more distant Galvanizing Line 2 (orders/day)
$r_{cB}(kT)$ nominal portion of the coil output rate of Company B that always is delivered by train to more distant Galvanizing Line 1 (orders/day)
K_c coil delivery adjustment parameter used by Companies A and B (days^{-1})
D_c delay in making and implementing coil delivery adjustments (days)
D_k delay in delivering coils by truck (days)
D_n delay in delivering coils by train (days)
$w_1(t)$ coil inventory of Galvanizing Line 1 (orders)
$w_2(t)$ coil inventory of Galvanizing Line 2 (orders)
$r_{i1}(t)$ coil input rate to Galvanizing Line 1 (orders/day)
$r_{i2}(t)$ coil input rate to Galvanizing Line 2 (orders/day)
$r_{o1}(t)$ coil processing rate of Galvanizing Line 1 (orders/day)
$r_{o2}(t)$ coil processing rate of Galvanizing Line 2 (orders/day)

As shown in the block diagram in Figure 7.5, both companies use the same coil delivery adjustment strategy:

$$\begin{bmatrix} R_{B1}(z) \\ R_{A2}(z) \end{bmatrix} = K_c \begin{bmatrix} -1 & 1 \\ 1 & -1 \end{bmatrix} \begin{bmatrix} W_1(z) \\ W_2(z) \end{bmatrix} + \begin{bmatrix} R_{cB}(z) \\ R_{cA}(z) \end{bmatrix}$$

The model for coil inventory in the galvanizing lines is

$$\begin{bmatrix} W_1(z) \\ W_2(z) \end{bmatrix} = \begin{bmatrix} \mathcal{Z}\left\{ \dfrac{\mathrm{e}^{-D_k s}}{s} R_A(s) \right\} \\ \mathcal{Z}\left\{ \dfrac{\mathrm{e}^{-D_k s}}{s} R_B(s) \right\} \end{bmatrix} - \begin{bmatrix} \mathcal{Z}\left\{ \dfrac{1}{s} R_{o1}(s) \right\} \\ \mathcal{Z}\left\{ \dfrac{1}{s} R_{o2}(s) \right\} \end{bmatrix}$$

$$+ \begin{bmatrix} \mathcal{Z}\left\{ \dfrac{\left(1-\mathrm{e}^{-Ts}\right)}{s} \dfrac{\mathrm{e}^{-D_c s}\mathrm{e}^{-D_n s}}{s} \right\} & -\mathcal{Z}\left\{ \dfrac{\left(1-\mathrm{e}^{-Ts}\right)}{s} \dfrac{e^{-D_c s}\mathrm{e}^{-D_k s}}{s} \right\} \\ -\mathcal{Z}\left\{ \dfrac{\left(1-\mathrm{e}^{-Ts}\right)}{s} \dfrac{\mathrm{e}^{-D_c s}\mathrm{e}^{-D_k s}}{s} \right\} & \mathcal{Z}\left\{ \dfrac{\left(1-\mathrm{e}^{-Ts}\right)}{s} \dfrac{\mathrm{e}^{-D_c s}\mathrm{e}^{-D_n s}}{s} \right\} \end{bmatrix} \begin{bmatrix} R_{B1}(z) \\ R_{A2}(z) \end{bmatrix}$$

Combining the models for the companies and the galvanizing lines yields

$$\begin{bmatrix} W_1(z) \\ W_2(z) \end{bmatrix}$$

$$= \left(\mathrm{I} - \begin{bmatrix} \mathcal{Z}\left\{ \dfrac{\left(1-\mathrm{e}^{-Ts}\right)}{s} \dfrac{\mathrm{e}^{-D_{cB} s}\mathrm{e}^{-D_n s}}{s} \right\} & -\mathcal{Z}\left\{ \dfrac{\left(1-\mathrm{e}^{-Ts}\right)}{s} \dfrac{\mathrm{e}^{-D_{cA} s}\mathrm{e}^{-D_k s}}{s} \right\} \\ -\mathcal{Z}\left\{ \dfrac{\left(1-\mathrm{e}^{-Ts}\right)}{s} \dfrac{\mathrm{e}^{-D_{cB} s}\mathrm{e}^{-D_k s}}{s} \right\} & \mathcal{Z}\left\{ \dfrac{\left(1-\mathrm{e}^{-Ts}\right)}{s} \dfrac{\mathrm{e}^{-D_{cA} s}\mathrm{e}^{-D_n s}}{s} \right\} \end{bmatrix} K_c \begin{bmatrix} 1 & -1 \\ -1 & 1 \end{bmatrix} \right)^{-1}$$

$$\times \left(\begin{bmatrix} \mathcal{Z}\left\{ \dfrac{\mathrm{e}^{-D_k s}}{s} R_A(s) \right\} \\ \mathcal{Z}\left\{ \dfrac{\mathrm{e}^{-D_k s}}{s} R_B(s) \right\} \end{bmatrix} - \begin{bmatrix} \mathcal{Z}\left\{ \dfrac{1}{s} R_{o1}(s) \right\} \\ \mathcal{Z}\left\{ \dfrac{1}{s} R_{o2}(s) \right\} \end{bmatrix} \right.$$

$$\left. + \begin{bmatrix} \mathcal{Z}\left\{ \dfrac{\left(1-\mathrm{e}^{-Ts}\right)}{s} \dfrac{\mathrm{e}^{-D_{cB} s}\mathrm{e}^{-D_n s}}{s} \right\} & -\mathcal{Z}\left\{ \dfrac{\left(1-\mathrm{e}^{-Ts}\right)}{s} \dfrac{\mathrm{e}^{-D_{cA} s}\mathrm{e}^{-D_k s}}{s} \right\} \\ -\mathcal{Z}\left\{ \dfrac{\left(1-\mathrm{e}^{-Ts}\right)}{s} \dfrac{\mathrm{e}^{-D_{cB} s}\mathrm{e}^{-D_k s}}{s} \right\} & \mathcal{Z}\left\{ \dfrac{\left(1-\mathrm{e}^{-Ts}\right)}{s} \dfrac{\mathrm{e}^{-D_{cA} s}\mathrm{e}^{-D_n s}}{s} \right\} \end{bmatrix} \begin{bmatrix} R_{cB}(z) \\ R_{cA}(z) \end{bmatrix} \right)$$

The goal is to reduce the difference between the coil inventories of the two galvanizing lines:

$$\Delta W(z) = \begin{bmatrix} 1 & -1 \end{bmatrix} \begin{bmatrix} W_1(z) \\ W_2(z) \end{bmatrix}$$

The companies' coil output rates $r_A(t)$ and $r_B(t)$ vary due to seasonality or other factors that cause demand for steel coils to fluctuate. The frequency of this fluctuation can be expected to significantly affect dynamic behavior and hence the efficacy of coil delivery adjustment in reducing the difference between coil inventories. Adjustments should be effective in reducing fluctuations in the difference between inventories when the period of fluctuation is relatively long, but should not be expected to be effective when the period of fluctuation is relatively short. This dynamic behavior can be assessed by analyzing the frequency response of the production network.

Program 7.2 shows how the time and frequency response of the galvanizing line inventories and the difference between the inventories can be calculated for this

Program 7.2 Calculation of galvanizing line inventory frequency response

```
T7=7;  % adjustment period (days)
Dc7=7;  % delay in implementing transportation adjustments (days)
Kc7=0.0083;  % chosen to avoid oscillatory time response (days^-1)

T1=1;  % adjustment period (days)
Dc1=1;  % delay in implementing transportation adjustments (days)
Kc1=0.044;  % chosen to avoid oscillatory time response (days^-1)

Dk=1/24;  % delay for transportation by truck (days)
Dn=1;  % delay for transportation by train (days)

p=[240, 120, 60]';  % periods of variation of interest (days)
omegaP=2*pi./p;  % frequencies of interest (radians/day)
A=40;  % amplitude of variation in Company A coil output (coils/day)

s=tf('s','TimeUnit','days');

% transportation transfer function matrices for T=7 days
Gk7=absorbDelay(c2d(exp(-Dc7*s)*exp(-Dk*s)/s,T7));  % delivery by truck
Gn7=absorbDelay(c2d(exp(-Dc7*s)*exp(-Dn*s)/s,T7));  % delivery by train

% transportation transfer function matrices for T=1 day
Gk1=absorbDelay(c2d(exp(-Dc1*s)*exp(-Dk*s)/s,T1));  % delivery by truck
Gn1=absorbDelay(c2d(exp(-Dc1*s)*exp(-Dn*s)/s,T1));  % delivery by train

% transfer function matrix for W(z)/WAB(z) with T=7 days
Gc7=ss(Kc7*[tf(-1), tf(1); tf(1), tf(-1)]);  % use state space
Gkn7=[Gn7, -Gk7; -Gk7, Gn7];
Gw7=inv((eye(2)-Gkn7*Gc7),'min'); Gw7=minreal(Gw7);
damp(Gw7)
```

Pole	Magnitude	Damping	Frequency (rad/days)	Time Constant (days)
5.66e-01	5.66e-01	1.00e+00	8.14e-02	1.23e+01
4.96e-01	4.96e-01	1.00e+00	1.00e-01	9.98e+00
-6.16e-02	6.16e-02	6.64e-01	6.00e-01	2.51e+00

```
% transfer function matrix for W(z)/WAB(z) with T=1 day
Gc1=ss(Kc1*[tf(-1), tf(1); tf(1), tf(-1)]); % use state space
Gkn1=[Gn1, -Gk1; -Gk1, Gn1];
Gw1=inv((eye(2)-Gkn1*Gc1),'min'); Gw1=minreal(Gw1);
damp(Gw1)
```

```
      Pole        Magnitude     Damping     Frequency     Time Constant
                                            (rad/days)      (days)
    6.70e-01     6.70e-01     1.00e+00     4.01e-01      2.49e+00
    5.70e-01     5.70e-01     1.00e+00     5.61e-01      1.78e+00
   -2.40e-01     2.40e-01     4.14e-01     3.45e+00      7.01e-01
```

```
p120=120; % plot sinusoidal responses for 120-day period
omega120=2*pi/p120; % corresponding frequency

t7=(0:T7:4*p120)'; % 4*120 days
wA7=[-A*cos(omega120*(t7-Dk))/omega120, zeros(size(t7))]; % integration
w7=lsim(Gw7,wA7,t7); % response to sinusoid when T=7 days

t1=(0:T1:4*p120)'; % 4*120 days
wA1=[-A*cos(omega120*(t1-Dk))/omega120, zeros(size(t1))]; % integration
w1=lsim(Gw1,wA1,t1); % response to sinusoid when T=1 day

stairs(t7,[1, -1]*w7','r'); hold on % T=7 days - Figure 7.6a
stairs(t7,w7(:,1),'k'); stairs(t7,w7(:,2),'b'); hold off
legend('\Deltaw(kT)','w_1(kT)','w_2(kT)')
xlabel('time kT (days)'); ylabel('inventory variation (coils)')

stairs(t1,[1, -1]*w1','r'); hold on % T=7 days - Figure 7.6b
stairs(t1,w1(:,1),'k'); stairs(t1,w1(:,2),'b'); hold off
legend('\Deltaw(kT)','w_1(kT)','w_2(kT)')
xlabel('time kT (days)'); ylabel('inventory variation (coils)')

% frequency response with T=7 days
[Mdw7omega,Pdw7omega,omega]=bode([1 -1]*Gw7,{0.01 1});
[Mdw7omegaP]=bode([1, -1]*Gw7,omegaP);
A7omega=A*squeeze(Mdw7omega(1,1,:))./omega; % include integration
A7omegaP=A*squeeze(Mdw7omegaP(1,1,:))./omegaP;

% frequency response with T=1 day
[Mdw1omega]=bode([1, -1]*Gw1,omega);
[Mdw1omegaP]=bode([1, -1]*Gw1,omegaP);
A1omega=A*squeeze(Mdw1omega(1,1,:))./omega; % include integration
A1omegaP=A*squeeze(Mdw1omegaP(1,1,:))./omegaP;

loglog(omega,A./omega,'r'); hold on % Kc=0 - Figure 7.7
loglog(omega,A7omega,'k'); loglog(omega,A1omega,'b')
loglog(omegaP,A./omegaP,'xr') % no coil delivery adjustments
loglog(omegaP,A7omegaP,'xk'); loglog(omegaP,A1omegaP,'xb'), hold off
legend('no coil delivery adjustments',...
  'T=7 days, D_c=7 days, K_c=0.0115 days^{-1}',...
  'T=1 day, D_c=1 day, K_c=0.0622 days^{-1}')
xlabel('frequency \omega (radians/day)')
ylabel('inventory variation (coils)')

[omegaP, A./omegaP, A7omegaP, A1omegaP]' % amplitude data - Table 7.2
```

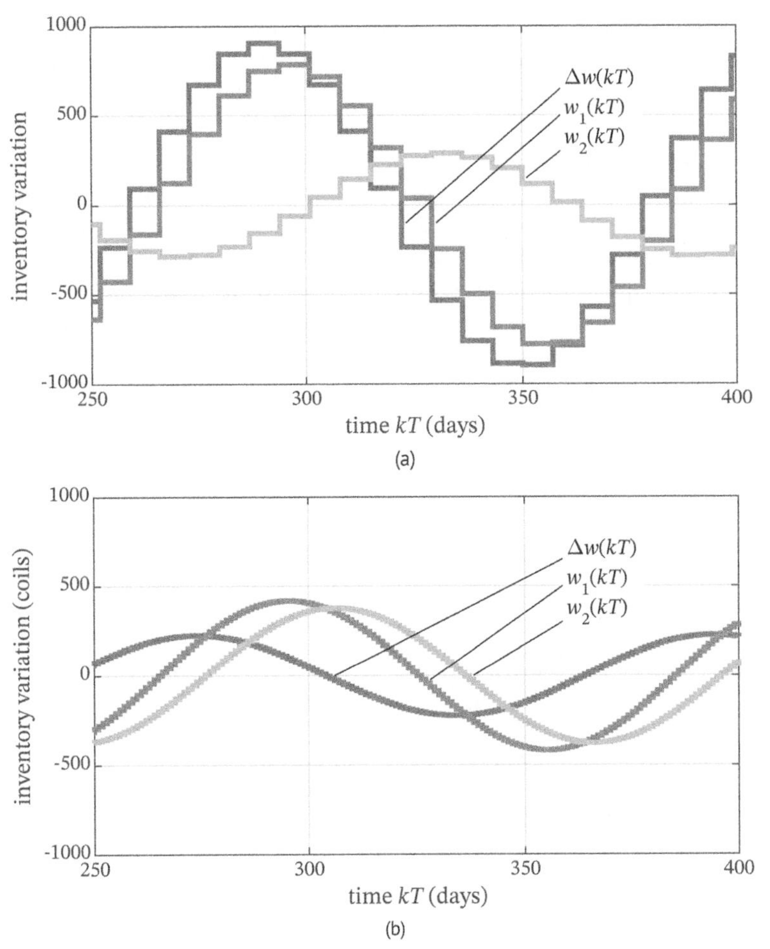

Figure 7.6 Variation in galvanizing line coil inventories when the period of variation in the coil output of Company A is 120 days and the amplitude of variation is 40 coils/day (variation in Company B coil output is assumed to be zero). (a) T = 7 days, D_c = 7 days, D_k = 1 hour, D_n = 1 day. (b) T = 1 day, D_c = 1 day, D_k = 1 hour, D_n =1 day.

production network. Of particular interest is the response for periods of variation in company coil output rate of 60, 120, and 240 days. The corresponding frequencies of variation ω are $2\pi/60$, $2\pi/120$, and $2\pi/240$ radians/day (frequencies 1/60, 1/120, and 1/240 cycles/day). The time response of galvanizing line coil inventory to variations in the coil output rate of Company A when the period of variation is 120 days and the amplitude of variation is 40 coils/day is shown in Figure 7.6 for two cases. In the first case, the company coil delivery adjustment period is $T = 7$ days, the delay in implementing adjustments is $D_c = 7$ days, and $K_c = 0.0115$ days^{-1}. In the second case, the company coil delivery adjustment period is $T = 1$ day, the delay in implementing adjustments is $D_c = 1$ days, and $K_c = 0.0622$ days^{-1}. In both cases, the delay in delivering coils by truck is $D_k = 1/24$ day (1 hour), the delay in delivering coils by train is $D_n = 1$ day, and K_c is chosen to produce rapid response while avoiding significant oscillations in time response. K_c can be higher when adjustment period T is shorter and the upper limit of K_c, beyond which dynamic behavior of company coil delivery adjustment becomes undesirable, also depends on delays D_c, D_k, and D_n.

The relationship between amplitude of variation in the difference between galvanizing lined coil inventories and the frequency of variation in company coil output rate is shown in Figure 7.7 for both cases as well as for the case with no coil delivery adjustment ($K_c = 0$). The results for the periods of interest are summarized in Table 7.2. The magnitude of the frequency response is multiplied by 40 coils/day in Program 7.2 to obtain the amplitude of variation in the difference between galvanizing line coil inventories at frequency ω radians/day. In Program 7.2, magnitude of the frequency response is calculated for input

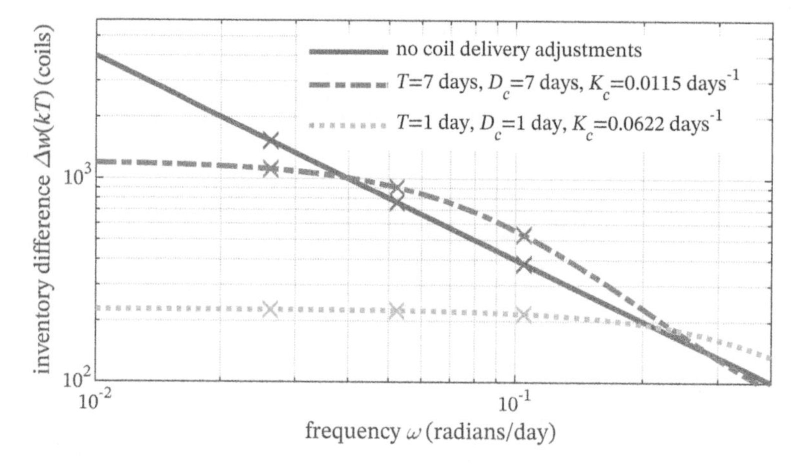

Figure 7.7 Amplitude of variation of difference between galvanizing line inventories in response to 40 coils/day sinusoidal variation in Company A coil output rate ("x" marks the frequencies of interest that correspond to periods of variation of 240, 120, and 60 days).

Table 7.2 Amplitude of variation of difference in inventory in response to 40 coils/day amplitude of variation in Company A coil output rate for periods of variations of 240, 120, and 60 days (variation in Company B coil output rate assumed to be zero).

period (frequency)	**240 days (0.0262 radians/day)**	**120 days (0.0524 radians/day)**	**60 days (0.1047 radians/day)**
no coil delivery adjustments ($K_c = 0$)	1528	764	382
$T = 7$ days, $D_c = 7$ days, $K_c = 0.0083$ days^{-1}	1112	906	530
$T = 1$ day, $D_c = 1$ day, $K_c = 0.044$ days^{-1}	227	224	216

$$\begin{bmatrix} W_A(z) \\ W_B(z) \end{bmatrix} = \begin{bmatrix} \mathcal{Z}\left\{\dfrac{e^{-D_k s}}{s} R_A(s)\right\} \\ \mathcal{Z}\left\{\dfrac{e^{-D_k s}}{s} R_B(s)\right\} \end{bmatrix}$$

Then, the effect of integration of company coil output rate inputs $r_A(t)$ and $r_B(t)$ to obtain inputs $w_A(t)$ and $w_B(t)$ is included by dividing the magnitude by frequency ω radians/day. (Delay D_k in delivering coils by truck does not affect the magnitude of the frequency response.)

For the case with the shorter coil delivery adjustment period $T = 1$ day, the results shown in Figure 7.7 and Table 7.2 imply that galvanizing line coil inventory sharing and company coil delivery adjustment is effective; however, for the case with the longer coil delivery adjustment period $T = 7$ days, the results show that galvanizing line coil inventory sharing and company coil delivery adjustment is detrimental at the shorter periods of company coil output rate variation. The efficacy of coil inventory sharing and company coil delivery adjustment therefore can be expected to be more significant when the period between adjustments and the delay in making adjustments are both relatively short.

7.3 Effect of Order Flow Information Sharing on the Dynamic Behavior of a Production Network

In this application example the potential benefits of sharing order flow information within a production network are quantitatively evaluated using an analysis of the network's dynamic behavior [3]. This behavior becomes more complicated when order flows in the network are omnidirectional, disturbances such as rush orders and equipment failures propagate from work system to work system, and work systems individually adjust their capacity to compensate for variations in order flows. The effects of these variations can be mitigated if up-to-date order flow information is shared within the network. This information can be shared either by centralized production planning and control or by distribution by individual work systems in the network.

An example of a network of production work systems is shown in Figure 7.8 where 659 orders flow through a network with five work systems during a 186-day period. The total number of orders that flow between each pair of work systems is shown in Figure 7.8 along with the total number of orders processed by each work system during the 186-day period. Different types of orders require different sequences of processing steps, and this is indicated in an aggregated fashion by the number of orders that entered each work system from outside the network, the number of orders that left the network from each work system, and the internal order flows within the network. Order flow is unidirectional between Work Systems 1, 2, 3 and 4 (shearing-sawing, ring rolling, drop forging and heat treatment), and order flow is omnidirectional between Work Systems 4 and 5 (heat treatment and quality control). In a network such as that shown in Figure 7.8, local capacity adjustments or disturbances in individual work systems result in fluctuations in order flows to other work systems. This affects the work in progress (WIP) in the other work systems and they react by adjusting their capacities. If order flow information can be shared within this network, there is an opportunity to proactively anticipate the adjustments in capacity that are needed to locally regulate WIP, which can assist in designing the network to have consistent, desirable dynamic behavior.

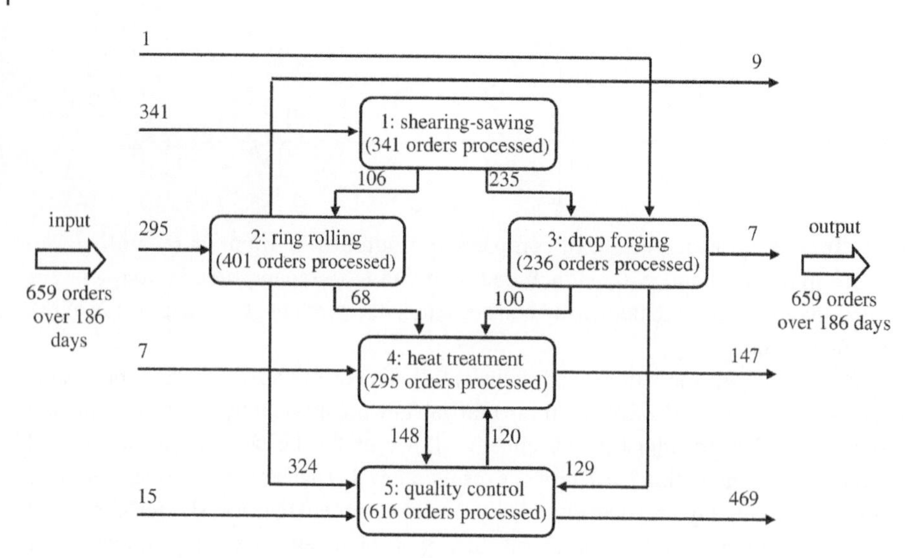

Figure 7.8 Example of a production network with five work systems.

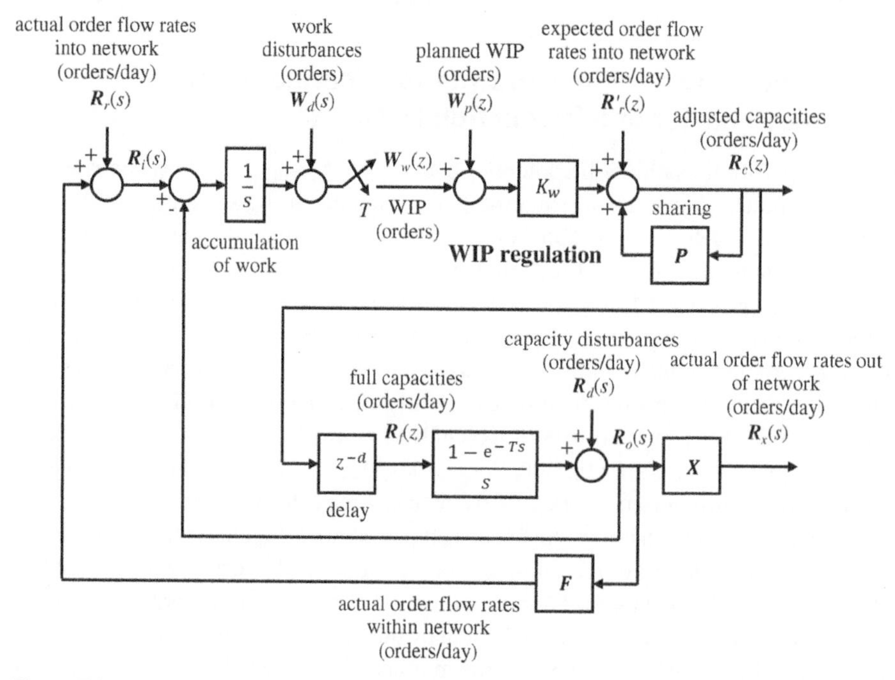

Figure 7.9 Block diagram for production network of N work systems with sharing of order flow information.

A model of the entire production network is shown in the block diagram in Figure 7.9, and the model of each work system in the network is shown in Figure 7.10, which includes flow of orders and sharing of information between work systems. It is assumed that the production capacities of all work systems are adjusted with period T days, and all adjustments are made at the same time. The variables and parameters in the model are defined as follows:

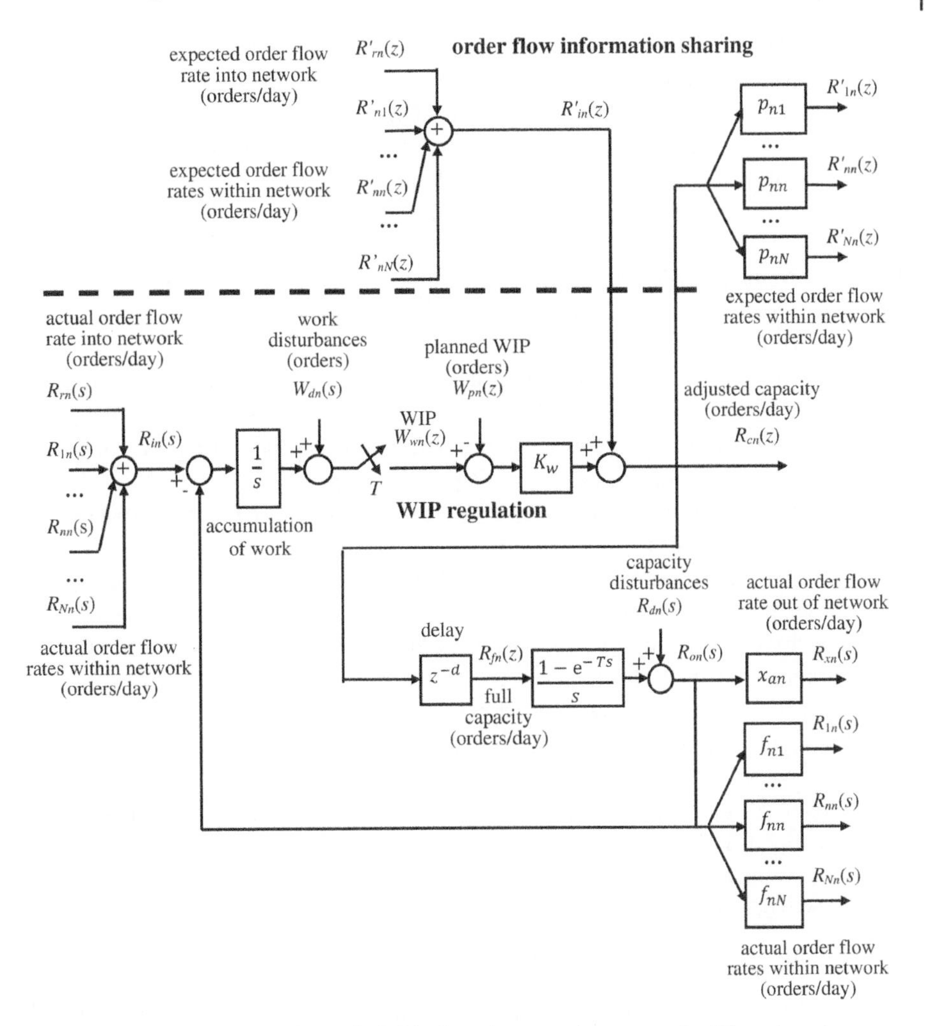

Figure 7.10 Block diagram for an individual work system in a network of N work systems with sharing of order flow information.

N	number of work systems in the production network
d	non-negative integer dT days delay in implementing production rate adjustments)
$\boldsymbol{K_w}$	WIP regulation parameter, assumed to be the same for all work systems (days^{-1})
$\boldsymbol{F}$	matrix of actual fractions of order flow rates into work systems in the production network where f_{nm} is the actual fraction of order flow rate out of the mth work system that flows into the nth work system
$\boldsymbol{X}$	diagonal matrix of actual fractions of order flow rates out of the production network where x_n is the actual fraction of order flow rate out of the nth work system that flows to outside the network
$\boldsymbol{P}$	matrix of expected fractions of order flow rates into work systems in the production network where p_{nm} is the expected fraction of order flow rate out of the mth work system that flows into the nth work system

$\boldsymbol{R}_i(s)$ vector of actual order flow rates into work systems in the network where $r_{in}(t)$ is the actual order flow rate into the nth work system (orders/day)

$\boldsymbol{R}_o(s)$ vector of actual order flow rates out of work systems in the network where $r_{on}(t)$ is the actual order flow rate out of the nth work system (orders/day)

$\boldsymbol{R}_r(s)$ vector of actual order flow rates to work systems in the network from outside the network where $r_{rn}(t)$ is the actual order flow rate from outside the network to the nth work system (orders/day)

$\boldsymbol{R}_x(s)$ vector of order flows from work systems in the network to outside the network where $r_{xn}(t)$ is the order flow rate from the nth work system to outside the network (orders/day)

$\boldsymbol{R}_c(z)$ vector of adjusted work system capacities, before delay in implementation where $r_{cn}(kT)$ is the adjusted capacity of the nth work system (orders/day)

$\boldsymbol{R}_f(z)$ vector of full work system capacities where $r_{fn}(kT)$ is the full capacity of the nth work system (orders/day)

$\boldsymbol{R}_d(z)$ vector of work system capacity disturbances where $r_{dn}(kT)$ is the capacity disturbances in the nth work system (orders/day)

$\boldsymbol{R}'_i(z)$ vector of expected order flow rates into the work systems in the network where $r'_{in}(kT)$ is the expected order flow rate into the nth work system (orders/day)

$\boldsymbol{R}'_r(z)$ vector of expected order flow rates to the work systems in the network from outside the network where $r'_{rn}(kT)$ is the expected order flow rate from outside the network to the nth work system (orders/day)

$\boldsymbol{W}_w(z)$ vector of WIP in work systems in the network where $w_{wn}(kT)$ is the WIP in the nth work system (orders)

$\boldsymbol{W}p(z)$ vector of planned WIP in work systems in the network where $w_{pn}(kT)$ is the planned WIP in the nth work system (orders)

$\boldsymbol{W}_d(z)$ vector of work disturbances in work systems in the network where $w_{dn}(kT)$ is the work disturbances in the nth work system (orders)

Additionally,

$r_{nm}(t)$ actual order flow rate within the network from the mth work system to the nth work system (orders/day)

$r'_{nm}(t)$ expected order flow rate within the network from the mth work system to the nth work system (orders/day)

The actual order flow rates into the N work systems in the production network are

$$\begin{bmatrix} r_{i1}(t) \\ r_{i2}(t) \\ \vdots \\ r_{iN}(t) \end{bmatrix} = \begin{bmatrix} r_{r1}(t) \\ r_{r2}(t) \\ \vdots \\ r_{rN}(t) \end{bmatrix} + \begin{bmatrix} f_{11} & f_{21} & \cdots & f_{N1} \\ f_{12} & f_{22} & \cdots & f_{N2} \\ \vdots & \vdots & \ddots & \vdots \\ f_{1N} & f_{2N} & \cdots & f_{NN} \end{bmatrix} \begin{bmatrix} r_{o1}(t) \\ r_{o2}(t) \\ \vdots \\ r_{oN}(t) \end{bmatrix}$$

or

$$\boldsymbol{R}_i(s) = \boldsymbol{R}_r(s) + \boldsymbol{F}\boldsymbol{R}_o(s)$$

where the actual structure of order flows within the production network is represented by matrix $\boldsymbol{F}$.

The actual order flow rates out of the network are

$$\begin{bmatrix} r_{x1}(t) \\ r_{x2}(t) \\ \vdots \\ r_{xN}(t) \end{bmatrix} = \begin{bmatrix} x_1 & 0 & \cdots & 0 \\ 0 & x_2 & \cdots & 0 \\ \vdots & \vdots & \ddots & \vdots \\ 0 & 0 & \cdots & x_N \end{bmatrix} \begin{bmatrix} r_{o1}(t) \\ r_{o2}(t) \\ \vdots \\ r_{oN}(t) \end{bmatrix}$$

or

$$\boldsymbol{R}_x(s) = \boldsymbol{X}\boldsymbol{R}_o(s)$$

where the actual structure of order flows out of the production network is represented by matrix $\boldsymbol{X}$. Order flows are conserved in the network, and for the output of the nth work system in the network

$$\sum_{m=1}^{N} f_{mn} + x_n = 1$$

The expected order flow rates into the N work systems in the production network are calculated and shared at instants in time separated by period T:

$$\begin{bmatrix} r'_{i1}(kT) \\ r'_{i2}(kT) \\ \vdots \\ r'_{iN}(kT) \end{bmatrix} = \begin{bmatrix} r'_{r1}(kT) \\ r'_{r2}(kT) \\ \vdots \\ r'_{rN}(kT) \end{bmatrix} + \begin{bmatrix} p_{11} & p_{21} & \cdots & p_{N1} \\ p_{12} & p_{22} & \cdots & p_{N2} \\ \vdots & \vdots & \ddots & \vdots \\ p_{1N} & p_{2N} & \cdots & p_{NN} \end{bmatrix} \begin{bmatrix} r_{c1}(kT) \\ r_{c2}(kT) \\ \vdots \\ r_{cN}(kT) \end{bmatrix}$$

or

$$\boldsymbol{R}'_i(z) = \boldsymbol{R}'_r(z) + \boldsymbol{P}\boldsymbol{R}_c(z)$$

where the expected structure of order flows within the production network is represented by matrix $\boldsymbol{P}$.

Work system capacity is adjusted to regulate WIP:

$$R_{cn}(z) = K_w\left(W_{wn}(z) - W_{pn}(z)\right) + R'_{in}(z)$$

or

$$\boldsymbol{R}_c(z) = K_w\left(\boldsymbol{W}_w(z) - \boldsymbol{W}_p(z)\right) + \boldsymbol{R}'_i(z)$$

where WIP regulation parameter K_w is chosen to obtain desired dynamic behavior. The WIP is

$$\boldsymbol{W}_w(z) = \mathcal{Z}\left\{\frac{1}{s}\boldsymbol{R}_r(s)\right\} + \boldsymbol{W}_d(z) - (\mathbf{I} - \boldsymbol{F})\left(\mathcal{Z}\left\{\frac{1-\mathrm{e}^{-Ts}}{s}\frac{1}{s}\right\}z^{-d}\boldsymbol{R}_c(z) + \mathcal{Z}\left\{\frac{1}{s}\boldsymbol{R}_d(s)\right\}\right)$$

and with WIP regulation as defined above,

$$\mathbf{W}_w(z) = \left(\mathbf{I} + K_w(\mathbf{I}-\mathbf{F})(\mathbf{I}-\mathbf{P})^{-1}\frac{Tz^{-1}}{1-z^{-1}}z^{-d}\right)^{-1}\left(Z\left\{\frac{1}{s}\mathbf{R}_r(s)\right\} + \mathbf{R}_d(z)\right.$$
$$\left. + (\mathbf{I}-\mathbf{F})\left(\frac{Tz^{-1}}{1-z^{-1}}z^{-d}(\mathbf{I}-\mathbf{P})^{-1}\left(K_w\mathbf{W}_p(z) - \mathbf{R}_r'(z)\right) + Z\left\{\frac{1}{s}\mathbf{R}_d(s)\right\}\right)\right)$$

The fundamental dynamic behavior of a network of production systems modeled as shown in Figure 7.10 is a function of the roots of the characteristic equation

$$\det\left(\mathbf{I} + K_w(\mathbf{I}-\mathbf{F})(\mathbf{I}-\mathbf{P})^{-1}\frac{Tz^{-1}}{1-z^{-1}}z^{-d}\right) = 0$$

If the expected order flow structure $\mathbf{P}$ is shared by the work systems and accurately represents the actual order flow structure, then $\mathbf{F} = \mathbf{P}$ and the fundamental dynamic behavior is not affected by the flow of orders between work systems. In this case, the work systems have independent fundamental dynamic behavior that is a function of the roots of characteristic equation

$$z^{d+1} - z^d + K_w T = 0$$

On the other hand, if there is no order flow information sharing in the network then

$$\mathbf{P} = \mathbf{0}$$

and the work systems interact with fundamental dynamic behavior that is characterized by the roots of

$$\det\left(\mathbf{I} + K_w(\mathbf{I}-\mathbf{F})\frac{Tz^{-1}}{1-z^{-1}}z^{-d}\right) = 0$$

These results can be used to quantitatively evaluate the effect of order flow information sharing on the fundamental dynamic behavior of the network of production systems. For the 5-work-system example shown in Figure 7.8, the actual order flow structure matrices are

$$\mathbf{F} = \begin{bmatrix} 0 & 0 & 0 & 0 & 0 \\ 106/341 & 0 & 0 & 0 & 0 \\ 235/341 & 0 & 0 & 0 & 0 \\ 0 & 68/401 & 100/236 & 0 & 120/616 \\ 0 & 324/401 & 129/236 & 148/295 & 0 \end{bmatrix}$$

$$\mathbf{X} = \begin{bmatrix} 0 & 0 & 0 & 0 & 0 \\ 0 & 9/401 & 0 & 0 & 0 \\ 0 & 0 & 7/236 & 0 & 0 \\ 0 & 0 & 0 & 147/295 & 0 \\ 0 & 0 & 0 & 0 & 496/616 \end{bmatrix}$$

Program 7.3 can be used to calculate the response of this network to a 1-order work disturbance in the Work System 4 (heat treatment), a rush order for example. Capacity adjustments are made daily ($T = 1$ day) and there is a delay of 1 day ($d = 1$) in implementing these adjustments. The WIP regulation parameter is chosen as $K_w = 0.25$ days^{-1} for all five work systems.

Program 7.3 Calculation of fundamental dynamic characteristics and time response of a production network without and with order flow information sharing

```
T=1;  % daily adjustments
N=5;  % number of work systems in network
d=1;  % delay dT days in making capacity adjustments
Kw=(d^d/(d+1)^(d+1)); % WIP regulation parameter for two equal roots

% order flow structure
F=[0, 0, 0, 0, 0; 106/341, 0, 0, 0, 0; 235/341, 0, 0, 0, 0;…
  0, 68/401, 100/236, 0, 120/616; 0, 324/401, 129/236, 148/295, 0]
```

```
F =
        0      0         0         0         0
   0.3109      0         0         0         0
   0.6891      0         0         0         0
        0      0.1696    0.4237    0         0.1948
        0      0.8080    0.5466    0.5017    0
```

```
X=[0, 0, 0, 0, 0; 0, 9/401, 0, 0, 0; 0, 0, 7/236, 0, 0;…
  0, 0, 0, 147/295, 0; 0, 0, 0, 0, 496/616]
```

```
X =
        0         0         0         0         0
        0    0.0224         0         0         0
        0         0    0.0297         0         0
        0         0         0    0.4983         0
        0         0         0         0    0.8052
```

```
z=tf('z',T,'TimeUnit','days');
Gi=minreal(T*z^-1/(1-z^-1));
Gi=ss(Gi);  % convert to state space to improve subsequent precision

% transfer function Ww(z)/Wd(z) without order flow information sharing
GwdI=minreal((eye(N)+Kw*Gi*(eye(N)-F)*z^-d)^-1);
zpk(GwdI(:,4))
```

```
ans =

 From input to output…
  1: 0
  2: 0
  3: 0
                     z (z-0.5)^2 (z-1)
  4: ---------------------------------------
      (z-0.7796) (z-0.2204) (z^2 - z + 0.3282)
                     0.12542 z (z-1)
  5: ---------------------------------------
      (z-0.7796) (z-0.2204) (z^2 - z + 0.3282)
```

```
Sample time: 1 days
Discrete-time zero/pole/gain model.
```

```
damp(zpk(GwdI(4,4)))  % heat treatment dynamic characteristics
```

```
         Pole              Magnitude      Damping      Time Constant
                                                          (days)
   7.80e-01                7.80e-01       1.00e+00      4.02e+00
   5.00e-01 + 2.80e-01i    5.73e-01       7.38e-01      1.79e+00
   5.00e-01 - 2.80e-01i    5.73e-01       7.38e-01      1.79e+00
   2.20e-01                2.20e-01       1.00e+00      6.61e-01
```

```
damp(zpk(GwdI(5,4)))  % quality control dynamic characteristics
```

```
         Pole              Magnitude      Damping      Time Constant
                                                          (days)
   7.80e-01                7.80e-01       1.00e+00      4.02e+00
   5.00e-01 + 2.80e-01i    5.73e-01       7.38e-01      1.79e+00
   5.00e-01 - 2.80e-01i    5.73e-01       7.38e-01      1.79e+00
   2.20e-01                2.20e-01       1.00e+00      6.61e-01
```

```
step(GwdI(4,4),GwdI(5,4),25)  % 1-order work disturbance - Figure 7.11
xlabel('time kT (days)'); ylabel('change in WIP W_w(kT) (orders)')
legend('heat treatment','quality control')
P=F;  % expected order flow structure same as actual order flow

% transfer function Ww(z)/Wd(z) with order flow information sharing
GwdP=minreal((eye(N)+Kw*Gi*(eye(N)-F)*((eye(N)-P)^-1)*z^-d)^-1);
zpk(GwdP(:,4))
```

```
ans =
 From input to output...
  1: 0
  2: 0
  3: 0
       z (z-1)
  4: ---------
      (z-0.5)^2
  5: 0
Sample time: 1 days
Discrete-time zero/pole/gain model.
```

```
damp(zpk(GwdP(4,4)))  % heat treatment dynamic characteristics
```

```
   Pole        Magnitude      Damping     Time Constant
                                             (days)
 5.00e-01      5.00e-01      1.00e+00      1.44e+00
 5.00e-01      5.00e-01      1.00e+00      1.44e+00
```

```
damp(zpk(GwdP(5,4)))  % quality control dynamic characteristics
```

```
System does not have any poles.
```

```
step(GwdP(4,4),GwdP(5,4),25)  % 1-order work disturbance - Figure 7.12
xlabel('time kT (days)'); ylabel('change in WIP W_w(kT) (orders)')
legend('heat treatment','quality control')

% 37 fewer orders than expected sent to quality control
P1=F; P1(5,4)=(148+37)/295; P1(4,5)=120/(616+37);
X1=X; X1(4,4)=(147-37)/295; X1(5,5)=(469+37)/(616+37);

% transfer function Ww(z)/Wd(z) with order flow information sharing
GwdP1=minreal((eye(N)+Kw*Gi*(eye(N)-F)*((eye(N)-P1)^-1)*z^-d)^-1);
zpk(GwdP1(:,4))
ans =
 From input to output…
```

```
  1: 0
  2: 0
  3: 0

                    z (z-1) (z^2 - z + 0.2565)
  4: -----------------------------------------------
      (z^2 - 0.8766z + 0.1982) (z^2 - 1.123z + 0.3216)

             -0.03544 z (z-1)
  5: ------------------------------------------------
     (z^2 - 0.8766z + 0.1982) (z^2 - 1.123z + 0.3216)
Sample time: 1 days
Discrete-time zero/pole/gain model.
```

```
damp(zpk(GwdP1(4,4)))  % heat treatment dynamic characteristics
```

Pole	Magnitude	Damping	Time Constant (days)
5.62e-01 + 7.80e-02i	5.67e-01	9.72e-01	1.76e+00
5.62e-01 - 7.80e-02i	5.67e-01	9.72e-01	1.76e+00
4.38e-01 + 7.80e-02i	4.45e-01	9.77e-01	1.24e+00
4.38e-01 - 7.80e-02i	4.45e-01	9.77e-01	1.24e+00

```
damp(zpk(GwdP1(5,4)))  % quality control dynamic characteristics
```

Pole	Magnitude	Damping	Time Constant (days)
5.62e-01 + 7.80e-02i	5.67e-01	9.72e-01	1.76e+00
5.62e-01 - 7.80e-02i	5.67e-01	9.72e-01	1.76e+00
4.38e-01 + 7.80e-02i	4.45e-01	9.77e-01	1.24e+00
4.38e-01 - 7.80e-02i	4.45e-01	9.77e-01	1.24e+00

```
step(GwdP1(4,4),GwdP1(5,4),25)  % 1-order work disturbance - Figure 7.14
xlabel('time kT (days)'); ylabel('change in WIP W_w(kT) (orders)')
legend('heat treatment','quality control')
```

Table 7.3 Summary of results for cases without and with order flow information sharing.

case	without order flow information sharing ($P = 0$)	with order flow information sharing: expected order flow structure = actual order flow structure ($P = F$)		expected order flow structure ≠ actual order flow structure ($P \neq F$)	
	heat treatment and quality control	heat treatment	quality control	heat treatment	quality control
time constants (days)	4.02	1.44	no response	1.76	1.76
	0.61	1.44		1.76	1.76
	0.79			1.24	1.24
	0.79			1.24	1.24
damping ratios	1.00	1.00	no response	0.97	0.97
	1.00	1.00		0.97	0.97
	0.74			0.98	0.98
	0.74			0.98	0.98

Results of analysis of fundamental dynamic behavior without order flow information sharing are obtained with $\boldsymbol{P} = \boldsymbol{0}$. The transfer functions for WIP in the Work System 1 (shearing-sawing), Work System 2 (ring rolling) and Work System 3 (drop forging) in response to a work disturbance in Work System 4 (heat treatment) are zero because of the unidirectional flow of orders in this portion of the production network: disturbances in the heat treatment work system do not affect the shearing-sawing, ring rolling and drop forging work systems. However, the flow of orders is omnidirectional between Work System 4 (heat treatment) and Work System 5 (quality control). The transfer functions for WIP in response to a work disturbance in Work System 4 (heat treatment) therefore are not zero. The characteristic equation is the same for these two work systems, and the time constants that correspond to its roots are 4.02, 1.79, and 0.66 days. The associated damping ratios are 1.00, 0.74, and 1.00; these results are summarized in Table 7.3. The time responses of change in WIP in these two work systems to a 1-order work disturbance in the heat treatment work system are shown in Figure 7.11. These results illustrate the propagation of the disturbance in the heat treatment work system to the quality control work system when there is no order flow information sharing. The model predicts that it will take a relatively long time for the effects of the disturbance to disappear and WIP to return to planned WIP, approximately 20 days.

Results of analysis of fundamental dynamic behavior with order flow information sharing when the expected order flow structure is the same as the actual order flow structure are obtained with $\boldsymbol{P} = \boldsymbol{F}$. In this case, the transfer functions for WIP in all work systems except Work System 4 (heat treatment) in response to a work disturbance in Work System 4 are zero: disturbances in the heat treatment work system do

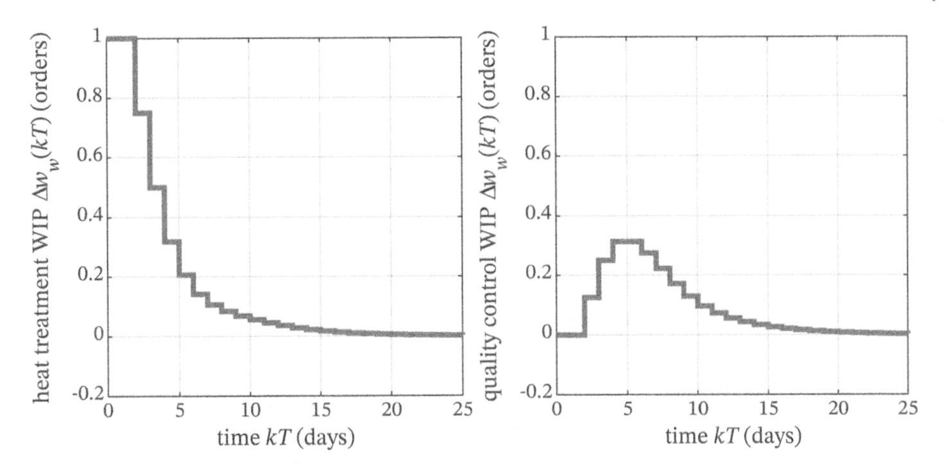

Figure 7.11 Change in WIP in response to a rush order in the heat treatment work system without order flow information sharing: $\boldsymbol{P} = \boldsymbol{0}$.

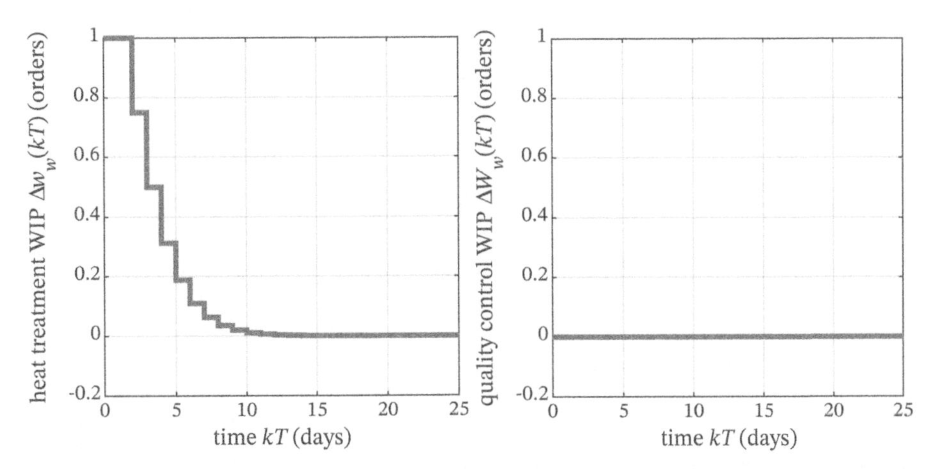

Figure 7.12 Change in WIP in response to a rush order in the heat treatment work system with order flow information sharing and $\boldsymbol{P} = \boldsymbol{F}$ (the expected order flow structure is the same as the actual order flow structure).

not affect the shearing-sawing, ring rolling, drop forging and quality control work systems. The transfer function for WIP in Work System 4 (heat treatment) in response to a work disturbance in Work System 4 is not zero, but the characteristic equation obtained from this transfer function has two equal real roots as expected with the choice $K_w = 0.25$ days^{-1}. These roots correspond to a time constant of 1.44 days and damping ratio 1.0; these results are summarized in Table 7.3. The time responses of the heat treatment and quality control work systems to a 1-order work disturbance in the heat treatment work system are shown for this case in Figure 7.12. There is no change in WIP in the quality control work system in this case, and WIP in heat treatment work system returns to planned WIP in approximately 12 days.

Results of analysis of fundamental dynamic behavior with order flow information sharing when the expected order flow structure is not the same as the actual order flow structure are obtained with $\boldsymbol{P} \neq \boldsymbol{F}$. A case was considered where approximately 25% fewer orders than expected leave the heat treatment work system and enter the quality

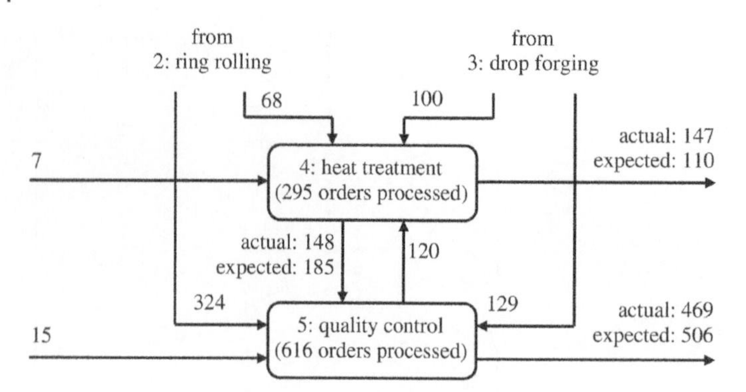

Figure 7.13 Case where 37 fewer orders than expected go from heat treatment to quality control.

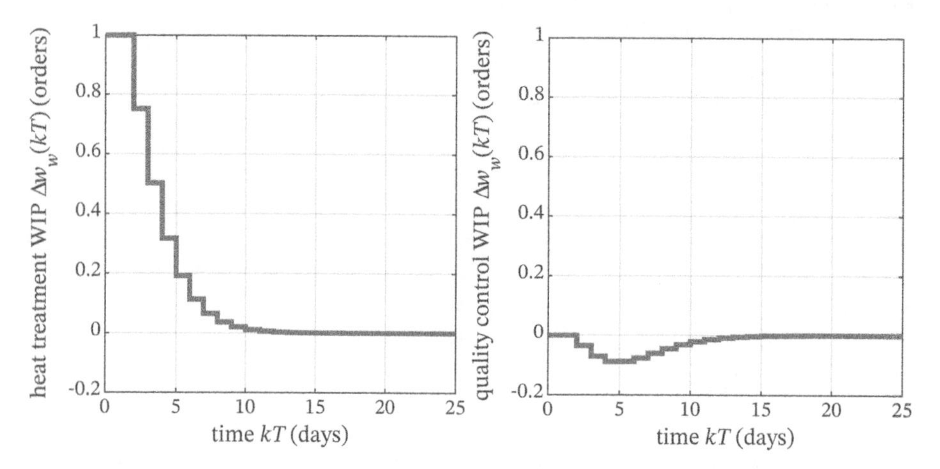

Figure 7.14 Change in WIP in response to a rush order in the heat treatment work system with information sharing and $\boldsymbol{P} \neq \boldsymbol{F}$ (37 fewer orders than expected leave the heat treatment work system and enter the quality control work system).

control work system; these orders leave the production network without quality control. As shown in Figure 7.13, 37 orders that were expected to go to quality control do not. The fractions that affect dynamic behavior then are $F(5,4) = 148/295$, $P(5,4) = (148 + 37)/295$, $F(4,5) = 120/616$, and $P(4,5) = 120/(616 + 37)$. Again, work disturbances in the heat treatment work system do not affect the shearing-sawing, ring rolling, and drop forging work systems because the flow of orders is unidirectional in that portion of the production network and the corresponding transfer functions are zero. However, the transfer function for WIP in Work System 5 (quality control) in response to a work disturbance in Work System 4 (heat treatment) is not zero, and the roots of the characteristic equation correspond to time constants 1.44 and 1.47 days and corresponding damping ratios 1.0 and 0.97; these results are summarized in Table 7.3. The time responses of these two work systems to a 1-order work disturbance in the heat treatment work system are shown in Figure 7.14. There is a change in WIP in the quality control work system in this case because of the difference between the expected and actual order flow structures. WIP returns to planned WIP in approximately 14 days.

7.4 Adjustment of Cross-Trained and Permanent Worker Capacity

In this application example it is shown that frequency response analysis can be a valuable tool in designing a decision-making approach for adjusting the portions of the capacity of a production system that are provided by cross-trained and permanent workers, taking into account differences in both the range of frequencies of fluctuation in demand that adjustments in cross-trained worker capacity and permanent worker capacity should address and how often adjustments are made for these two types of workers [4]. The desired combined range of frequencies of fluctuation in demand that adjustments in capacity should address can be defined first. Then, the portion of fluctuations to be responded to by adjustments in permanent worker capacity can be determined. The remaining portion is responded to by adjustments in cross-trained worker capacity. The frequency response perspective provides a single, convenient framework for designing these three aspects of capacity decision-making for the production system.

There are many ways in which production capacity can be adjusted including overtime/short-time work, hiring/dismissing contract or permanent workers, relocation of cross-trained workers, and outsourcing/insourcing. The cost of adjustments can be a function of frequency and can be prohibitive, particularly for rapid adjustments. Figure 7.15 shows how cost and frequency could be related to several types of adjustment of capacity in a production system. Adjustment of capacity using overtime at both high and low frequencies can be expensive and limited by work agreements or law. Adjustment of capacity using permanent workers may become increasingly expensive with frequency due to the administrative effort involved and challenges in rapidly hiring and training new workers. On the other hand, adjustment of capacity using cross-trained workers may have a relatively low, fixed cost that is not significantly affected by how often or for how long cross-trained workers are reassigned. Only permanent and cross-trained workers are adjusted in this application example, and cost will not be directly considered; however, adjustments in capacity using permanent workers will be restricted to lower frequencies, and adjustments in capacity using cross-trained workers will be used to adapt capacity over a broader range of frequencies of fluctuation in demand.

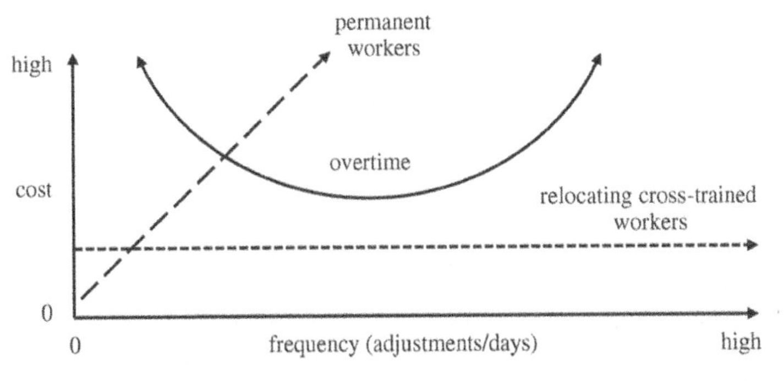

Figure 7.15 Frequency range and cost of capacity adjustments for various types of workers.

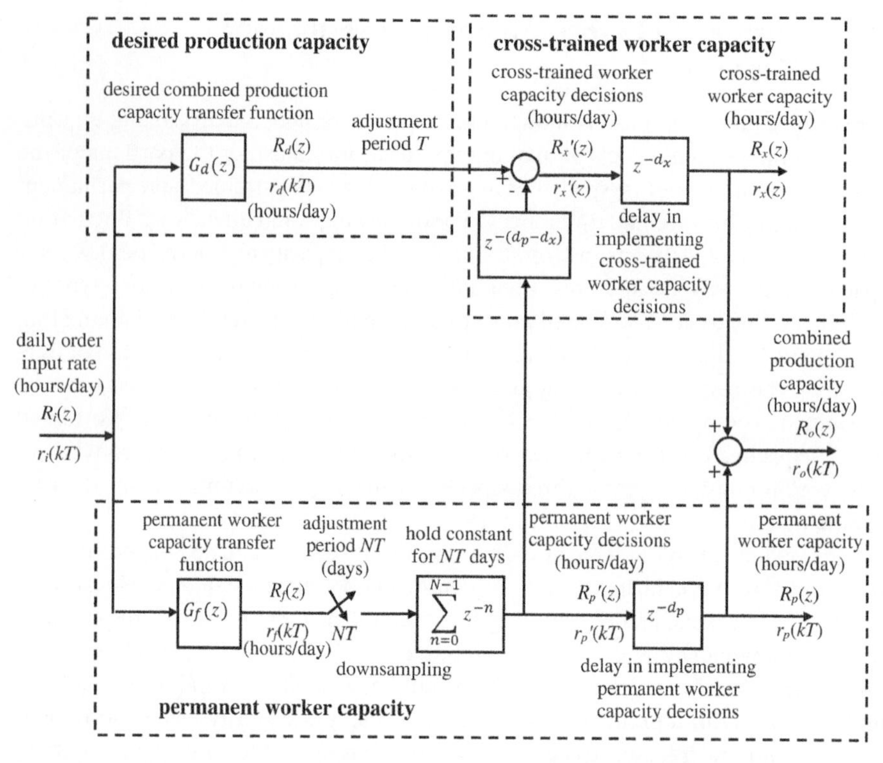

Figure 7.16 Block diagram for adjusting cross-trained worker and permanent worker capacities.

A block diagram for an approach for using permanent and cross-trained workers to adjust capacity in a production system is shown in Figure 7.16. Transfer function $G_d(z)$ specifies the desired dynamic behavior of production capacity that is to result from the combination of both types of capacity adjustments. This transfer function is used to define the range of frequencies over which combined production capacity is to follow fluctuations in order input rate. Response to higher frequency fluctuations is not desired, and the frequency response magnitude at higher frequencies ω is ideally $M_d(\omega) = 0$, while at intermediate and lower frequencies ω the frequency response magnitude is ideally $M_d(\omega) = 1$.

Transfer function $G_f(z)$ is used to make permanent worker capacity adjustment decisions that are focused on the lower-frequency portion of that frequency range. Decisions in response to intermediate and higher frequency fluctuations in order input rate are not desired, and the frequency response magnitude at these frequencies ω is ideally $M_f(\omega) = 0$, while at lower frequencies ω the frequency response magnitude is ideally $M_d(\omega) = 1$.

Cross-trained worker capacity adjustment decisions are focused on the portion of the desired dynamic behavior of production capacity that is not provided by adjustments in permanent worker capacity. This portion is not directly defined using a transfer function; instead, it is what remains after permanent worker capacity is subtracted from the desired production capacity. Little response of cross-trained worker capacity therefore is expected at lower and higher frequency fluctuations in order input rate. At some intermediate frequencies ω the frequency response magnitude for cross-trained workers may be greater than one due to differences between the frequency response phase of desired production capacity and permanent worker capacity.

The system's order input rate is assumed to change daily at instants in time separated by period T days. Decisions regarding hiring/dismissing of permanent workers are made with longer period NT days, where N is a positive integer, and there is a delay d_pT days in implementing permanent worker adjustments where d_p is a nonnegative integer. Decisions made regarding relocating cross-trained workers are made daily, and there is a delay d_xT days in implementing cross-trained worker adjustments where d_x is a nonnegative integer. The variables in the model in Figure 7.16 are as follows:

- $r_i(kT)$ rate of flow of orders into the system (orders/day)
- $r_d(kT)$ desired combined worker capacity (orders/day)
- $r_f(kT)$ permanent worker capacity decisions made before downsampling with longer period NT (orders/day)
- $r_p'(kT)$ permanent worker capacity decisions after downsampling with longer period NT (orders/day)
- $r_p(kT)$ implemented permanent worker capacity adjustments (orders/day)
- $r_x'(kT)$ cross-trained worker capacity decisions (orders/day)
- $r_x(kT)$ implemented cross-trained worker capacity adjustments (orders/day)
- $r_o(kT)$ combined production capacity (orders/day)

If desired production capacity transfer function $G_d(z)$ is chosen to be 1st-order with time constant τ_d days and permanent worker capacity decision filter transfer function $G_f(z)$ is chosen to be 1st-order with time constant τ_f days, then the relationships between variables in the model in Figure 7.16 are

$$r_d(kT) = e^{-T/\tau_d} r_d((k-1)T) + \left(1 - e^{-T/\tau_d}\right) r_i(kT)$$

$$r_f(kT) = e^{-T/\tau_f} r_f((k-1)T) + \left(1 - e^{-T/\tau_f}\right) r_i(kT)$$

$$r_x'(kT) = r_d(kT) - r_p'((k - d_d + d_x)T)$$

$$r_x(kT) = r_x'((k - d_x)T)$$

$$r_p(kT) = r_p'((k - d_d)T)$$

$$r_o(kT) = r_x(kT) + r_p(kT)$$

where $k = 0, 1, 2, \ldots$. The downsampling (slower sampling) with period NT of the output of filter $G_f(z)$ and the holding of each downsampled value for N sample periods T is modeled using

$$r_p'(kT) = r_p'((k+1)T) = r_p'((k+2)T) = \cdots = r_p'((k+N-1)T) = r_f(kT)$$

where $k = 0, N, 2N, \ldots$

The transformed relationships between variables in the model in Figure 7.16 then are

$$G_d(z) = \frac{1 - e^{-T/\tau_d}}{1 - e^{-T/\tau_d} z^{-1}}$$

$$G_f(z) = \frac{1 - e^{-T/\tau_f}}{1 - e^{-T/\tau_f} z^{-1}}$$

$$R_d(z) = G_d(z) R_i(z)$$

$$R_f(z) = G_f(z) R_i(z)$$

$$R_p'(z) = \sum_{n=0}^{N-1} z^{-n} \mathcal{Z}_N \{R_f(z)\}$$

$$R_x'(z) = G_d(z) R_i(z) - z^{-(d_p - d_x)} R_p'(z)$$

$$R_i(z) = z^{-d_p} R_p'(z) - z^{-d_x} R_x'(z)$$

For a sampled function $f(kT)$, The z transform of permanent worker capacity decisions $r_f(kT)$ with sample period T is

$$R_f(z) = r_f(0) + r_f(T) z^{-1} + r_f(2T) z^{-2} + \cdots + r_f(NT) z^{-N} + \cdots + r_f(2NT) z^{-2N} + \cdots$$

The corresponding z transform when downsampled with period NT is

$$\mathcal{Z}_N\{R_f(z)\} = r_f(0) + 0z^{-1} + 0z^{-2} + \cdots + 0z^{-(N-1)} + r_f(NT) z^{-N} + 0z^{-(N+1)} + \cdots$$

or

$$\mathcal{Z}_N\{R_f(z)\} = r_f(0) + r_f(NT) z^{-N} + r_f(2NT) z^{-2N} + r_f(3NT) z^{-3N} + \cdots$$

The relationship between transformed order input rate $R_i(z)$ and transformed permanent worker capacity decisions $R_p'(z)$ cannot be modeled by a transfer function that is valid for all order input rate sequences $r_i(kT)$ because of downsampling and holding. However, the frequency response, which specifically describes the relationship for sinusoidal inputs, is

$$G_p\left(e^{Tj\omega}\right) = \frac{R_p'\left(e^{Tj\omega}\right)}{R_i\left(e^{Tj\omega}\right)} = \frac{G_f\left(e^{Tj\omega}\right)}{N} \sum_{n=0}^{N-1} e^{-nTj\omega}$$

The frequency response of the relationship between transformed order input rate $R_i(z)$ and transformed cross-trained worker capacity decisions $R_x'(z)$ is then

$$G_x\left(e^{Tj\omega}\right) = \frac{R_x'\left(e^{Tj\omega}\right)}{R_i\left(e^{Tj\omega}\right)} = G_d\left(e^{Tj\omega}\right) - G_p\left(e^{Tj\omega}\right) e^{-(d_p - d_x)Tj\omega}$$

and the frequency response of the relationship between transformed order input rate $R_i(z)$ and transformed combined production capacity $R_0(z)$, including delays in implementation, is

$$G_o\left(e^{Tj\omega}\right) = \frac{R_o\left(e^{Tj\omega}\right)}{R_i\left(e^{Tj\omega}\right)} = G_d\left(e^{Tj\omega}\right)e^{-d_x Tj\omega}$$

Program 7.4 illustrates how the above relationships can be used to design and predict the dynamic behavior of permanent worker capacity, cross-trained worker capacity, and combined capacity. The calculation is done using the real and imaginary parts of frequency response rather than magnitude and phase. The adjustment

Program 7.4 Adjustment of permanent and cross-trained worker capacity

```
T=1;  % period between cross-trained worker capacity decisions (days)
z=tf('z',T,'TimeUnit','days');

N=10;  % period NT days between cross-trained worker capacity decisions
dp=5;  % 5-day delay in permanent worker capacity adjustments
dx=1;  % 1-day delay in cross-trained worker capacity adjustments

ff=1/100;  % cutoff for permanent worker adjustments (cycles/day)
fd=1/20;  % cutoff for combined capacity adjustments (cycles/day)

load RiData;  % load daily order input rate data into Ri
t=(0:length(Ri)-1)*T;  % time (days)

% define transfer functions
tauf=1/ff/(2*pi);  % time constant for permanent worker capacity
Gf=(1-exp(-T/tauf))/(1-exp(-T/tauf)/z)  % permanent worker filter
```

```
Gf =

   0.0609 z
  ----------
  z - 0.9391

Sample time: 1 days
Discrete-time transfer function.
```

```
taud=1/fd/(2*pi);  % time constant for desired capacity
Gd=(1-exp(-T/taud))/(1-exp(-T/taud)/z)  % combined capacity filter
```

```
Gd =

   0.2696 z
  ----------
  z - 0.7304

Sample time: 1 days
Discrete-time transfer function.
```

```
Ghold=tf(ones(1,N),1,T,'TimeUnit','days')/z^(N-1)  % hold (NT days)
```

```
Ghold =

 z^9 + z^8 + z^7 + z^6 + z^5 + z^4 + z^3 + z^2 + z + 1
 -----------------------------------------------------
                          z^9

Sample time: 1 days
Discrete-time transfer function.
```

```
% calculate frequency response
W=logspace(-4,pi,2000);  % logarithmically spaced frequencies
[Rep1,Imp1]=nyquist(Gf*Ghold/N,W);  % permanent worker decisions
FRp1=squeeze(complex(Rep1,Imp1));
[Rex1,Imx1]=nyquist((Gd-Gf*Ghold/N/z^(dp-dx)),W);  % cross-trained
FRx1=squeeze(complex(Rex1,Imx1));
[Redp,Imdp]=nyquist(z^-dp,W);  % delay dpT
FRdp=squeeze(complex(Redp,Imdp));
[Redx,Imdx]=nyquist(z^-dx,W);  % delay dxT
FRdx=squeeze(complex(Redx,Imdx));
FRo=FRx1.*FRdx+FRp1.*FRdp;  % combined production capacity after delays

% plot frequency response magnitudes - Figure 7.17
F=W/(2*pi); % frequency (cycles/day)
semilogx(F,abs(FRp1),F,abs(FRx1),F,abs(FRo)) % permanent, cross-, total
xlabel('frequency (cycles/day)'); ylabel('magnitude (-/-)')
legend('|G_p(e^{Tj\omega})|','|G_x(e^{Tj\omega})|',...
    '|G_o(e^{Tj\omega})|')

% calculate time response
Rf=lsim(Gf,Ri,t);  % output of permanent worker capacity filter
for m=1:length(Rf)  % downsample with period NT, hold for N samples
    m1=floor((m-1)/N)*N+1;
    Rp1(m,1)=Rf(m1);  % permanent worker capacity decisions
end
Rp=Rp1;
for m=dp+1:length(Rp)
    Rp(m)=Rp1(m-dp);  % adjusted permanent worker capacity
end
Rd=lsim(Gd,Ri,t);  % desired combined capacity filter
Rx1=Rd;
for m=dp-dx+1:length(Rf)
    Rx1(m)=Rx1(m)-Rp1(m-(dp-dx));  % cross-trained worker decisions
end
Rx=Rx1;
for m=dx+1:length(Rx)
    Rx(m)=Rx1(m-dx);  % adjusted cross-trained worker capacity
end
Ro=Rp+Rx;  % combined production capacity

stairs(t,Ri); hold on  % order rate input - Figure 7.18
stairs(t,Rp1)  % permanent worker capacity decisions
stairs(t,Rx1); hold off  % cross-trained worker capacity decisions
xlabel('time itkT (days)'); ylabel('(orders/day)')
legend('r_i(kT)','r_p''(kT)','r_x''(kT)')

stairs(t,Ri); hold on  % order input rate - Figure 7.19
stairs(t,Ro); hold off  % combined production capacity
xlabel('time {\itkT} (days)'); ylabel('(orders/day)')
legend('r_i(kT)','r_o(kT)')
```

period for cross-trained worker capacity is $T = 1$ day, while the adjustment period for permanent workers is $NT = 10$ days ($N = 10$). The delay in implementing cross-trained worker capacity decisions is $d_xT = 1$ day ($d_x = 1$) and the delay in implementing permanent worker capacity decisions is $d_pT = 5$ days ($d_p = 5$).

It is desired to focus adjustments in permanent worker capacity on frequencies below 0.01 cycles/day (period 100 days/cycle); hence, it is desired to reduce the amplitude of adjustments in permanent worker capacity for frequencies above 0.01 cycles/day. The time constant chosen for permanent worker filter transfer function $G_f(z)$ therefore is $\tau_f = 100/2\pi$ days.

It is desired to focus adjustments in cross-trained worker capacity on frequencies between 0.01 cycles/day (period 100 days/cycle) and 0.05 cycles/day (period 20 days/cycle); hence, it is desired to reduce the amplitude of adjustments for frequencies above 0.05 cycles/day. The time constant chosen for desired transfer function $G_d(z)$ therefore is $\tau_d = 20/2\pi$ days.

The resulting frequency response magnitude for adjustments in permanent worker capacity decisions and cross-trained worker capacity decisions are shown in Figure 7.17 together with the frequency response for combined adjustments in worker capacity. Several features should be noted:

- The frequency response magnitude for permanent worker capacity decisions and the frequency response for combined production capacity focus on the desired ranges of lower frequencies. The magnitude is unity at lower frequencies and is significantly reduced at higher frequencies.
- The frequency response magnitude for cross-trained worker capacity decisions covers the expected range of intermediate frequencies. The magnitude is greater than unity over portions of this range because of phase relationships: cross-trained worker capacity decisions compensate for additional time delay in permanent worker capacity decisions.

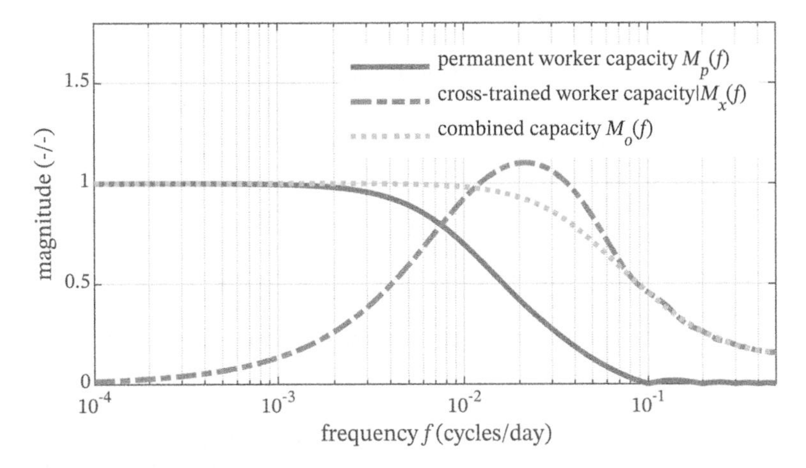

Figure 7.17 Magnitude of frequency response of permanent worker capacity decisions, cross-trained worker capacity decisions, and combined production capacity.

- The longer period between permanent worker capacity decisions could result in unexpected response[2] to higher frequencies above ½ the downsampling rate of $1/NT$ days^{-1}; however, this is unlikely when the amplitude of the frequency content of input $r_i(kT)$ is low at these higher frequencies and the magnitude of the frequency response for permanent worker capacity decisions is low at these higher frequencies. The latter is aided by the choice

$$\frac{1}{2\pi\tau_f} \ll \frac{1}{NT}$$

The permanent worker filter $G_f(z)$ is implemented with period T before downsampling rather than with period NT after downsampling.

An example of order input rate $r_i(kT)$ and the resulting time response of permanent worker capacity decisions $r_p'(kT)$ and cross-trained worker capacity decisions $r_x'(kT)$ are shown in Figure 7.18 for the parameter values described above. These results illustrate how longer-term fluctuations in order input rate are reflected in permanent worker capacity decisions and also illustrate the longer period between these decisions. The results also illustrate how shorter-term fluctuations in order input rate are reflected in cross-trained worker capacity decisions and the shorter period between these decisions. (In practice, a nominal number of cross-trained workers would be assigned so that cross-trained worker capacity is not negative.) The resulting combined production capacity after adjustment delays d_pT and d_sT is shown in Figure 7.19 and follows all but the higher frequency fluctuations in order input rate.

In this example the value of the frequency response perspective was illustrated both in defining how decision-making in a production system should react to fluctuating conditions and in distributing responses to these fluctuating conditions among available types of manipulations. Specifically, the desired overall frequency response of

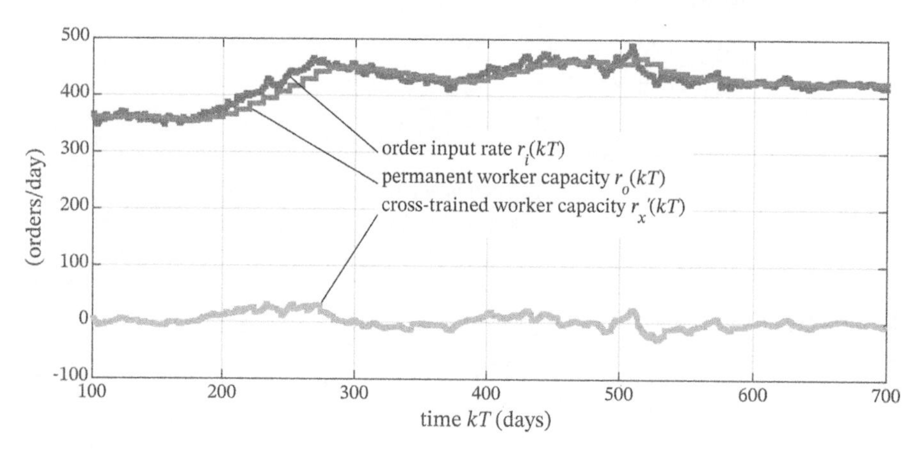

Figure 7.18 Example of order input rate and resulting permanent worker and cross-trained worker capacity decisions.

---

2 Sampling a signal that has content at frequencies above ½ the sampling frequency can result in these higher frequencies to being incorrectly responded to as if they are lower frequencies. This is an effect of aliasing, which is described in Section 5.2.1. The reader is referred to the Bibliography and many other publications on signal processing for more information on this topic.

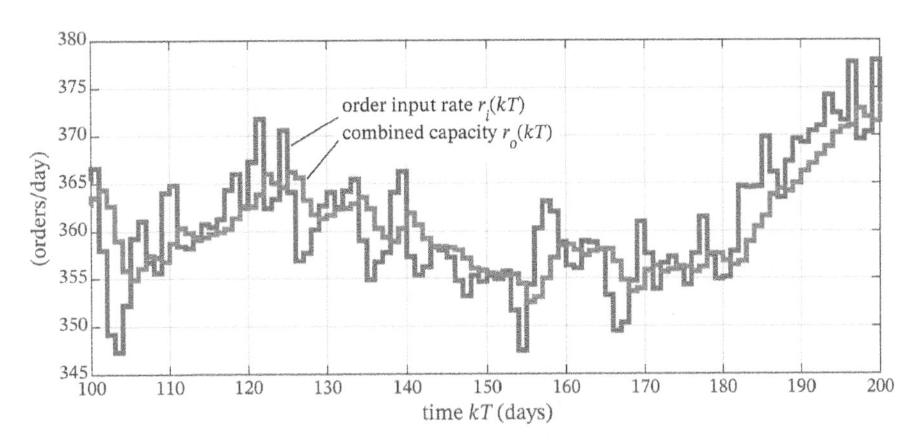

Figure 7.19 Example of order input rate and resulting combined production capacity.

production capacity was defined and a lower-frequency portion was addressed by decisions regarding permanent worker capacity. Then the remaining portion was addressed by decisions regarding cross-trained worker capacity, taking into account delays in implementing decisions and the dynamic characteristics of desired production capacity and permanent worker capacity.

7.5 Closed-Loop, Multi-Rate Production System with Different Adjustment Periods for WIP and Backlog Regulation

Various types of decisions are made and implemented on an ongoing basis in a production system, and they often interact via the physical and logistical variables that they affect. This can make it difficult to design each decision-making component individually, and also can make it difficult to predict the overall dynamic behavior that will result and ensure that it is favorable. Analysis of dynamic behavior can be even more difficult when some decisions are made and implemented relatively often, and others relatively less often. In this application example, different decision-making components adjust the production system to regulate backlog and work in progress (WIP) [5], with the latter being made relatively more often, daily for example, and the former being made relatively less often, weekly for example. These decisions interact, and the decision-making components must be designed with this in mind. This application example shows how the methods that have been presented in earlier chapters can be used to analyze and design the dynamic behavior of planning and scheduling in a production system that has multiple interacting decision-making components that manipulate different portions of the production system with different goals and different periods between adjustments.

In this example, the production system includes both backlog and WIP regulation as illustrated in Figure 1.3. As shown in the block diagram in Figure 7.20, decisions are made at the planning level regarding production capacity, which is adjusted with the goal of eliminating backlog, which here is the difference between the amount of work that has been planned to be completed and the amount of work that actually has been completed by the work system. At the scheduling level, order release rate is adjusted with the goal of eliminating any difference between actual WIP and planned WIP. Backlog and WIP regulation decisions interact in the production system because they both affect the release of orders to production. Furthermore, response to capacity and work disturbances can result in interacting adjustments in both production capacity and order release rate.

The model shown in the block diagram in Figure 7.20 can be used to better understand the fundamental dynamic behavior of this closed-loop, multi-rate production system and design the decision-making that is used to adjust production capacity and order release rate. Analysis of this system is complicated by the interactions between the inner WIP regulation loop and the outer backlog regulation loop, which have different adjustment periods because the rate at which work is released to the work system can be adjusted significantly more rapidly than production capacity. The following variables and parameters are used in the model in Figure 7.20:

T period between capacity adjustments (days)
N positive integer (T/N is the period between order release rate adjustments)
$r_p(t)$ planned rate of order input to the production work system (orders/day)
$w_b(t)$ backlog (orders)

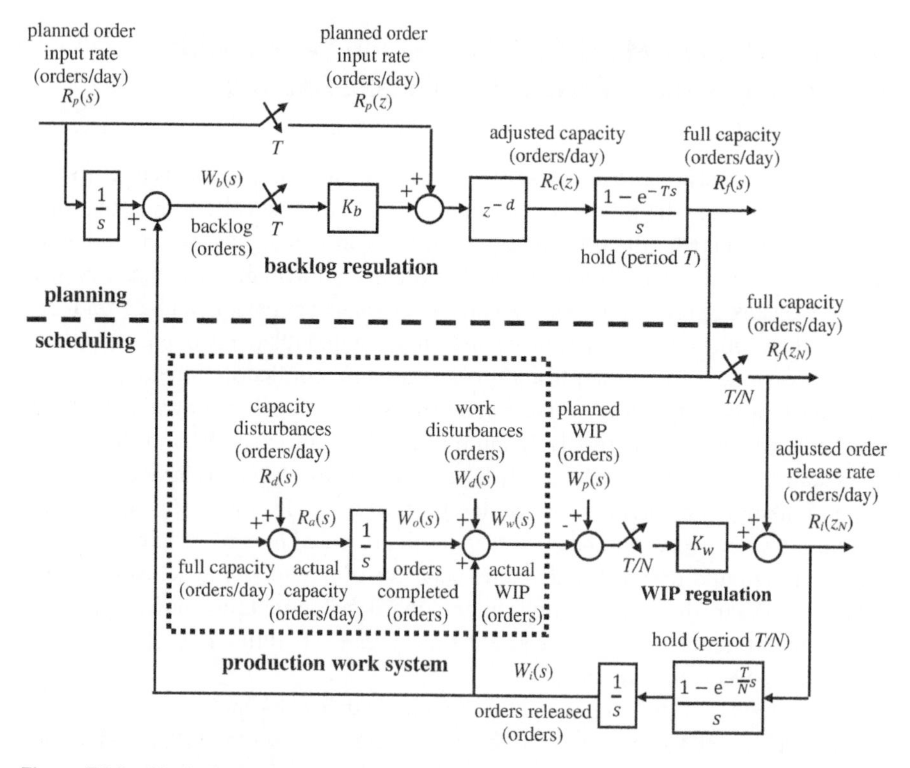

Figure 7.20 Block diagram with adjustment period T for production capacity and adjustment period T/N for order release rate.

K_b backlog regulation parameter (days^{-1})
d non-negative integer, delay dT days in implementing capacity adjustments
$r_c(kT)$ adjusted capacity (orders/day)
$r_f(t)$ full capacity (orders/day)
$r_d(t)$ capacity disturbances (orders/day)
$r_a(t)$ actual capacity (orders/day)
$w_o(t)$ orders completed (orders)
$w_d(t)$ work disturbances (orders)
$w_w(t)$ actual WIP (orders)
$w_p(t)$ planned WIP (orders)
K_w WIP regulation parameter (days^{-1})
$r_i(kT)$ order release rate (orders/day)
$w_i(t)$ orders released (orders)

The backlog is

$$w_b(t) = \int_0^t r_p(t)\,dt - w_i(t)$$

Backlog is measured and production capacity is adjusted with period T days, and there is a delay of dT days in implementing the adjustments. Proportional control is used in backlog regulation, and the relationship between planned order input rate, backlog, and adjusted capacity is

$$r_c(kT) = r_p((k-d)T) + K_b w_b((k-d)T)$$

The WIP in the production work system is

$$w_w(t) = w_i(t) + w_d(t) - \int_0^t (r_f(t) + r_d(t))\,dt$$

where work disturbances $w_d(t)$ include rush orders and cancelled orders. The rate at which orders are released is adjusted with period T/N days. Proportional control is used in WIP regulation, and the relationship between full capacity, planned WIP, actual WIP, and order release rate is

$$r_i\left(k\frac{T}{N}\right) = r_f\left(k\frac{T}{N}\right) + K_w\left(w_p\left(k\frac{T}{N}\right) - w_w\left(k\frac{T}{N}\right)\right)$$

There is no delay in implementing adjustments in order release rate.

The two regulation loops need to be designed by choosing the values of parameters K_b and K_w. It is convenient to design the two regulation loops independently, which is possible if the inner WIP regulation loop is designed to be significantly faster-acting than the outer backlog regulation loop. This is feasible because the adjustment period for backlog regulation is integer N times longer than the adjustment period for WIP regulation and the delay in production capacity adjustment further limits achievable backlog regulation performance.

WIP regulation has adjustment period T/N. and the Z transform for order release rate adjustment with this period is

$$R_i(z_N) = R_f(z_N) + K_w\left(\mathcal{Z}_N\{W_p(s)\} - \mathcal{Z}_N\left\{\left(\frac{1-e^{-\frac{T}{N}s}}{s}\right)\frac{1}{s}\right\}R_i(z_N) - \mathcal{Z}_N\{W_d(s)\} + \mathcal{Z}_N\left\{\frac{1}{s}R_f(s)\right\} + \mathcal{Z}_N\left\{\frac{1}{s}R_d(s)\right\}\right)$$

where

$$z_N = e^{\left(T/N\right)s}$$

and $\mathcal{Z}_N\{F(s)\}$ is the Z transform, with sample period T/N, of continuous-time function $f(t)$:

$$\mathcal{Z}_N\{F(s)\} = F(z_N) = f(0) + f\left(\frac{T}{N}\right)z_N^{-1} + f\left(2\frac{T}{N}\right)z_N^{-2} + f\left(3\frac{T}{N}\right)z_N^{-3} + \cdots$$

The zero-order hold with discrete input $R_i(z_N)$ holds its output constant over each period T/N; hence

$$\mathcal{Z}_N\left\{\left(\frac{1-e^{-\frac{T}{N}s}}{s}\right)\frac{1}{s}\right\}R_i(z_N) = \frac{\frac{T}{N}z_N^{-1}}{1-z_N^{-1}}R_i(z_N)$$

Then

$$R_i(z_N) = \frac{1-z_N^{-1}}{1-\left(1-K_w\frac{T}{N}\right)z_N^{-1}}\left(R_f(z_N) + K_w\left(\mathcal{Z}_N\{W_p(s)\} - \mathcal{Z}_N\{W_d(s)\} + \mathcal{Z}_N\left\{\frac{1}{s}R_f(s)\right\} + \mathcal{Z}_N\left\{\frac{1}{s}R_d(s)\right\}\right)\right)$$

K_w should be chosen such that $0 < K_w \leq N/T$, and choosing $K_w = N/T$ days^{-1} results in complete response of adjustments in order release rate in one period T/N. With this choice, the dynamic behavior of the inner loop can be ignored if the outer loop is designed for a response time that is significantly greater than T/N.

Ignoring disturbances,

$$W_w(s) \approx W_p(s)$$

$$W_i(s) \approx \frac{1}{s}R_f(s)$$

and a simplified model for adjustment of production capacity is

$$R_c(z) \approx \left(R_p(z) + K_b \left(\mathcal{Z}\left\{ \frac{1}{s} R_p(s) \right\} - \mathcal{Z}\left\{ \left(\frac{1-e^{-Ts}}{s} \right) \frac{1}{s} \right\} R_c(z) \right) \right) z^{-d}$$

$$R_c(z) \approx \frac{1-z^{-1}}{1-z^{-1}+K_b T z^{-(1+d)}} \left(R_p(z) + K_b \mathcal{Z}\left\{ \frac{1}{s} R_p(s) \right\} \right) z^{-d}$$

The design methods described in Chapter 6 are among those that can be used to choose K_b. One approach is to find the value of K_b that results in a characteristic equation with two equal real roots, anticipating that higher values of K_b will result in oscillatory response, while lower values of K_b will result in slower response. The characteristic equation is

$$z^{d+1} - z^{d} + K_b T = 0$$

Because the desired value of K_b results in a pair of equal roots, the first derivative of the characteristic equation also has this root:

$$\left(z - \frac{d}{d+1} \right) z^{d-1} = 0$$

Substituting the non-zero root into the characteristic equation and solving for K_b yields

$$K_b = \frac{d^d}{T(d+1)^{d+1}}$$

The combined dynamic behavior of backlog and WIP regulation is likely to resemble the designed independent behavior when $K_b \ll K_w$.

A combined Z transform model for the complete system in Figure 7.20 will facilitate verification of the actual dynamic behavior that results from choices of K_b and K_w including choices made using the simplifying assumptions described above. However, there are two different adjustment periods and the models of backlog and WIP regulation that were used above cannot be directly combined because the Z transforms have different variables: z and z_N. One approach for obtaining a combined model is to decompose the fast samplers with period T/N into N slow samplers with period T as shown in Figure 7.21. Each slow sampler is preceded by a time advance and followed by a corresponding time delay. All sample sequences then have the same instants in time separated by period T.

Variables of interest include

$$R_f(z_N) = \sum_{n=0}^{N-1} z_N^{-n} R_c(z)$$

$$W_o(z_N) = \frac{\frac{T}{N} z_N^{-1}}{1-z_N^{-1}} R_f(z_N) + \mathcal{Z}_N\left\{ \frac{1}{s} R_d(s) \right\}$$

$$W_i(z_N)=\frac{\frac{T}{N}z_N^{-1}}{1-\left(1-K_w\frac{T}{N}\right)z_N^{-1}}R_f(z_N)-\frac{K_w\frac{T}{N}z_N^{-1}}{1-\left(1-K_w\frac{T}{N}\right)z_N^{-1}}\left(\mathcal{Z}_N\{W_d(s)\}-\mathcal{Z}_N\{W_p(s)\}\right)$$

$$W_w(z_N)=W_i(z_N)+\mathcal{Z}_N\{W_d(s)\}-W_o(z_N)$$

$$R_i(z_N)=R_f(z_N)+K_w\left(-W_a(z_N)+\mathcal{Z}_N\{W_p(s)\}\right)$$

The orders released is

$$W_i(z_N)=\frac{\frac{T}{N}z_N^{-1}}{1-\left(1-K_w\frac{T}{N}\right)z_N^{-1}}\left(R_f(z_N)+K_w\left(W_o(z_N)-\mathcal{Z}_N\{W_d(s)\}+\mathcal{Z}_N\{W_p(s)\}\right)\right)$$

Decomposing each sampler with period T/N into N samplers with period T as shown in Figure 7.21,

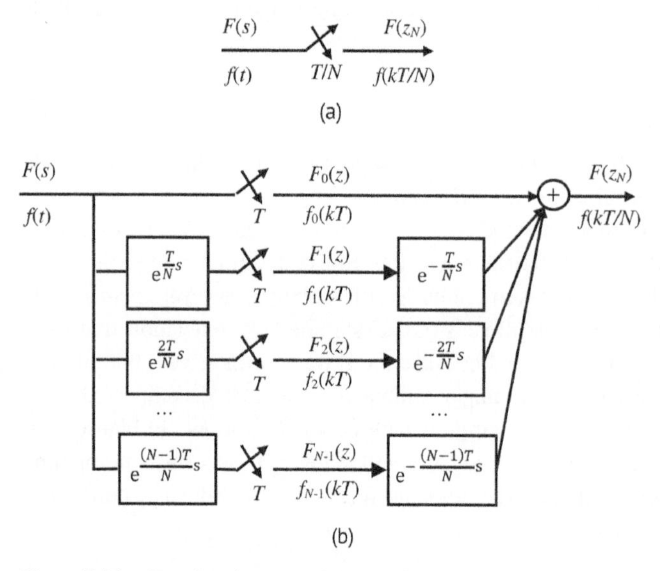

Figure 7.21 Sampler decomposition. (a) Fast sampler with period T/N. (b) Fast sampler decomposed into N slow samplers.

$$W_i(z_N) = \frac{\frac{T}{N} z_N^{-1}}{1 - \left(1 - K_w \frac{T}{N}\right) z_N^{-1}} \left(\sum_{n=0}^{N-1} z_N^{-n} R_c(z) + K_w \left(\sum_{n=0}^{N-1} z_N^{-n} \left(\frac{T\left(\frac{n}{N} + \frac{N-n}{N} z^{-1}\right)}{1 - z^{-1}} R_c(z) + \mathcal{Z}\left\{ e^{n\frac{T}{N}s} \frac{1}{s} R_d(s) \right\} \right) - \sum_{n=0}^{N-1} z_N^{-n} \mathcal{Z}\left\{ e^{n\frac{T}{N}s} W_d(s) \right\} + \sum_{n=0}^{N-1} z_N^{-n} \mathcal{Z}\left\{ e^{n\frac{T}{N}s} W_p(s) \right\} \right) \right)$$

Downsampling from period T/N to period T,[3]

$$\begin{aligned} W_i(z) = {} & \mathcal{Z}\left\{ \frac{\frac{T}{N} z_N^{-1}}{1 - \left(1 - K_w \frac{T}{N}\right) z_N^{-1}} \left(\sum_{n=0}^{N-1} z_N^{-n} \right) \right\} R_c(z) \\ & + \sum_{n=0}^{N-1} \mathcal{Z}\left\{ \left(\frac{K_w \frac{T}{N} z_N^{-1}}{1 - \left(1 - K_w \frac{T}{N}\right) z_N^{-1}} \right) z_N^{-n} \right\} \left(\frac{T\left(\frac{n}{N} + \frac{N-n}{N} z^{-1}\right)}{1 - z^{-1}} R_c(z) \right. \\ & \left. + \mathcal{Z}\left\{ e^{n\frac{T}{N}s} \frac{1}{s} R_d(s) \right\} \right) - \sum_{n=0}^{N-1} \mathcal{Z}\left\{ \left(\frac{K_w \frac{T}{N} z_N^{-1}}{1 - \left(1 - K_w \frac{T}{N}\right) z_N^{-1}} \right) z_N^{-n} \right\} \mathcal{Z}\left\{ e^{n\frac{T}{N}s} W_d(s) \right\} \\ & + \sum_{n=0}^{N-1} \mathcal{Z}\left\{ \left(\frac{K_w \frac{T}{N} z_N^{-1}}{1 - \left(1 - K_w \frac{T}{N}\right) z_N^{-1}} \right) z_N^{-n} \right\} \mathcal{Z}\left\{ e^{n\frac{T}{N}s} W_p(s) \right\} \end{aligned}$$

3 One way to obtain this result is to expand the transfer function in z_N, which includes delay, into the sum of a transfer function in z_N without delay and a polynomial in z_N. The result of downsampling from period T/N to period T then is

$$\begin{aligned} \mathcal{Z}\left\{ \frac{b_0 + b_1 z_N^{-1} + b_2 z_N^{-2} + \cdots}{\left(1 - r z_N^{-1}\right)} \right\} &= \mathcal{Z}\left\{ \frac{a}{\left(1 - r z_N^{-1}\right)} + c_0 + c_1 z_N^{-1} + c_2 z_N^{-2} + \cdots + c_N z^{-1} + \ldots \right\}. \\ &= \frac{a}{\left(1 - r^N z^{-1}\right)} + c_0 + c_N z^{-1} + c_{2N} z^{-2} + \cdots \end{aligned}$$

and then applying the summation formula for geometric series,

$$W_i(z) = \frac{T}{N}\left(\frac{\left(1-\left(1-K_w\frac{T}{N}\right)^N\right)z^{-1}}{K_w\frac{T}{N}\left(1-\left(1-K_w\frac{T}{N}\right)^N z^{-1}\right)}\right)R_c(z)$$

$$+\sum_{n=0}^{N-1}K_w\frac{T}{N}\frac{\left(1-K_w\frac{T}{N}\right)^{N-n-1}z^{-1}}{1-\left(1-K_w\frac{T}{N}\right)^N z^{-1}}\frac{T\left(\frac{n}{N}+\frac{N-n}{N}z^{-1}\right)}{1-z^{-1}}R_c(z)$$

$$+\sum_{n=0}^{N-1}K_w\frac{T}{N}\frac{\left(1-K_w\frac{T}{N}\right)^{N-n-1}z^{-1}}{1-\left(1-K_w\frac{T}{N}\right)^N z^{-1}}\mathcal{Z}\left\{e^{n\frac{T}{N}s}\frac{1}{s}R_d(z)\right\}$$

$$-\sum_{n=0}^{N-1}K_w\frac{T}{N}\frac{\left(1-K_w\frac{T}{N}\right)^{N-n-1}z^{-1}}{1-\left(1-K_w\frac{T}{N}\right)^N z^{-1}}\mathcal{Z}\left\{e^{n\frac{T}{N}s}W_d(s)\right\}$$

$$+\sum_{n=0}^{N-1}K_w\frac{T}{N}\frac{\left(1-K_w\frac{T}{N}\right)^{N-n-1}z^{-1}}{1-\left(1-K_w\frac{T}{N}\right)^N z^{-1}}\mathcal{Z}\left\{e^{n\frac{T}{N}s}W_p(s)\right\}$$

Also,

$$R_c(z) = \left(K_b W_b(z) + \mathcal{Z}\left\{R_p(s)\right\}\right)z^{-d}$$

$$W_b(z) = \mathcal{Z}\left\{\frac{1}{s}R_p(s)\right\} - W_i(z)$$

The backlog then is

$$W_b(z) = \Bigg(1+\left(\frac{\frac{T}{N}\left(1-\left(1-K_w\frac{T}{N}\right)^{N)}\right)z^{-1}}{K_w\frac{T}{N}\left(1-\left(1-K_w\frac{T}{N}\right)^N z^{-1}\right)}\right.$$

$$
\left.\left.+\sum_{n=0}^{N-1}K_w\frac{T}{N}\frac{\left(1-K_w\frac{T}{N}\right)^{N-n-1}z^{-1}}{1-\left(1-K_w\frac{T}{N}\right)^{N}z^{-1}}\frac{T\left(\frac{n}{N}+\frac{N-n}{N}z^{-1}\right)}{1-z^{-1}}\right)K_b z^{-d}\right]^{-1}\left(\mathcal{Z}\left\{\frac{1}{s}R_p(s)\right\}\right.
$$

$$
-\left(\frac{\frac{T}{N}\left(1-\left(1-K_w\frac{T}{N}\right)^{N)}\right)z^{-1}}{K_w\frac{T}{N}\left(1-\left(1-K_w\frac{T}{N}\right)^{N}z^{-1}\right)}\right.
$$

$$
\left.+\sum_{n=0}^{N-1}K_w\frac{T}{N}\frac{\left(1-K_w\frac{T}{N}\right)^{N-n-1}z^{-1}}{1-\left(1-K_w\frac{T}{N}\right)^{N}z^{-1}}\frac{T\left(\frac{n}{N}+\frac{N-n}{N}z^{-1}\right)}{1-z^{-1}}\right)z^{-d}\mathcal{Z}\left\{R_p(s)\right\}
$$

$$
-\sum_{n=0}^{N-1}K_w\frac{T}{N}\frac{\left(1-K_w\frac{T}{N}\right)^{N-n-1}z^{-1}}{1-\left(1-K_w\frac{T}{N}\right)^{N}z^{-1}}\mathcal{Z}\left\{e^{n\frac{T}{N}s}\frac{1}{s}R_d(z)\right\}
$$

$$
+\sum_{n=0}^{N-1}K_w\frac{T}{N}\frac{\left(1-K_w\frac{T}{N}\right)^{N-n-1}z^{-1}}{1-\left(1-K_w\frac{T}{N}\right)^{N}z^{-1}}\mathcal{Z}\left\{e^{n\frac{T}{N}s}W_d(s)\right\}
$$

$$
\left.-\sum_{n=0}^{N-1}K_w\frac{T}{N}\frac{\left(1-K_w\frac{T}{N}\right)^{N-n-1}z^{-1}}{1-\left(1-K_w\frac{T}{N}\right)^{N}z^{-1}}\mathcal{Z}\left\{e^{n\frac{T}{N}s}W_p(s)\right\}\right)
$$

Program 7.5 can be used to investigate the dynamic behavior of a production system in which capacity is adjusted with period $T = 5$ days, there is a delay $dT = 5$ days ($d = 1$) in implementing these adjustments, and order release rate is adjusted daily ($N = 5$). Using the backlog and WIP regulation design approach described above, $K_b = 0.05$ days^{-1} and $K_w = 1$ days^{-1}. Because the inner WIP regulation loop is designed for complete (deadbeat) response in one day, its dynamic properties do not appear in the characteristic equation for backlog regulation response. This characteristic equation has the expected pair of roots at 0.5; each corresponds to a time constant of 7.21 days. The change in backlog and change in capacity that result from a rush order (a 1-order work disturbance) are shown in Figure 7.22a when there is no capacity disturbance, planned order input is constant, and planned WIP is constant. The change in WIP and change in order release rate are shown in Figure 7.22b. As expected, the deviation in WIP is removed in one day, while the deviation in backlog is removed with a response that is

Program 7.5 Calculation of fundamental dynamic properties and time response for a production system with backlog and WIP Regulation and two different adjustment periods

```
T=5;  % slow sample period (days)
N=5;  % fast sampling period is T/N (days)
d=1;  % delay dT in implementing capacity adjustments
Kb=(d^d/(d+1)^(d+1))/T  % equal real roots for backlog regulation
```

```
Kb = 0.0500
```

```
Kw=N/T  % deadbeat response for WIP regulation
```

```
Kw = 1
```

```
[wb,rc,wwN,riN,t,tN]=TwoLoops(T,N,d,Kb,Kw);  % characteristics
```

```
  Pole        Magnitude    Damping    Time Constant
                                         (days)
 5.00e-01     5.00e-01    1.00e+00     7.21e+00
 5.00e-01     5.00e-01    1.00e+00     7.21e+00
```

```
stairs(t,wb); hold on  % plot time response - Figure 7.22
stairs(t,rc); hold off
xlabel('time kT (days)')
legend('Δw_b(kT) (orders)','Δr_c(kT) (orders/day)')

stairs(tN,wwN); hold on
stairs(tN,riN); hold off
xlabel('time kT (days)')
legend('Δw_w(kT/N) (orders)','Δr_i(kT/N) (orders/day)')

Kw=Kw/2  % slower WIP regulation response
```

```
Kw = 0.5000
```

```
Kb=2*Kb  % oscillatory backlog regulation response
```

```
Kb = 0.1000
```

```
[wb2,rc2,wwN2,riN2,t,tN]=TwoLoops(T,N,d,Kb,Kw);  % characteristics
```

```
          Pole              Magnitude     Damping    Time Constant
                                                        (days)
5.00e-01 + 5.00e-01i        7.07e-01     4.04e-01     1.44e+01
5.00e-01 - 5.00e-01i        7.07e-01     4.04e-01     1.44e+01
3.13e-02                    3.13e-02     1.00e+00     1.44e+00
```

```
stairs(t,wb2); hold on  % plot time response - Figure 7.23
stairs(t,rc2); hold off
xlabel('time kT (days)')
legend('backlog Δw_b(kT) (orders)','capacity Δr_c(kT) (orders/day)')
```

```
stairs(tN,wwN2); hold on
stairs(tN,riN2); hold off
xlabel('time kT (days)')
legend('WIP Δw_w(kT/N) (orders)','order release rate Δr_i(kT/N) (orders/
day)')

% calculate transfer functions, characteristics, and time response
function [wb,rc,wwN,riN,t,tN]=TwoLoops(T,N,d,Kb,Kw);
z=tf('z',T,'TimeUnit','days');
zN=tf('z',T/N,'TimeUnit','days');
a=1-Kw*T/N;

if a==0
  Gic=1/z;
else
  Gic=minreal((T/N)*((1-a^N)/(1-a))/z/(1-a^N/z));
end
for n=0:N-1
  Gin(1,n+1)=minreal(Kw*(T/N)*(a^(N-n-1)/z/(1-a^N/z)));
end
for n=0:N-1
  Gnc(n+1,1)=minreal((T*(n/N+(N-n)/N/z)/(1-1/z)));
end

Gb=minreal(Gin*feedback(1,Kb*(z^-d)*(ss(Gic)+ss(Gin)*ss(Gnc))));
damp(Gb)

t=(0:T:99);  % need complete sets of N days
tN=(0:T/N:99);

wd=ones(length(t),N);  % 1-order work disturbance occuring at time t=0
wb=lsim(Gb,wd,t);  % change in backlog
rc=lsim(Kb*(z^-d),wb,t);  % change in adjusted capacity

rfN=zeros(length(tN),1);  % preallocate change in full capacity
for m=1:length(t)
  for n=0:N-1
    rfN((m-1)*N+1+n,1)=rc(m);  % change in full capacity
  end
end

GifN=minreal((T/N)/zN/(1-a/zN));
wfN=lsim(GifN,rfN,tN);

GofN=minreal(T/N/zN/(1-1/zN));
woN=lsim(GofN,rfN,tN);  % corresponding change in orders completed

GiodN=minreal(Kw*(T/N)/zN/(1-a/zN));
wdN=ones(size(woN));  % 1-order work disturbance occuring at time t=0
wodN=lsim(GiodN,woN-wdN,tN);

wiN=wfN+wodN;  % change in orders released
wwN=wiN+wdN-woN;  % change in actual WIP

riN=-Kw*wwN+rfN;  % change in adjusted order release rate
end
```

well behaved but delayed by both the relatively long 5-day capacity adjustment period and the 5-day delay in implementing capacity adjustments. Reducing either or both of these would allow backlog and WIP regulation to be redesigned so that the effects of the rush order on backlog could be rejected more quickly.

If the backlog regulation parameter is increased by a factor of two to $K_b = 0.1$ days^{-1} and the WIP regulation parameter is reduced by a factor of two to $K_w = 0.5$ days^{-1}, the results demonstrate the interaction between backlog regulation and WIP regulation. In this case the characteristic equation for backlog regulation response has a real root that corresponds to a time constant of 1.44 days, which is closely associated with the inner WIP regulation loop, and a pair of complex conjugate roots that correspond to a damping ratio of 0.4 and a time constant of 14.4 days. The change in backlog and change in production capacity that result from a rush order are shown in Figure 7.23a. In this case, the response of backlog to the rush order is oscillatory and the settling time is longer than in Figure 7.22a. As shown in Figure 7.23b, oscillations in

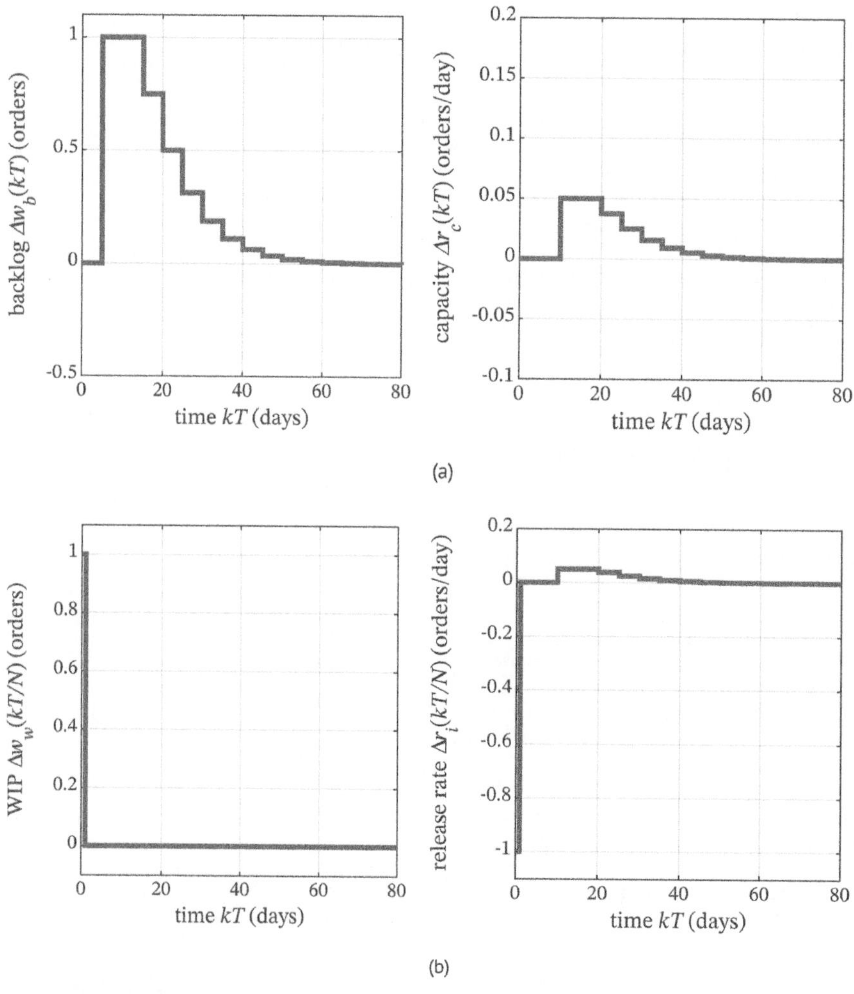

Figure 7.22 Response to a rush order when $K_b = 0.05$ days^{-1} and $K_w = 1$ days^{-1}. (a) backlog regulation. (b) WIP regulation.

production capacity, which is adjusted by backlog regulation, result in oscillations in order release rate, which is adjusted by WIP regulation.

The dynamic behavior in Figure 7.23 is significantly less desirable than the behavior in Figure 7.22. The ability to model and analyze this production system, with its two adjustment rates, is valuable in both understanding the complexities of the system's behavior and designing decision-making for backlog and WIP regulation that results in favorable dynamic behavior.

7.6 Summary

The application examples that have been presented in this chapter further illustrate how control theoretical modeling and analyses can lead to better understanding of the dynamic behavior of production systems and how this dynamic behavior can be

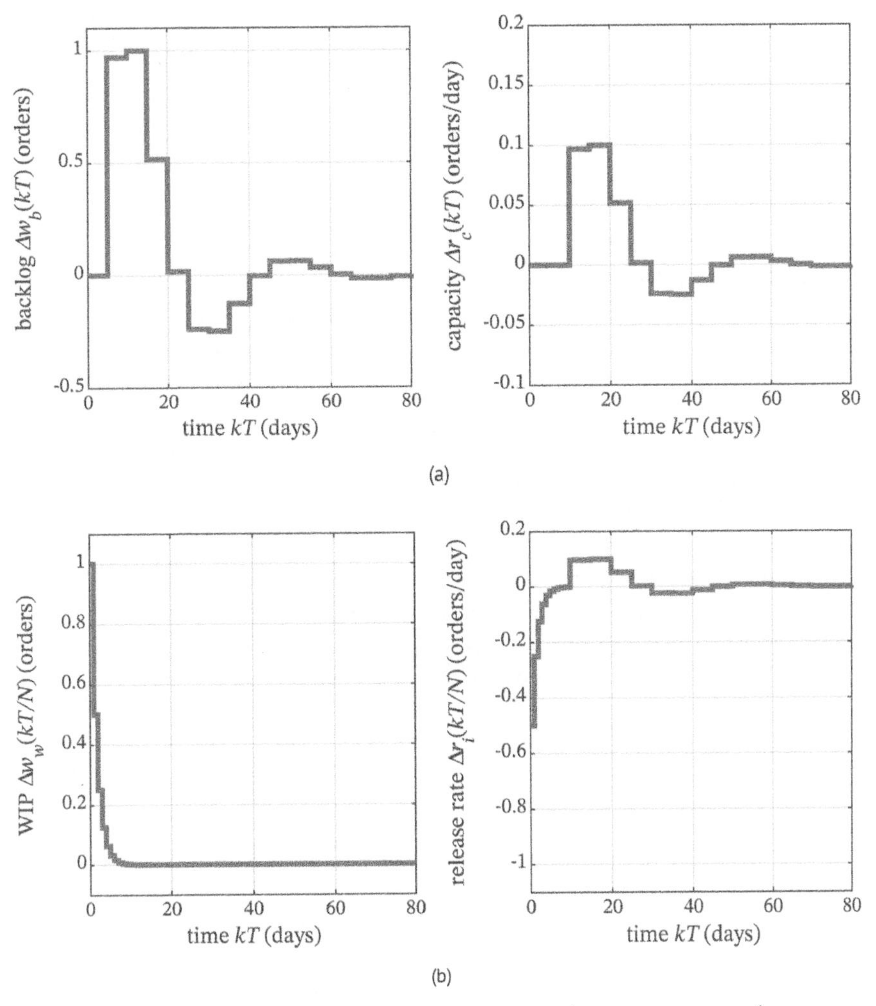

Figure 7.23 Response to a rush order when $K_b = 0.1$ days^{-1} and $K_w = 0.5$ days^{-1}. (a) backlog regulation. (b) WIP regulation.

designed from both time and frequency perspectives. The examples presented in previous chapters have been focused on basic aspects of control theoretical analysis and design of dynamic production systems. The examples presented here illustrated approaches for analysis and design when there are more complex aspects such as multiple inputs and outputs, network structures, information sharing, matrices of transfer functions, multiple closed loops, multiple delays, and multi-rate decision-making. A program has been included with each application example, although sometimes lengthy, to inform the interested reader of possible techniques for computation using control system engineering software.

The application examples in this chapter are by no means exhaustive, and significant opportunities exist for control theoretical analysis and design of future dynamic production systems that are changeable, agile, and respond rapidly and effectively to fluctuations in demand and production disturbances. Many more tools of control system engineering are available including state-space methods, internal model control, optimal control, nonlinear control, and stochastic control; however, even the most basic control theoretical methods for modeling and analysis can be important additions to the productions system engineer's toolbox for understanding and designing dynamic production systems.

References

1. Duffie, N., Bendul, J., and Knollmann, M. (2017). Prediction and avoidance of unfavorable dynamic behavior of lead time in production work systems. *SME Journal of Manufacturing Systems* 45: 273–285. doi: 10.1016/j.jmsy.2017.10.001.
2. Duffie, N. and Freitag, M. (2020). Frequency response analysis of inventory variation in production networks with information sharing. *Procedia CIRP* 93: 765–770. doi: 10.1016/j.procir.2020.03.058.
3. Duffie, N.A. and Shi, L. (2010). Dynamics of WIP regulation in large production networks of autonomous work systems. *IEEE Transactions on Automation Science & Engineering* 7 (3): 665–670. doi: 10.1109/TASE.2009.2036374.
4. Echsler Minquillon, F., Schömer, J., Stricker, N., Lanza, G., and Duffie, N. (2019). Planning for changeability and flexibility using a frequency perspective. *CIRP Annals-Manufacturing Technology* 68 (1): 427–430. doi: 10.1016/j.cirp.2019.03.006.
5. Duffie, N. and Falu, I. (2002). Control-theoretic analysis of a closed-loop PPC system. *CIRP Annals-Manufacturing Technology* 51 (1): 379–382.

Bibliography

Astolfi, A., Karagiannis, D., and Ortega, R. (2008). *Nonlinear and Adaptive Control with Applications*. Springer.

Äström, K.A. and Murray, R.M. (2021). *Feedback Systems: An Introduction for Scientists and Engineers, 2nd Edition*. Princeton University Press.

Black, H.S. (1977). Inventing the negative feedback amplifier. *IEEE Spectrum* 14 (12): 55–60.

Bode, H.W. (1960). Feedback—the history of an idea. *Proceedings of the Symposium on Active Networks and Feedback Systems*. New York: Polytechnic Institute of Brooklyn.

Bolton, W. (2019). *Mechatronics: Electronic Control Systems in Mechanical and Electrical Engineering, 7th Edition*. Pearson.

Dorf, R.C. and Bishop, H.B. (2022). *Modern Control Systems, 14th Edition*. Pearson.

Dullerud, G.E. and Paganini, F. (2000). *A Course in Robust Control Theory: A Convex Approach*. Springer.

Franklin, G.F., Powell, J.D., and Emami-Naeini, A.F. (2019). *Feedback Control of Dynamic Systems, 8th Edition*. Pearson.

Groover, M. (2019). *Automation, Production Systems and Computer-Integrated Manufacturing, 5th Edition*. Pearson.

Hespanha, J.P. (2018). *Linear System Theory, 2nd Edition*. Princeton University Press.

Liu, K.-Z. and Yao, Y. (2016). *Robust Control: Theory and Applications*. Wiley.

Lödding, H. (2013). *Handbook of Manufacturing Control*. Springer.

Mayr, O. (1970). The origins of feedback control. *Scientific American* 223 (10): 110–118.

Nise, N.S. (2019). *Control Systems Engineering, 8th Edition*. Wiley.

Nyhuis, P. and Wiendahl, H.-P. (2009). *Fundamentals of Production Logistics: Theory, Tools and Applications*. Springer.

Ogata, K. (2008). *MATLAB for Control Engineers*. Pearson.

Oppenheim, A.V. and Schafer, R.W. (2010). *Discrete-Time Signal Processing, 3rd Edition*. Pearson.

Phillips, C.L., Nagle, H.T., and Chakrabortty, A. (2015). *Digital Control System Analysis and Design, 4th Edition*. Pearson.

Russell, S. and Norvig, P. (2020). *Artificial Intelligence: A Modern Approach, 4th Edition*. Pearson.

Skogestad, S. and Postlethwaite, I. (2005). *Multivariable Feedback Control: Analysis and Design, 2nd Edition*. Wiley.

Control Theory Applications for Dynamic Production Systems: Time and Frequency Methods for Analysis and Design, First Edition. Neil A. Duffie.

Soroush, M., Baldea, M., and Edgar, T.F. (eds.) (2020). *Smart Manufacturing: Concepts and Methods, 1st Edition*. Elsevier.

Williams, R. and Lawrence, D. (2007). *Linear State-Space Control Systems*. Wiley.

Xue, D. and Chen, Y. (2015). *Modeling, Analysis and Design of Control Systems in MATLAB and Simulink*. World Scientific.

Zeigler, B.P., Muzy, A., and Kofman, E. (2018). *Theory of Modeling and Simulation: Discrete Event & Iterative System Computational Foundations, 3rd Edition*. Academic Press.

Index

Control Theory Applications for Dynamic Production Systems: Time and Frequency Methods for Analysis and Design, First Edition. Neil A. Duffie.

f

g